Computing Supplementum 2

Fundamentals of Numerical Computation (Computer-Oriented Numerical Analysis)

Edited by G. Alefeld and R. D. Grigorieff

in cooperation with
R. Albrecht, U. Kulisch, and F. Stummel

Springer-Verlag
Wien New York

Prof. Dr. Götz Alefeld
Prof. Dr. Rolf Dieter Grigorieff
Fachbereich 3 – Mathematik
Technische Universität Berlin

With 34 Figures

Library of Congress Cataloging in Publication Data. Main entry under title: Fundamentals of numerical computation. (Computing: Supplementum; 2.) "Mainly a collection of the invited lectures which were given during a conference ... held in June 5–8, 1979, on the occasion of the centennial of the Technical University of Berlin." 1. Numerical analysis – Data processing – Addresses, essays, lectures. I. Alefeld, G., 1941–. II. Grigorieff, R. D., 1938–. III. Series. QA297.F84.519.4.80–12131

ISSN 0344-8029
ISBN-13:978-3-211-81566-3 e-ISBN-13:978-3-7091-8577-3
DOI: 10.1007/978-3-7091-8577-3

Preface

This volume contains mainly a collection of the invited lectures which were given during a conference on "Fundamentals of Numerical Computation", held in June, 5–8, 1979, on the occasion of the centennial of the Technical University of Berlin. About hundred scientists from several countries attended this conference.

A preceding meeting on "Fundamentals of Computer-Arithmetic" was held in August, 1975, at the "Mathematisches Forschungsinstitut Oberwolfach". The lectures of this conference have been published as Supplementum 1 of Computing (Editors R. Albrecht, U. Kulisch).

After a period of four years of active research the purpose of the Berlin-Conference was to give a broad survey of the present status of the closely connected topics Interval Analysis, Mathematical Foundation of Computer Arithmetic, Rounding Error Analysis and Stability of Numerical Algorithms and to give prospects of future activities in these fields. Besides the invited lectures 35 short communications, each of 20 minutes length, were given.

We gratefully acknowledge the support of the President of the Technical University and of his Aussenreferat as well as of the Department of Mathematics. Besides these institutions financial support was given by AEG-Telefunken, Berlin, Allianz Lebensversicherungs A.G., Stuttgart, CDC, Hamburg/Berlin, DATA 100, München, Gesellschaft von Freunden der TU Berlin e.V., Berlin and Siemens AG., Berlin. Finally we express our thanks to Mrs. G. Froehlich and Mrs. B. Trajanović, who managed the paper work before, during and after the conference.

Berlin, February 1980 G. Alefeld and R. D. Grigorieff

Contents

Adams, E.: On Methods for the Construction of the Boundaries of Sets of Solutions for Differential Equations or Finite-Dimensional Approximations with Input Sets 1
Albrecht, R.: Roundings and Approximations in Ordered Sets. 17
Kaucher, E.: Interval Analysis in the Extended Interval Space $\mathbb{IR}$ 33
Kulisch, U. W., Miranker, W. L.: Arithmetic Operations in Interval Spaces 51
Markov, S. M.: Some Applications of Extended Interval Arithmetic to Interval Iterations. 69
Matula, D. W., Kornerup, P.: Foundations of Finite Precision Rational Arithmetic 85
Moore, R. E.: Interval Methods for Nonlinear Systems. 113
Oberaigner, W.: Algorithms for Multiplication with Given Precision . . . 121
Olver, F. W. J.: Unrestricted Algorithms for Generating Elementary Functions 131
Rall, L. B.: Applications of Software for Automatic Differentiation in Numerical Computation 141
Rump, S. M., Kaucher, E.: Small Bounds for the Solution of Systems of Linear Equations 157
Scherer, R., Zeller, K.: Shorthand Notation for Rounding Errors 165
Stummel, F.: Rounding Error Analysis of Elementary Numerical Algorithms 169
Ullrich, Ch.: Iterative Methods in the Spaces of Rounded Computations . 197
Yohe, J. M.: Portable Software for Interval Arithmetic. 211

Computing, Suppl. 2, 1 – 16 (1980)

On Methods for the Construction of the Boundaries of Sets of Solutions for Differential Equations or Finite-Dimensional Approximations with Input Sets*

E. Adams, Karlsruhe

Abstract

Collections of linear or nonlinear operator equations $Au = f$ are considered which may represent (i) differential or integral equations or (ii) finite-dimensional approximations. Input sets of coefficients a or data f are admitted. The envelope of the set of solutions is to be constructed where this boundary refers (i) to the range of values of the solutions or (ii) to a finite-dimensional space. The construction employs either topological boundary mapping or truncated Taylor expansions. Estimates of the local procedural errors are due to suitable a priori sets and interval mathematics. The relation between local and global error estimates is due to boundary mapping or an auxiliary inverse-monotone operator $\hat{B}$. The operator $\hat{B}$ is constructed for the case of arbitrary linear ordinary differential equations with boundary or initial conditions, provided the admitted A satisfy a mild condition.

1. Outline of the Problem

Collections of operator equations

$$Au = f \tag{1.1}$$

are considered which may represent (i) differential or integral equations with the usual side conditions or (ii) finite-dimensional approximations. The following types of input sets are admitted in (1.1): (*a*) sets S_f of data f and (*b*) sets S_a of coefficients a where "coefficient" refers to any input in A. An envelope ∂S_u of the set of solutions is to be approximated. The envelopes ∂S_a, ∂S_f, and ∂S_u either (*i*) refer to sets in finite-dimensional spaces if this is true for (1.1) or (*ii*) represent the upper and the lower envelopes of the ranges of the respective functions in (1.1). The existence of the solution of (1.1) will not be discussed.

The consideration of input sets S_a or S_f is motivated as follows:

Case (I) (Applied Mathematics). $A = A_0$ and $f = f_0$ are fixed:

(*1*) The execution of numerical methods requires that the solution u_0 of $A_0 u = f_0$ is well-conditioned with respect to neighboring coefficients and data;

(*2*) if a neighboring problem $A_v w = f_v$ can be solved, ∂S_u is of interest if S_a contains both a and a_v and if S_f contains both f and f_v;

(*3*) sets may appear in the analysis due to error estimates.

---

* This paper is dedicated to Prof. Dr. H. Görtler on the occasion of his 70th birthday. – The research was supported by the NATO Senior Fellowship Award SA.5-2-03B(112)961(78)MDL.

Case (II) (Applications of Mathematics). S_a and S_f are prescribed:

(*4*) The range of the solutions is to be bracketed as, e.g., for the case of a collection of loads in problems in civil engineering;

(*5*) mathematical models of real world problems have imprecisely known input.

With the possible exception of (4), input sets are usually "small", e.g., an input interval possesses a small span. The motivations (1)–(4) generally admit input sets with fixed deterministic envelopes; stochastic envelopes in the case of (5) will not be discussed here.

2. Ordinary Linear Initial Value Problems with Initial Sets

The solution of the linear ordinary ivp (initial value problem)

$$u' - A(t)u = g(t) \quad \text{for} \quad t \in J := (0, T], u(0) \in \mathbb{R}^n,$$
$$u: J \to \mathbb{R}^n, \quad A: J \to L(\mathbb{R}^n), \quad A, g \in C(\bar{J}) \tag{2.1}$$

can be represented as follows [11, p. 139–141]:

$$u(t) = X(t)\left[u(0) + \int_0^t (X(s))^{-1} g(s)\, ds\right] \quad \text{for } t \in J, \tag{2.2}$$

where X is the fundamental matrix of $u' - Au = 0$. A compact initial set $E(0) \subset \mathbb{R}^n$ is admitted. Then, (2.2) represents a bijective affine mapping with parameter t of $E(0)$ onto $E(t)$, the set of solutions at any $t \in J$, such that $\partial E(0)$ is mapped onto $\partial E(t)$. For the case of (2.1), this *Boundary Mapping* was recognized independently in 1977 by K. Nickel [18] and R. Lohner [14], [15].

The theory of differential inequalities ([13] or [23]) may be employed to construct an interval $I(t) = [\underline{u}(t), \bar{u}(t)] \subset \mathbb{R}^n$ such that $E(0) \subseteq I(0)$. Provided the off-diagonal elements of A are nonnegative, the ivp are inverse-monotone, i.e.,

$$u' - Au \geqslant w' - Aw \quad \text{for } t \in J \qquad \text{and} \qquad u(0) \geqslant w(0) \Rightarrow u(t) \geqslant w(t) \quad \text{for } t \in J. \tag{2.3}$$

Then, $\underline{u}, \bar{u}$ such that $\underline{u} \leqslant u \leqslant \bar{u}$ for $t \in J$ and every solution of (2.1) with $E(0)$ are solutions of $\bar{u}' - A\bar{u} = \underline{u}' - A\underline{u} = g$ on J and $\underline{u} \leqslant u \leqslant \bar{u}$ for $t = 0$ and every $u(0) \in E(0)$. If A does not possess this property, then

$$\left\{\begin{aligned} \bar{u}_i' - a_{ii}\bar{u}_i - \operatorname*{Max}_{u \in \hat{I}(t)} \sum_{\substack{j=1 \\ j \neq i}}^{n} a_{ij}u_j = g_i(t) \\ \underline{u}_i' - a_{ii}\underline{u}_i - \operatorname*{Min}_{u \in \hat{I}(t)} \sum_{\substack{j=1 \\ j \neq i}}^{n} a_{ij}u_j = g_i(t) \end{aligned}\right\} \quad \begin{aligned} &\text{for } t \in J, \underline{u}_i(0) \leqslant u_i(0) \leqslant \bar{u}_i(0), \\ &i = 1(1)n. \end{aligned} \tag{2.4}$$

Due to Max or Min, $\hat{I}(t) := [\underline{u}(t), \bar{u}(t)]$ generally exceeds the set $E(t)$.

Example 2.1. If

$$E(0) = [0.9, 1.1] \times [-0.1, 0.1] \subset \mathbb{R}^2 \quad \text{and} \quad A = \begin{pmatrix} 0 & 1 \\ -1 & 0 \end{pmatrix}, g = 0, \tag{2.5}$$

then $\bar{u}_i(t) - \underline{u}_i(t) = 0.2\,e^t$ for $i = 1$ or 2 even though $E(t)$ is uniformly bounded with respect to every $t \in J$ for $T \to \infty$.

Since I or $\hat{I}$ at t completely determine I or $\hat{I}$ at $t + dt$, (2.3) and (2.4) are interval methods. The overestimates due to (2.4) make it desirable to look for a constructive execution of boundary mapping in the case of (2.1).

Example 2.2. The ivp (2.1), (2.5) is reconsidered. The superscript $k = 1(1)4$ denotes the four corners of the set $E(0)$. By use of $v := u_1$ and $w := u_2$ and the trapezoidal rule, the representation of the ivp by Volterra integral equations can be discretized as follows:

$$v_{j+1}^{(k)} = v_j^{(k)} + (h/2)[w_j^{(k)} + w_{j+1}^{(k)}] - (h^3/12)w''(\alpha_j)$$
$$\text{for } j = 0(1)N-1,\ h = T/N,\ \alpha_j \in [t_j, t_{j+1}],$$
$$w_{j+1}^{(k)} = \cdots \tag{2.6}$$

Since $v'' + v = w'' + w = 0$, a suitably selected a priori interval $\tilde{I}_{j+1}^{(k)} = \tilde{V}_{j+1}^{(k)} \times \tilde{W}_{j+1}^{(k)} \subset \mathbb{R}^2$ may be employed to estimate the remainder term in the truncated Taylor expansion

$$w^{(k)}(t) = w_j^{(k)} - (t - t_j)v_j^{(k)} - \tfrac{1}{2}(t - t_j)^2 w_j^{(k)} - \tfrac{1}{6}(t - t_j)^3 \underbrace{v_j^{(k)}(\beta_j)}_{\in \tilde{V}_{j+1}^{(k)}} \overset{!}{\in} \tilde{W}_{j+1}^{(k)}$$
$$\text{for } t \in [t_j, t_{j+1}],\ \beta_j \in [t_j, t_{j+1}], \tag{2.7}$$

of the function $w^{(k)}$ where ! denotes a condition. If the corresponding condition $v^{(k)}(t) \overset{!}{\in} \tilde{V}_{j+1}^{(k)}$ for $t \in [t_j, t_{j+1}]$ is also satisfied, $w''(\alpha_j)$ may be replaced by $\tilde{W}_{j+1}^{(k)}$ in (2.6). Then (2.6) is a linear algebraic interval system with a fixed matrix whose solution is a parallelogram, $P_{j+1}^{(k)}$, which can be enclosed by the smallest interval $I_{j+1}^{(k)}$. The four intervals $I_{j+1}^{(k)}$ with $k = 1(1)4$ then can be enclosed by the smallest quadrilateral $\tilde{E}(t_{j+1})$ which is an outer approximation of $E(t_{j+1})$ such that the four corners of $\tilde{E}(t_{j+1})$ are the starting vectors for the continuation of the construction. Due to the numerical results by E. Gerdon (Karlsruhe), $\partial E(t_j)$ and $\partial \tilde{E}(t_j)$ deviate by less than $1.5(10^{-6})t_j$ for $t_j \in (0, 20\pi]$ where $h = 10^{-2}$ was used and the CPU-time was 56 sec on the UNIVAC 1108 at the University of Karlsruhe.

Remark. As compared with Example 2.1, here the "interval-coarsening" is restricted to the small local procedural error and its transfer to the corresponding global error due to the outer approximation.

Remark. By use of the domain invariance theorem of topology ([12] or [19]), boundary mapping holds for nonlinear ordinary ivp provided a uniqueness condition is satisfied, e.g., [12] or [19]. Boundary mapping also holds for suitable linear operators in Banach spaces with infinite dimensions, e.g., [8].

3. Ordinary Linear Boundary Value Problems with Sets of Data and Coefficients

The following collection of linear ordinary Sturm-Liouville bvp (boundary value problem) is considered:

$$Au = f \Leftrightarrow \begin{cases} Pu := u'' - a(x)u = -g(x) \quad \text{for } x \in I := (0,1), \\ Ru := u = 0 \quad \text{for } x \in \partial I; \quad u \in U := C^2(\bar{I}); \; \gamma := a \text{ or } g, \\ \forall \gamma \in S_\gamma := [\underline{\gamma}, \bar{\gamma}] \cap C(\bar{I}). \end{cases} \tag{3.1}$$

It is assumed that there exists $a_0 \in S_a$ such that Green's function G_0 is explicitly known for the operator A_0 pertaining to a_0. Each individual admitted bvp can be represented equivalently by

$$Bu := u(x) - \int_0^1 K(x,\xi)u(\xi)\,d\xi = -\int_0^1 G_0(x,\xi)g(\xi)\,d\xi =: g^*(x) \quad \text{for } x \in \bar{I},$$

$$\forall \gamma \in S_\gamma, \; K(x,\xi) := G_0(x,\xi)[a(\xi) - a_0(\xi)] \quad \text{for } (x,\xi) \in \bar{I} \times \bar{I}. \tag{3.2}$$

Auxiliary kernels are introduced:

$$K^+(x,\xi) := K(x,\xi) \quad \text{if } K(x,\xi) \geqslant 0 \text{ locally and } K^+(x,\xi) = 0 \text{ otherwise,}$$

$$K^-(x,\xi) := K(x,\xi) - K^+(x,\xi) \quad \text{on } \bar{I} \times \bar{I}. \tag{3.3}$$

The values of the set of solutions of $Bu = g^*$ can be bracketed by use of an interval $[\underline{u}, \bar{u}](x)$ for $x \in \bar{I}$ which solves the following interval extension of $Bu = g^*$

$$[\underline{u}, \bar{u}](x) = \int_0^1 ([\underline{K}^-, \bar{K}^-] + [\underline{K}^+, \bar{K}^+])(x,\xi)[\underline{u}, \bar{u}](\xi)\,d\xi = \int_0^1 [\underline{G_0 g}, \overline{G_0 g}]\,d\xi$$

for $x \in \bar{I}$, where $\underline{K}^- := \operatorname{Min}_{a \in S_a} K^-$ and $\underline{G_0 g} := \operatorname{Min}_{g \in S_g} G_0 g$ locally at every

$$(x,\xi) \in \bar{I} \times \bar{I}, \text{ etc.} \tag{3.4}$$

Generally, the relation between the input a, g and the corresponding solution u is lost as $Bu = g^*$ is replaced by (3.4). By use of the rule of interval multiplication, rearrangement of terms in (3.4) yields the "extended operator equation" with

$$\hat{B}\hat{u} := \bar{u} - \int_0^1 \left[\begin{pmatrix} (\operatorname{Max}_{a \in S_a} K^+)\bar{u} \text{ if } \bar{u} \geqslant 0 \text{ locally} \\ (\operatorname{Min}_{a \in S_a} K^+)\bar{u} \text{ if } \bar{u} < 0 \text{ locally} \end{pmatrix} + \begin{pmatrix} |\operatorname{Max}_{a \in S_a} K^-|(-\underline{u}) \text{ if } \underline{u} \geqslant 0 \text{ locally} \\ |\operatorname{Min}_{a \in S_a} K^-|(-\underline{u}) \text{ if } \underline{u} < 0 \text{ locally} \end{pmatrix} \right] d\xi,$$

$$:= (-\underline{u}) - \int_0^1 \left[\begin{pmatrix} (\operatorname{Min}_{a \in S_a} K^+)(-\underline{u}) \text{ if } \underline{u} \geqslant 0 \text{ locally} \\ (\operatorname{Max}_{a \in S_a} K^+)(-\underline{u}) \text{ if } \underline{u} < 0 \text{ locally} \end{pmatrix} + \begin{pmatrix} |\operatorname{Min}_{a \in S_a} K^-|\bar{u} \text{ if } \bar{u} \geqslant 0 \text{ locally} \\ |\operatorname{Max}_{a \in S_a} K^-|\bar{u} \text{ if } \bar{u} < 0 \text{ locally} \end{pmatrix} \right] d\xi,$$

$$\text{for } x \in \bar{I} \text{ with } \hat{u}(x) := (\bar{u}(x), -\underline{u}(x)) \text{ and } \hat{u}: \bar{I} \to \mathbb{R}^2. \tag{3.5}$$

The execution of the operators Max or Min yields (1) $\mathrm{Min}_{a\in S_a} K^+ = \mathrm{Max}_{a\in S_a} K^- = 0$ if $a_0(x)\in(\underline{a}(x), \bar{a}(x))$ for every $x\in\bar{I}$ and (2) discontinuous functions $a_x(\xi)$ whose values for any fixed $x\in\bar{I}$ are either $\underline{a}(\xi)$ or $\bar{a}(\xi)$; $\hat{B}$ is nonlinear unless $\underline{a} = \bar{a} = a_0$.

In order to show that $\hat{B}$ is inverse-monotone with respect to $\hat{u}$, it is required that there exists a test element $\hat{v} = (v, v)$ with $v\in C(\bar{I})$ and $v(x) > 0$ on $\bar{I}$ such that $\hat{B}\hat{v} > 0$ on $\bar{I}$. This inequality is trivially satisfied by the choice of $v(x) = 1$ for $x\in\bar{I}$ provided there holds

$$1 - \int_0^1 [(\mathrm{Max}_{a\in S_a} K^+) + |\mathrm{Min}_{a\in S_a} K^-|](x,\xi)\,d\xi \geqslant 1 - (\|\bar{K}^+\|_\infty + \|\underline{K}^-\|_\infty) \overset{!}{>} 0. \quad (3.6)$$

Theorem 3.1. *The operator $\hat{B}$ is inverse-monotone with respect to u if there exists a test element $\hat{v}\in C(\bar{I})$ such that $\hat{v} > 0$ and $\hat{B}\hat{v} > 0$ on $\bar{I}$.*

The *proof* as presented in [2] follows the one in [1].

Theorem 3.2. *The system of Fredholm integral equations*

$$\begin{aligned}\hat{B}\hat{u} = \hat{f} :&= \int_0^1 \mathrm{Max}_{g\in S_g}(-G_0 g)\,d\xi = -\int_0^1 G_0 \begin{Bmatrix}\underline{g}\,d\xi \ \textit{if}\ G_0 \geqslant 0\ \textit{locally}\\ \bar{g}\,d\xi\ \textit{if}\ G_0 < 0\ \textit{locally}\end{Bmatrix}\\
&=: \int_0^1 G_0 \breve{g}_x(\xi)\,d\xi,\\
&:= -\int_0^1 \mathrm{Min}_{g\in S_g}(-G_0 g)\,d\xi = \int_0^1 G_0 \begin{Bmatrix}\bar{g}\,d\xi \ \textit{if}\ G_0 \geqslant 0\ \textit{locally}\\ \underline{g}\,d\xi\ \textit{if}\ G_0 < 0\ \textit{locally}\end{Bmatrix}\\
&=: \int_0^1 G_0 \underset{\vee}{g}_x(\xi)\,d\xi, \end{aligned} \quad (3.7)$$

(1) possesses a unique solution $\underline{u}, \bar{u}\in C(\bar{I})$ such that

$$\underline{u}(x) \leqslant u(x) \leqslant \bar{u}(x)\ \textit{for}\ x\in\bar{I}\ \textit{and every solution}\ u\in C(\bar{I})\ \textit{of}\ (3.2), \quad (3.8)$$

provided (3.6) *is satisfied*; (*2*) *$\underline{u}, \bar{u}$ are sharp bounds if* (*i*) *$\underline{a} = \bar{a} = a_0$ or* (*ii*) *$G_0 \geqslant 0$ on $\bar{I}\times\bar{I}$ and $\underline{a} = a_0$ on $\bar{I}$.*

Proof. (*1*) This follows from Theorem 3.1 and the uniform convergence of the Neumann sequence of successive approximations; (*2*) this follows from a theorem by Arzela (e.g., [9, p. 772]) since the discontinuous functions $g_x(\xi)$ with values $\underline{g}(\xi)$ or $\bar{g}(\xi)$ for any fixed $x\in\bar{I}$ can be approximated with arbitrary accuracy by a sequence of functions $g_x^{(\nu)}(\xi)\in C(\bar{I})$, e.g., by use of the L_1-norm. □

For every collection (3.1) with $\|G_0\|_\infty\|\bar{a} - \underline{a}\|_\infty$ sufficiently small, (3.6) is satisfied; then, an interval $[\underline{u}, \bar{u}](x)$ can be constructed by use of the inverse-monotone extended operator $\hat{B}$. Hansen (e.g., [3, p. 232]) recognized the advantage of premultiplying a system of linear algebraic interval equations, $Au = f$, by A_0^{-1}.

Remark. The ivp (2.1) can be treated analogously.

Example 3.1. The operator equation $Au = f$ in (3.1) is reconsidered with a and g fixed. It is assumed that there exists a neighboring coefficient a_0 such that (*i*)

Green's function is (explicitly) known for (3.1) with a_0 instead of a and (*ii*) (3.6) is satisfied for functions $\underline{a}, \bar{a}$ such that $a, a_0 \in [\underline{a}, \bar{a}]$ on $\bar{I}$. Cubic Spline functions (e.g., [22, p. 81])

$$u_s := u_{sj} := u_j\left(\frac{x_{j+1} - x}{h}\right) + u_{j+1}\left(\frac{x - x_j}{h}\right)$$

$$+ \frac{M_j}{6} h^2\left[\left(\frac{x_{j+1} - x}{h}\right)^3 + \left(\frac{x - x_j}{h}\right) - 1\right] + M_{j+1}(h^2/6)[\cdots]$$

$$\text{for } x \in [x_j, x_{j+1}],\ x_j = jh,\ j = 0(1)n+1,\ h = (n+1)^{-1},\ x_0 = 0, \tag{3.9}$$

with any fixed $n \in \mathbb{N}$, are employed to construct an approximate solution of (3.1) for the case of fixed a and g. The composite Spline function $u_s \in C^2(\bar{I})$ yields a residual $r := Au_s - f$. A correction z is defined by use of $Az = -r$ such that $A[u_s + z] = f$. Analogous to (3.2) and (3.5), the equation $A_0^{-1}Az = A_0^{-1}r =: r^*$ gives rise to $\hat{B}\hat{z} = \hat{r}$. This yields the quantitative error estimates $z \in [\underline{z}, \bar{z}]$ and

$$u(x) \in [u_s(x) + \underline{z}(x), u_s(x) + \bar{z}(x)] \quad \text{for } x \in \bar{I}. \tag{3.10}$$

The standard error estimate of linear discrete analogies involves a constant $c \in \mathbb{R}^+$ which usually is not known quantitatively since c majorizes the norm of the inverse matrices for *every* $n \in \mathbb{N}$, [21, p. 9].

Remark. Linear elliptic bvp can be treated analogously by use of the Spline functions developed in [20] provided a Green's function G_0 is known.

4. Linear Ordinary Initial Value Problems with Sets of Coefficients

With reference to procedural or rounding errors, a complete conditioning (or sensitivity) analysis of a mathematical model requires a quantitative comparison of a "basic solution" $u(x, \gamma_0)$ with neighboring solutions $u(x, \gamma)$ where x stands for the independent variables and γ, γ_0 denote every data and coefficient. Interval mathematics *seems* to suggest that $\|u(x,\gamma) - u(x,\gamma_0)\|_\infty$ is majorized for $\gamma = \underline{\gamma}$ or for $\gamma = \bar{\gamma}$ provided $\gamma \in [\underline{\gamma}, \bar{\gamma}]$. This generally is *not true* unless the problem is inverse-monotone and γ stands for data only. The subsequent discussion of this topic cannot employ boundary mapping since each admitted data g defines a separate operator A of the ivp to be discussed.

Example 4.1. The following collection of ivp is considered:

$$u'' + u = g(t) \quad \text{for } t \in J := (0, T],\ u(0) = u_0,\ u'(0) = u_1 \text{ with fixed}$$

$$u_0, u_1 \in \mathbb{R},\ g \in S_g := [\underline{g}, \bar{g}] \cap C(\bar{J}). \tag{4.1}$$

For any fixed forcing function $g \in S_g$, the solution of (4.1) is as follows:

$$u(t) = \sigma(t) + \int_0^1 G(t,s)g(s)\,ds$$

$$\text{where } G := (\sin t)(\cos s) - (\cos t)(\sin s) = \sin(t - s),$$

$$\sigma := u_0 \cos t + u_1 \sin t. \tag{4.2}$$

Analogous to (3.7), the upper envelope of the values of the set of solutions of (4.1) is given by

$$\bar{u}(t) = \sigma(t) + \int_0^t \operatorname*{Max}_{g \in S_g} (G(t,s)g(s))\,ds = \sigma(t) + \int_0^t G(t,s)g_t(s)\,ds \quad \text{for } t \in J,$$

$$g_t(s) := \begin{cases} \bar{g}(s) \text{ if } G(t,s) \geqslant 0 \text{ locally} \\ \underline{g}(s) \text{ if } G(t,s) < 0 \text{ locally} \end{cases} \begin{cases} \text{for } t \in J \text{ fixed and} \\ \text{every } s \in [0,t]. \end{cases} \tag{4.3}$$

The lower envelope $\underline{u}$ is determined correspondingly; g_t may be interpreted as a control function whose choice majorizes the interval $[\underline{u}, \bar{u}](t)$. For the special choice of $\delta := -\underline{g} = \bar{g} \in \mathbb{R}^+$, obviously, $\bar{u} = \sigma + S$ and $\underline{u} = \sigma - S$ where $S := \delta \int_0^t |G|\,ds$. This yields

$$\bar{u}(m\pi) = \sigma(m\pi) + 2[\delta m] \quad \text{for } t = m\pi \text{ with every } m \in \mathbb{N}. \tag{4.4}$$

For comparison, (4.1) is considered for the "resonance case" of $g = \delta \sin t$:

$$u(m\pi) = \sigma(m\pi) + (\pi/2)[\delta(-1)^{m+1}m] \quad \text{for } t = m\pi \text{ with every } m \in \mathbb{N}. \tag{4.5}$$

Therefore, (*i*) the operator Max in (4.3) causes the selection of a forcing function g with the "resonance frequency", 1, (*ii*) the interval $[\underline{u}, \bar{u}](t)$ is almost covered by solutions of physical relevance, (*iii*) the interval $[\underline{u}, \bar{u}](t)$ is determined by functions $g_t(s) \notin S_g$ such that each g_t can be approximated with arbitrary accuracy by a sequence of functions $g_t^{(\nu)}(s) \in S_g$ whose values are *not* given by those of $\underline{g}$ and $\bar{g}$ only. Even if $\delta \in \mathbb{R}^+$ is arbitrarily small, (4.1) is unstable for $T \to \infty$ and ill-conditioned for any sufficiently large $T \in \mathbb{R}^+$. The unbounded increase of $|\bar{u}|$ and $|\underline{u}|$ can be avoided if there is a sufficiently large damping constant $b \in \mathbb{R}^+$ in the altered equation $u'' + bu' + u = g$ for $t \in J$.

As $t \to \infty$, a corresponding unbounded increase of $\bar{u}(t) - \underline{u}(t)$ occurs in the case of

$$u'' + a(t)u = 0 \quad \text{for } t \in J := (0, T];\ u(0), u'(0) \in \mathbb{R} \text{ fixed,}$$

$$a \in S_a := [\underline{a}, \bar{a}] \cap C(\bar{J}). \tag{4.6}$$

This "parameter-resonance" is well known in mechanics, e.g., [7, p. 225].

5. Partial Differential Equations with Input Sets

The discussions of ordinary (linear) ivp apply immediately to (linear) parabolic or hyperbolic pde with the usual side conditions and input sets, provided approximate solutions are constructed by use of the longitudinal line method, e.g., [23]. Elliptic pde with the usual side conditions and input sets can be treated approximately by use of discrete analogies, compare Section 6 and Example 3.1.

6. On Truncated Taylor Expansions for the Approximate Solution of Problems with Sets of Constant Input Properties

The collection of operator equations $Au = f$ in (1.1) is considered. In many cases, an algorithm for the exact or the approximate solution of (1.1) employs or yields a function from one finite-dimensional space into some other such space; e.g., the solution of a discrete analogy is such a function which depends on constant input parameters. Due to the motivations listed in Sect. 1, these real numbers may take values in certain input sets. The range of such a function then is to be determined precisely or to be approximated with sufficient accuracy. In many cases, interval methods are non-existent for these purposes or they yield unacceptable over-estimates.

This will be achieved by use of truncated Taylor expansions of the unknown function $\hat{F}$ representing the solution in terms of every constant input parameter (and the independent variables); $\hat{F}$ is assumed to be sufficiently smooth. Because of Sect. 7, first an interval polynomial will be treated.

Example 6.1. The following collection of functions F: $\mathbb{C} \to \mathbb{C}$ is considered:

$$F(z; a, b) := z^2 + 2az + b = 0, \quad z \in \mathbb{C}, \quad a \in [0.9, 1.1], \quad b \in [1.9, 2.1]. \quad (6.1)$$

Here, only the root $z = -a + i\sqrt{b - a^2}$ will be discussed further. For the special choice of $a_0 = 1$ and $b_0 = 2$, the root $z_0 := z(a_0, b_0) = -1 + i$ is obtained *from the nonlinear equation* (6.1). By use of $z = x + iy$, $\sigma = 1$, $\tau = i$, the root $z(a, b)$ of $F = 0$ defines the functions $\check{F}(a, b; \sigma, \tau) := F(x(a, b) + iy(a, b); a, b)$ and $\hat{F}(a, b) := \check{F}(a, b; \sigma, \tau)$ where $\hat{F}: D \to \mathbb{R}$ with $D := [0.9, 1.1] \times [1.9, 2.1] \subset \mathbb{R}^2$ and $\hat{F}(a, b) \equiv 0$. By use of recombining real quantities into complex quantities, differentiation of the last identity yields the following *linear decoupled equations* for the partial derivatives of z with respect to a or b:

$$\frac{\partial \hat{F}}{\partial a} \equiv 0 \Rightarrow v := \frac{\partial z}{\partial a} = \frac{-z}{z + a} \Rightarrow v_0 = \frac{-z_0}{z_0 + a_0} = -1 - i,$$

$$\frac{\partial \hat{F}}{\partial b} \equiv 0 \Rightarrow w := \frac{\partial z}{\partial b} = \frac{-1}{2(z + a)} \Rightarrow w_0 = \frac{-1}{2(z_0 + a_0)} = \frac{i}{2},$$

$$\frac{\partial^2 \hat{F}}{\partial a^2} \equiv 0 \Rightarrow \alpha := \frac{\partial^2 z}{\partial a^2} = \frac{-2v - v^2}{z + a}, \cdots \quad (6.2)$$

The derivatives of z of any order exist if $z + a \neq 0$. A linear truncated Taylor expansion of $z(a, b)$ with respect to (a_0, b_0) is introduced:

$$\tilde{z} := z_0 + (a - a_0)v_0 + (b - b_0)w_0 = -a + i[1 - (a - a_0) + (b - b_0)/2]. \quad (6.3)$$

By use of the a priori set

$$\tilde{S}_z := [-1.2, -0.8] + i[0.8, 1.1] \qquad \text{with } z_0 \in \tilde{S}_z, \quad (6.4)$$

the following estimates are obtained: $|\partial^2 z/\partial a^2| \leqslant 10.3$, $|\partial^2 z/\partial a\, \partial b| \leqslant 2.36$, $|\partial^2 z/\partial b^2| \leqslant 0.49$, and $|R_2| \leqslant 0.1551/2$ where R_2 denotes the remainder term of the truncated expansion (6.3). The following table compares results for $\tilde{z}$ and the exact root $z = -a + i\sqrt{b - a^2}$:

Table 1

a	b	$\tilde{z}$	z	a	b	$\tilde{z}$	z
1.1	2.1	$-1.1 + 0.95i$	$-1.1 + 0.942i$	0.9	2.1	$-0.9 + 1.15i$	$-0.9 + 1.100i$
1.1	1.9	$-1.1 + 0.85i$	$-1.1 + 0.830i$	0.9	1.9	$-0.9 + 1.05i$	$-0.9 + 1.000i$

In each case, $|\tilde{z} - z| < 0.1551/2$. These results and the monotonicity of $\tilde{z}(a, b)$ confirm that $\tilde{z} \in \tilde{S}_z$ is valid for every admitted $(a, b) \in \mathbb{R}^2$.

Remarks. If the intervals for a and b admit double roots $z(a, b)$, (6.3) is still valid if (a_0, b_0) is selected so as to ensure that $z(a_0, b_0) + a_0 \neq 0$. Since $v, w, \alpha, \ldots$ are determined from decoupled (linear) equations, interval polynomials of any degree may be treated in this way. The books by Alefeld and Herzberger [3] and by Moore [16] on interval mathematics do not discuss the subject of interval polynomials.

The following example treats truncated Taylor expansions for a problem where an input set appears only due to embedding the operator equation into a collection of such equations.

Example 6.2. A nonlinear system in $\mathbb{R}^n$ is considered:

$F(u) = 0$ with $F\colon D \to \mathbb{R}^n$ and $D \subset \mathbb{R}^n$; F possesses a continuous Frechet derivative $DF(u)$ on D such that $(DF(u))^{-1}$ exists. (6.5)

This system is embedded in the following collection:

$$F(u) = b \quad \text{with } b \in S_b \subset \mathbb{R}^n. \tag{6.6}$$

It is assumed that this system possesses a real or complex root $u(b)$ for every $b \in S_b$. A function $\hat{F}(b) := F(u(b)) - b$ is defined such that $\hat{F}\colon S_b \to \mathbb{R}^n$ and $\hat{F}(b) \equiv 0$, which implies that $\partial \hat{F}_i/\partial b_j \equiv 0$ for $i, j = 1(1)n$. The last identities yield the matrix $(\partial u_i/\partial b_j) \in L(\mathbb{R}^n)$. Due to

$$(\partial F_i/\partial u_k)(\partial u_k/\partial b_j) = I \Rightarrow (\partial u_k/\partial b_j) = (DF(u))^{-1}, \tag{6.7}$$

where I is the identity matrix. By use of a superscript k to be explained subsequently, a linear truncated Taylor expansion of $u(b)$ at $b^{(k)} = F(u^{(k)})$ is introduced, which employs a free vector $\tilde{b}^{(k+1)} \in \mathbb{R}^n$:

$$u^{(k+1)} = u^{(k)} + (\partial u_i/\partial b_j)(\tilde{b}^{(k+1)} - b^{(k)}) \quad \text{with } k + 1 \in \mathbb{N} \text{ and } b^{(k)} = F(u^{(k)}). \tag{6.8}$$

If $\tilde{b}^{(k+1)} = 0$, this is the classical Newton method, whose divergence can be avoided if $d_i^{(k+1)} := |\tilde{b}_i^{(k+1)} - b_i^{(k)}|$ is sufficiently small for $i = 1(1)n$. A check on the choice of the individual $\tilde{b}_i^{(k+1)}$ is possible by use of the $u_j^{(k+1)}$, as following from (6.8), and $|\tilde{b}_m^{(k+1)} - b_m^{(k+1)}|$ for $j, m = 1(1)n$ where $b^{(k+1)} = F(u^{(k+1)})$. For the special choice of $n = 2$ and

$$F_1 := u_1^2 u_2^3 - 0.5 \qquad \text{and} \qquad F_2 := u_1 u_2 - 1.0, \tag{6.9}$$

the classical Newton method diverges if $u_i^{(0)} = 2$ for $i = 1$ and 2. The choice of

$$\tilde{b}_1^{(k+1)} := \begin{cases} b_1^{(k)} - 0.1 & \text{if } |b_1^{(k)}| \geqslant 0.2, \\ 0 & \text{otherwise,} \end{cases}$$

$$\tilde{b}_2^{(k+1)} := \begin{cases} b_2^{(k)} - 0.01 & \text{if } |b_2^{(k)}| \geqslant 0.1, \\ 0 & \text{otherwise,} \end{cases} \tag{6.10}$$

yielded a slow linear convergence (of 307 iteration cycles) and, subsequently, a rapid quadratic convergence of only 3 iteration cycles when $\tilde{b}_i^{(k+1)} = 0$ was reached for both $i = 1$ and $i = 2$. The solution, thus approximated, is $u_1 = 2$ and $u_2 = 0.5$.

Remark. Modified Newton methods (e.g., [22, p. 208]) usually employ a parameter $\lambda_k \in \mathbb{R}$ in

$$u^{(k+1)} = u^{(k)} - \lambda_k (DF(u^{(k)}))^{-1} F(u^{(k)}) \quad \text{with } k+1 \in \mathbb{N}. \tag{6.11}$$

Even though convergence can be enforced under rather general conditions, the determination of the λ_k is rather involved.

Next, $Au = f$ in (1.1) is assumed to represent a collection of systems of linear algebraic equations with $f \in \mathbb{R}^n$ fixed and a set of matrices $A \in L(\mathbb{R}^n)$; (1.1) then may be the discrete analogy of a differential equation with side conditions and a set of coefficients. The smallest interval enclosing the set of solutions, S_u, of $Au = f$ can be determined if every admitted matrix is an M-matrix, e.g., [5, p. 156]. This condition is not required in the following example, where ∂S_u is approximated for the case of $n = 2$.

Example 6.3. The collection of systems

$$Au - f = 0 \Leftrightarrow \begin{pmatrix} a_{11} & a_{12} \\ a_{21} & a_{22} \end{pmatrix} \begin{pmatrix} u_1 \\ u_2 \end{pmatrix} = \begin{pmatrix} f_1 \\ f_2 \end{pmatrix} \quad \text{with } a_{ij} \in [\underline{a}_{ij}, \bar{a}_{ij}] \text{ and } f \in \mathbb{R}^2 \text{ fixed} \tag{6.12}$$

is considered. A special matrix $A_0 = (a_{ij}^{(0)})$ is selected from the admitted set such that A_0^{-1} exists. The subsequent results are optimal with respect to the choice of A_0 if there holds

$$a_{ij}^{(0)} := \underline{a}_{ij} + \tfrac{1}{2}(\bar{a}_{ij} - \underline{a}_{ij}) \quad \text{for } i,j = 1 \text{ or } 2. \tag{6.13}$$

An auxiliary vector $\beta := (a_{11}, a_{12}, a_{21}, a_{22}) \in S_\beta := \prod_{i,j=1}^{2} [\underline{a}_{ij}, \bar{a}_{ij}] \subset \mathbb{R}^4$ is introduced. A function $F(u, A) := Au - f$ is defined. Since a sufficient condition for the existence of one and only one solution $u(\beta)$ is implied in the following, it may be assumed that a unique solution $u(\beta)$ exists for every $\beta \in S_\beta$. The function $\hat{F}(\beta) := F(u(\beta), A(\beta))$ has the properties $\hat{F}\colon S_\beta \to \mathbb{R}^2$ and $\hat{F}(\beta) \equiv 0$. The following notations are introduced:

$$v_k^{(i,j)}(\beta) := \frac{\partial u_k(\beta)}{\partial a_{ij}} \quad \text{and} \quad w_k^{(i,j:\sigma,\tau)}(\beta) := \frac{\partial^2 u_k(\beta)}{\partial a_{ij}\, \partial a_{\sigma\tau}}; k, i, j, \sigma, \tau = 1 \text{ or } 2. \tag{6.14}$$

By use of differentiating the identity $\hat{F}(\beta) \equiv 0$, four systems of linear algebraic equations with the identical matrix $A(\beta)$ are obtained for the four vectors $v^{(i,j)} \in \mathbb{R}^2$. In these systems, β is now replaced by $\beta_0 := (a_{11}^{(0)}, \ldots, a_{22}^{(0)}) \in S_\beta$. This defines the special solution $u(\beta_0) = A_0^{-1} f$ and the Taylor coefficients $v_0^{(i,j)} \in \mathbb{R}^2$ in the following linear truncated Taylor expansion:

$$\tilde{u}(\beta) = u(\beta_0) + \sum_{i,j=1}^{2} (a_{ij} - a_{ij}^{(0)}) v_0^{(i,j)} \quad \text{where } A_0 v_0^{(1,1)} = \begin{pmatrix} -u_1(\beta_0) \\ 0 \end{pmatrix}, \text{ etc.,} \tag{6.15}$$

i.e., the existence of A_0^{-1} implies the existence of the $v_0^{(i,j)}$. The validity of the following matrix representation is implied by the Neumann series for $A^{-1}(\beta)$:

$$\tilde{u}(\beta) = A_0^{-1}[f - B(\beta)u(\beta_0)] \qquad \text{where } A_0^{-1} = A^{-1}(\beta_0)$$
$$\text{and} \quad B(\beta) := ((a_{ij} - a_{ij}^{(0)})) \in L(\mathbb{R}^2). \tag{6.16}$$

Since $f - B(\beta)u(\beta_0) =: I(\beta) \subset \mathbb{R}^2$ is an interval, $A_0^{-1}I(\beta)$ is a parallelogram $P \subset \mathbb{R}^2$ such that (*i*) $\partial I \leftrightarrow \partial P$ due to (affine bijective) boundary mapping and (*ii*) ∂P is an approximation of ∂S_u where S_u is the set of solutions of (6.12). The theory of the Neumann series implies that the terms of the second order,

$$\Gamma_2 := \frac{1}{2} \sum_{i,j,\sigma,\tau=1}^{2} (a_{ij} - a_{ij}^{(0)})(a_{\sigma\tau} - a_{\sigma\tau}^{(0)}) w^{(i,j:\sigma,\tau)}(\beta), \tag{6.17}$$

in an extension of the expansion (6.15), admit the following matrix-representation:

$$\Gamma_2 = A^{-1}(\beta)B(\beta)A^{-1}(\beta)B(\beta)u(\beta). \tag{6.18}$$

The 10 vectors $w^{(i,j:\sigma,\tau)} \in \mathbb{R}^2$ in (6.17) are the solutions of 10 systems with the identical matrix $A(\beta)$. By use of (6.18), the remainder term, R_2, of the linear truncated expansion (6.15) can be estimated as follows by use of a suitably selected a priori set $\tilde{S}_u \subset \mathbb{R}^2$:

$$\|R_2\|_\infty \leqslant \operatorname*{Max}_{\beta \in S_\beta} \operatorname*{Max}_{u \in \tilde{S}_u} \|\Gamma_2\|_\infty \leqslant [\operatorname*{Max}_{\beta \in S_\beta} \|B(\beta)\|_\infty^2]\gamma^2[\operatorname*{Max}_{u \in \tilde{S}_u} \|u\|_\infty] =: \rho,$$
$$\text{where } \gamma := \frac{\|A_0^{-1}\|_\infty}{1 - \|A_0^{-1}\|_\infty \|(\bar{a}_{ij} - a_{ij}^{(0)})\|_\infty} \text{ if } \|A_0^{-1}\|_\infty \|(\bar{a}_{ij} - a_{ij}^{(0)})\|_\infty \overset{!}{<} 1. \tag{6.19}$$

Here, γ is due to a well-known estimate in Numerical Analysis, e.g., [6, p. 65]. If the condition denoted by ! is satisfied, then every admitted matrix is invertible, compare (3.6). For ∂P to be a sufficiently accurate approximation of the set of solutions, S_u, of (6.12), ρ as defined in (6.19) should be very much smaller than a characteristic geometric extension of P. The selected a priori set, $\tilde{S}_u$, is sufficiently large if $\partial \tilde{S}_u$ is an outer approximation of ∂P whose distance exceeds ρ. This construction has been applied to the following system:

$$A(\varepsilon)u - \mu I u = f \quad \text{with } \mu = 1.5, f \in S_f := [0.8, 1.2] \times [0.8, 1.2] \subset \mathbb{R}^2,$$
$$\varepsilon = (\varepsilon_1, \varepsilon_2, \varepsilon_3, \varepsilon_4) \in S_\varepsilon \subset \mathbb{R}^4,$$
$$A(\varepsilon) :\in \begin{pmatrix} [-3 - \varepsilon_1, -3 + \varepsilon_1] & [1 - \varepsilon_2, 1 + \varepsilon_2] \\ [-4 - \varepsilon_3, -4 + \varepsilon_3] & [2 - \varepsilon_4, 2 + \varepsilon_4] \end{pmatrix} \text{ with } |\varepsilon_i| \leqslant \bar{\varepsilon} \text{ for } i = 1(1)4. \tag{6.20}$$

The admitted matrices are not M-matrices; $A(0)$ possesses (*a*) the eigenvalues -2 and 1 and (*b*) the eigenvectors (1, 1) and (1, 4). The numerical results in Fig. 1 reveal that the set of solutions possesses roughly the direction of the eigenvector (1, 4) since $\mu = 1.5$ is close to the eigenvalue 1. J. Steckelberg (Karlsruhe) has computed the numerical results in Fig. 1 which hold for the two fixed choices of $\bar{\varepsilon} = 0.01$ and $\bar{\varepsilon} = 0.05$. The dashed parallelogram represents the set of solutions $A^{-1}(0)S_f$. The solid quadrilateral encloses $3^4 = 81$ parallelograms $A^{-1}(\varepsilon)S_f$ with 81 different fixed choices of $\varepsilon \in S_\varepsilon$ such that $|\varepsilon_i| \leqslant \bar{\varepsilon}$. Therefore, the solid boundary of this quadrilateral encloses the set of solutions, $S_u(\bar{\varepsilon})$, with a high degree of approximation. The double solid lines for the case of $\bar{\varepsilon} = 0.01$ are due to an application of the estimate denoted

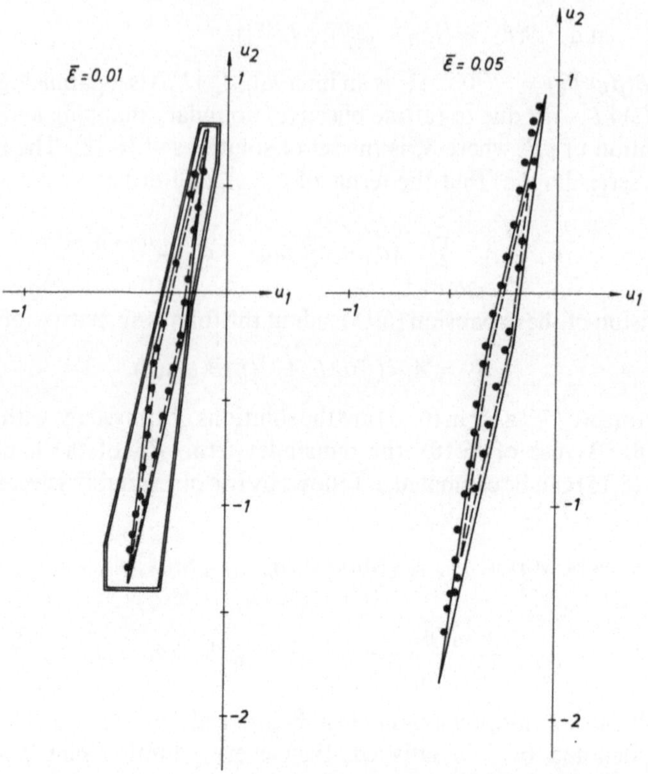

Fig. 1. The set of solutions for the collection (6.20) of systems of linear algebraic equations. The symbols are explained subsequent to (6.20)

by γ in (6.19) to selected fixed vectors $f \in S_f$. The linear truncated Taylor-expansion (6.15) was employed separately for (*i*) the two choices of $\bar{\varepsilon}$ and (*ii*) fixed choices of $f \in \partial S_f$ in a discretization of ∂S_f. The resulting parallelograms $P(\bar{\varepsilon}, f)$ are enclosed by the dots in Fig. 1 which define a quadrilateral $\partial S_T(\bar{\varepsilon})$. Due to (6.15), each $P(\bar{\varepsilon}, f)$ is determined by $A^{-1}(0)f$ and four vectors $v_0^{(i,j)} \in \mathbb{R}^2$ which are the solutions of systems with the identical matrix $A(0)$. Inspection of Fig. 1 reveals that $\partial S_T(\bar{\varepsilon})$ is a highly accurate approximation of the envelope $\partial S_u(\bar{\varepsilon})$ of the set of solutions for both $\bar{\varepsilon} = 0.01$ and $\bar{\varepsilon} = 0.05$. The number $\rho(\bar{\varepsilon})$ in (6.19) was determined by use of an a priori set $\tilde{S}_u(\bar{\varepsilon})$ with the following properties: (*i*) $S_T(\bar{\varepsilon}) \subset \tilde{S}_u(\bar{\varepsilon})$ and (*ii*) the distance of $\partial S_T(\bar{\varepsilon})$ and $\partial \tilde{S}_u(\bar{\varepsilon})$ exceeds a chosen number $d(\bar{\varepsilon})$:

$$\begin{aligned} \bar{\varepsilon} = 0.01\colon \quad & d(0.01) = 0.06 \quad \Rightarrow \rho(0.01) = 0.01619 \leqslant d(0.01), \\ \bar{\varepsilon} = 0.05\colon \quad & d(0.05) = 1.3266 \Rightarrow \rho(0.05) = 2.67 \quad \nleqslant d(0.05), \end{aligned} \tag{6.21}$$

i.e., the chosen a priori set $\tilde{S}_u(0.05)$ is too small. By construction, an expansion of each side of the dotted quadrilateral $\partial S_T(\bar{\varepsilon})$ by the distance $\rho(\bar{\varepsilon})$ yields an outer approximation of the set of solutions, $S_u(\bar{\varepsilon})$. For $\bar{\varepsilon} = 0.05$, the estimate $\rho \geqslant \|R_2\|_\infty$ exceeds considerably the distance between the solid quadrilateral and the dotted $\partial S_T(\bar{\varepsilon})$. Without a knowledge of the solid quadrilateral, a truncated Taylor expansion of the second order should be employed, with a remainder term of the third order.

Remarks. (*1*) Even though $S_T(\bar{\varepsilon})$ is convex in Fig. 1, this is not necessarily true in other examples. (*2*) If the intervals $[\underline{a}_{ij}, \bar{a}_{ij}]$ are not independent of each other, the truncated Taylor expansion is executed in terms of only the independent intervals; as opposed to interval mathematics, there is no overestimate due to this restriction. (*3*) A collection of systems $Au = f$ with $f \in \mathbb{R}^n$ fixed, $A = (a_{ij}) \in L(\mathbb{R}^n)$, and $a_{ij} \in [\underline{a}_{ij}, \bar{a}_{ij}]$ is considered. The linear truncated Taylor expansion then requires the computation of n^2 vectors $v^{(i,j)} \in \mathbb{R}^n$ which are the solutions of n^2 systems with the *same matrix* A_0. If A_0^{-1} is known, the numerical work in evaluating the estimate (6.19) of the remainder term increases only slightly with n. This remainder term yields a small number $\rho \in \mathbb{R}^+$ if the span of every interval $[\underline{a}_{ij}, \bar{a}_{ij}]$ is sufficiently small, compare Sect. 1. (*4*) The system $Au = f$ is again considered for the case of $f \in \mathbb{R}^n$ fixed and matrices $(a_{ij}) \in L(\mathbb{R}^n)$ with n^2 intervals $[\underline{a}_{ij}, \bar{a}_{ij}]$. The resulting interval $I := \prod_{i,j=1}^n [\underline{a}_{ij}, \bar{a}_{ij}] \subset R^{n^2}$ has 2^{n^2} corners. The boundary ∂S_u of the set of solutions then can be approximated by solving $Au = f$ for the choice of 2^{n^2} *matrices* with elements in the corners of the interval I, as follows from a theorem which has been derived independently (*a*) by use of interval mathematics in [4] and (*b*) by use of elementary methods in [17] and in [15]. The computational work here is prohibitive in comparison to the employment of only one matrix A_0 in applications of linear truncated Taylor expansions to this problem. (*5*) By use of the Neumann series for A^{-1}, (6.12) can be generalized to the case of $A \in L(\mathbb{R}^n)$ with any $n \in \mathbb{N}$.

7. Joint Application of the Laplace Transform and Truncated Taylor Expansions

According to [10, p. 23], the starting of an electric direct current motor can be described by the mathematical model

$$\Omega^{-2}y'' + 2d\Omega^{-1}y' + y = ku \quad \text{for } t \in J := (0, T];\ y(0) = y_0,\ y'(0) = y_1;$$

$$u := \begin{cases} 0 & \text{for } t < 0, \\ 1 & \text{for } t \geqslant 0, \end{cases} \tag{7.1}$$

with the five constant parameters

$$\text{the data } k \in \mathbb{R}^+ \text{ and } y_0, y_1 \in \mathbb{R}, \text{ and the coefficients } d, \Omega \in \mathbb{R}^+. \tag{7.2}$$

For the purpose of a quantitative sensitivity analysis, intervals are admitted:

$$[\underline{k}, \bar{k}], [\underline{y}_i, \bar{y}_i] \text{ for } i = 1 \text{ or } 2,\ [\underline{d}, \bar{d}], \text{ and } [\underline{\Omega}, \bar{\Omega}] \text{ with } \underline{k}, \underline{d}, \underline{\Omega} > 0,\ \bar{d} < 1. \tag{7.3}$$

The condition $\bar{d} < 1$ restricts the analysis to oscillatory solutions. An auxiliary vector is introduced: $\Gamma = (k, y_0, y_1, d, \Omega) \in S_\Gamma \subset \mathbb{R}^5$, where S_Γ is the interval in $\mathbb{R}^5$ following from (7.3). By treating the parameters (7.2) as if they were fixed, an application of the classical Laplace transform,

$$Y(s) = \int_0^\infty y(t)e^{-st}\,dt \qquad \text{and} \qquad U(s) = 1/s, \tag{7.4}$$

yields the following representation of (7.1), (7.2):

$$Y(s)[\Omega^{-2}s^2 + 2d\Omega^{-1}s + 1] = kU(s) + \Omega^{-2}[y_0 s + y_1] + 2d\Omega^{-1}y_0. \quad (7.5)$$

The expression in brackets is an interval polynomial; its roots s_1 and s_2 can be enclosed in complex intervals by use of a (first) application of linear truncated Taylor expansions, according to Example 6.1 (of course, the explicit representation of the roots is preferable here). Due to $\bar{d} < 1, s_1 \neq s_2$. In the following, s_1 and s_2 will be treated as if they were fixed. By use of standard tables, the execution of the inverse Laplace transform yields an explicit representation of the solution y which can be written as follows in terms of the data k, y_0, y_1 and the coefficients d, Ω:

$$y(t,\Gamma) = kA_1(d,\Omega) + y_0 A_2(d,\Omega) + y_1 A_3(d,\Omega). \quad (7.6)$$

A second application of linear truncated Taylor expansions yields the approximately valid representation

$$\tilde{y}(t,\Gamma) = \sum_{i=1}^{3} \gamma_i \left[A_i(\gamma_0) + \frac{\partial A_i(\Gamma_0)}{\partial d}(d - d_0) + \frac{\partial A_i(\Gamma_0)}{\partial \Omega}(\Omega - \Omega_0) \right],$$

$$\text{where } \gamma_1 := k,\ \gamma_2 := y_0,\ \gamma_3 := y_1, \text{ and } \Gamma_0 \in S_\Gamma \text{ is fixed,} \quad (7.7)$$

which depends explicitly on the five input parameters (7.2). It is therefore immediately possible to replace these parameters by the respective input intervals in (7.3). A resulting interval for $\tilde{y}(t,\Gamma)$ is obtained easily since (7.7) is a bilinear expression.

Remark. It is generally advantageous to replace constant parameters by sets only at the latest possible stage in an analysis.

8. Conclusions Concerning Operator Equations $Au = f$ with Input Sets S_a and S_f

I. Data Sets S_f Are Admitted, the Operator A Is Fixed

1. Interval mathematics, determining bounds $\underline{u}, \bar{u}$ by use of $\underline{f}, \bar{f}$ only, is not applicable for a construction of the smallest interval $[\underline{u}, \bar{u}]$ containing the values of the set of solutions, unless A is inverse-monotone.

2. The boundary mapping $S_f \to S_u$ is executed by use of individual solutions; i.e., standard error estimates are applicable in a numerical execution, and there is no overestimate if these solutions are determined exactly.

3. The local procedural (or rounding) error may be bracketed by use of standard estimates, a priori sets, and interval mathematics.

4. A combination of 2. and 3. yields a global outer approximation of the range of the set of solutions and, thus, a global error estimate.

II. Sets of Data and Coefficients Are Admitted

1. The auxiliary operator $\hat{B}$, which is the basis of global error estimates, is inverse-monotone under mild conditions.

2. Truncated Taylor expansions yield the range of F: $\mathbb{R}^n \rightarrow \mathbb{R}^m$ where F is either defined by an algorithm or given. These approximations are in particular useful with respect to accuracy or computational work if (*i*) $m, n \in \mathbb{N}$ are large and the "spans" of the input sets are small, (*ii*) if the input sets are not independent of each other.

3. In Example 2.2 and Example 6.2, algorithms have been derived by embedding a given equation $Au = f$ without input sets in a collection of such equations.

References

[1] Adams, E., Spreuer, H.: Uniqueness and stability for boundary value problems with weakly coupled systems of nonlinear integro-differential equations and application to chemical reactions. JMAA **49**, 393–410 (1975).

[2] Adams, E., Spreuer, H.: On the construction of an interval containing the set of solutions of non-inverse isotone linear problems with intervals admitted for both data and coefficients. Report **CAM 14** (1978). (Center for Appl. Math., The Univ. of Georgia, Athens, Georgia, U.S.A.)

[3] Alefeld, G. Herzberger, J.: Einführung in die Intervallrechnung. Mannheim-Wien-Zürich: Bibliographisches Institut 1974.

[4] Beeck, H.: Über Struktur und Abschätzungen der Lösungsmenge von linearen Gleichungssystemen mit Intervallkoeffizienten. Comp. **10**, 231–244 (1972).

[5] Beeck, H.: Zur Problematik der Hüllenbestimmung von Intervallgleichungssystemen, in: Lecture Notes in Computer Science, Vol. 29. Berlin-Heidelberg-New York: Springer 1975.

[6] Blum, K. E.: Numerical Analysis and Computation Theory and Practice. Reading, Mass.-Menlo Park, Calif.-London-Don Mills, Ont.: Addison-Wesley 1972.

[7] Bogoljubow, N. N., Mitropolski, J. A.: Asymptotische Methoden in der Theorie der nichtlinearen Schwingungen. Berlin: Akademie-Verlag 1965. (Translation from the Russian.)

[8] Deimling, K.: Nichtlineare Gleichungen und Abbildungsgrade. Berlin-Heidelberg-New York: Springer 1974.

[9] Fichtenholz, G. M.: Differential- und Integralrechnung, Vol. II, 6. Aufl. Berlin: VEB Deutscher Verlag der Wissenschaften 1974. (Translation from the Russian.)

[10] Föllinger, O.: Laplace- und Fourier-Transformation. Berlin: Elitera-Verlag 1977.

[11] Kamke, E.: Differentialgleichungen, Vol. I, 6. Aufl. Leipzig: Akad. Verlagsgesellschaft Geest & Portig K.G. 1969.

[12] Kasriel, R. H.: Undergraduate Topology. Philadelphia-London-Toronto: W. B. Saunders Co. 1971.

[13] Lakshmikantham, V., Leela, S.: Differential and Integral Inequalities, Theory and Applications, Vol. I and II. New York-London: Academic Press 1969.

[14] Lohner, R., Adams, E.: On initial value problems in R^N with intervals for both initial data and a parameter in the differential equation. Report **CAM 8** (1978). (Center for Appl. Math., The Univ. of Georgia, Athens, Georgia, U.S.A.)

[15] Lohner, R.: Anfangswertaufgaben im $\mathbb{R}^n$ mit kompakten Mengen für Anfangswerte und Parameter. Diplomarbeit, Karlsruhe, 1978.

[16] Moore, R. E.: Interval Analysis. Englewood Cliffs, N. J.: Prentice-Hall 1966.

[17] Nickel, K.: Die Überschätzung des Wertebereichs einer Funktion in der Intervallrechnung mit Anwendung auf lineare Gleichungssysteme. Comp. **18**, 15–36 (1977).

[18] Nickel, K.: Schranken für die Lösungsmenge von Funktional-Differentialgleichungen. Freiburger Intervall-Berichte **79/4** (1979).

[19] Ortega, J. M., Rheinboldt, W. C.: Iterative Solution of Nonlinear Equations in Several Variables. New York-San Francisco-London: Academic Press 1970.

[20] Spreuer, H.: Konvergente numerische Schranken für partielle Randwertaufgaben von monotoner Art, in: Lecture Notes in Computer Science, Vol. 29. Berlin-Heidelberg-New York: Springer 1975.

[21] Stetter, H. J.: Analysis of Discretization Methods for Ordinary Differential Equations. Berlin-Heidelberg-New York: Springer 1973.

[22] Stoer, J.: Einführung in die Numerische Mathematik, Vol. I. Berlin-Heidelberg-New York: Springer 1972.
[23] Walter, W.: Differential and Integral Inequalities. Berlin-Heidelberg-New York: Springer 1970.

Prof. Dr.-Ing. E. Adams
Institut für Angewandte Mathematik
Universität Karlsruhe
Kaiserstrasse 12
D-7500 Karlsruhe
Federal Republic of Germany

Computing, Suppl. 2, 17–31 (1980)

Roundings and Approximations in Ordered Sets

R. Albrecht, Innsbruck

Abstract

The topological aspect of rounding was introduced by the author in an earlier article [1] and applied to the theory of a rounded algebraic law of composition. For example common rounding in $\mathbb{R}$ as well as interval arithmetic on $\mathbb{R}$ are covered by this general theory. In the present paper the development of the topological theory is continued by introducing the notions of adherence, co-adherence, and limit of a rounding, by a uniform topologization of the space of filter- and ideal-basis on an ordered set with a rounding, by giving a general contraction principle, by a comparison of two roundings on the same ordered set, and by extending the theory of consistency and stability of approximations to mapping problems.

1. Roundings and Non-Classical Topologies

Considered are non-empty sets $V, X, Y, \ldots$ on each of which a (partial) order relation $\leqslant_V, \leqslant_X, \leqslant_Y, \ldots$ is defined. If no confusion is possible we suppress the subscripts. Without loss of generality it is assumed that each of the ordered sets $(V, \leqslant_V), \ldots$ has a least element and a greatest element. On any subset of an ordered set the induced ordering is introduced.

The relations $\in$ and $\subseteq$ are used for sets as well as for families of elements without changing the notations. For any pair (U, W) of subsets $U, W \subseteq V$ respectively of families $U := (u_i)_{i \in I}$, $W := (w_j)_{j \in J}$ on V we define the following ordering:

$$U \leqslant W :\Leftrightarrow \bigwedge_{u \in U} \bigwedge_{w \in W} u \leqslant w.$$

If $f\colon X \to Y$ and $U \subseteq X$, respectively $U := (u_i)_{i \in I}$, $\bigwedge_{i \in I} u_i \in X$, then $f(U)$ or fU means $\{f(u) | u \in U\}$, respectively $(f(u_i))_{i \in I}$.

$\sqcup$ respectively $\sqcap$ means the lattice theoretical join respectively meet.

Dual cases in definitions, theorems and proofs are labelled by (a) and (b).

Definition 1.1. Given $(V, \leqslant)$ and $T\colon V \to V$ with

(1) $\bigwedge_{u, v \in V} u \leqslant v \Rightarrow Tu \leqslant Tv,$

(2) $\bigwedge_{v \in V} TTv = Tv.$

Then T is a "one-stage rounding" on $(V, \leqslant)$ [1], [3]. If in addition holds

(3) $\bigwedge_{v \in V} v \leqslant Tv \vee Tv \leqslant v$

then T is named a "one-stage, directed rounding" or a "one-stage, two-sided, non-classical topology" on $(V, \leqslant)$. $(V, \leqslant, T)$ denotes the set V together with the ordering $\leqslant$ and the mapping T.

Remarks. 1. If in the previous definition (2) is generalized to $\bigwedge_{v \in V} T^{m+1}v \leqslant Tv$, $m \in \mathbb{N}$, then T is an m-stage rounding.

2. If T equals the identity mapping, then T is also called a "discrete" rounding Δ.

For convenience sake we use in the following the word "rounding" for a one-stage rounding and the word "topology" for a one-stage, two-sided, non-classical topology.

Definition 1.2. Let T be a topology on $(V, \leqslant)$.

(a) $V_{T\uparrow} := \{v \mid v \in V \wedge v \leqslant Tv\}$,

$T\uparrow := T \mid V_{T\uparrow}$ (i.e. T restricted to $V_{T\uparrow}$),

(b) $V_{T\downarrow} := \{v \mid v \in V \wedge Tv \leqslant v\}$,

$T\downarrow := T \mid V_{T\downarrow}$.

Theorem 1.1.
1. $V = V_{T\uparrow} \cup V_{T\downarrow}$,
2. $TV = V_{T\uparrow} \cap V_{T\downarrow} \neq \emptyset$.

The proof follows easily from Definition 1.1. Further consequences of this definition are:

1. T is a $\leqslant$-endomorphism of V and $\bigwedge_{u,v \in V} u \underset{T}{\sim} v :\Leftrightarrow Tu = Tv$ is an equivalence-relation on $(V, \leqslant, T)$.

2. T is isotone on $V\uparrow$, $T\downarrow$ is antitone on $V\downarrow$.

3. If T is a topology in our sense, then

(a) $T\uparrow$ on $(V_{T\uparrow}, \leqslant)$ and

(b) $T\downarrow$ on $(V_{T\downarrow}, \overset{-1}{\leqslant})$

are one-stage, non-classical topologies in the usual sense [4].

Examples. 1. Let $(\mathbb{R}, \leqslant)$ be the set of reals with its natural ordering, and let $\mathbb{Z}$ be the set of integers. Then $T\colon \mathbb{R} \to \mathbb{Z}$ defined by $\bigwedge_{x \in \mathbb{R}} Tx = \text{entier}\,(x + 0.5)$, the usual rounding of $\mathbb{R}$ into $\mathbb{Z}$, is a topology.

2. Let $(L, \leqslant)$, $L \neq \emptyset$, be a complete lattice. $(\mathfrak{P}(L) \setminus \{\emptyset\}, \subseteq)$ is an ordered set with the set of intervals $\mathbb{I}(L) := \{\langle a, b \rangle \mid a, b \in L \wedge a \leqslant b\}$ as a subset. Then $T\colon \mathfrak{P}(L) \setminus \{\emptyset\} \to \mathbb{I}(L)$, defined by $\bigwedge_{X \in \mathfrak{P}(L) \setminus \{\emptyset\}} TX = \langle \inf X, \sup X \rangle$, is a topology with $T = T\uparrow$. A special case is $L := \mathbb{R} \cup \{-\infty, \infty\}$ with $\{-\infty\} \leqslant \mathbb{R} \leqslant \{\infty\}$.

In the following we consider (a) filter-bases (F-bases), respectively (b) ideal-bases (I-bases) on $(V, \leqslant)$, $(X, \leqslant)$, $(Y, \leqslant)$. We write them as families, e.g. $\mathfrak{R} = (v_i)_{i \in I}$, $I \neq \emptyset$, $\bigwedge_{i \in I} v_i \in V$, and we use the name "base" as an abbreviation for either an F-base or an

I-base. Let $\mathfrak{R}$ be a base. Then in case (a) $\mathfrak{F}(\mathfrak{R})$ denotes the filter generated by $\mathfrak{R}$, in case (b) $\mathfrak{I}(\mathfrak{R})$ denotes the ideal generated by $\mathfrak{R}$. As a notation for both, either $\mathfrak{F}(\mathfrak{R})$ or $\mathfrak{I}(\mathfrak{R})$, we use $\mathfrak{F}\mathfrak{I}(\mathfrak{R})$.

Theorem 1.2. *Let T be a rounding and let $\mathfrak{R}$ be a base on $(V, \leqslant)$. Then*

1. *$T\mathfrak{R}$ is a base on $(TV, \leqslant)$.*
2. *If T is a topology and*

   (*a*) *$\mathfrak{R}$ is an F-base on $(V_{T\uparrow}, \leqslant)$,*
   (*b*) *$\mathfrak{R}$ is an I-base on $(V_{T\downarrow}, \leqslant)$,*

   then $\mathfrak{R}$ is finer than $T\mathfrak{R}$.
3. *$T\mathfrak{R} \subseteq T\mathfrak{F}\mathfrak{I}(\mathfrak{R}) \subseteq T\mathfrak{F}\mathfrak{I}(T\mathfrak{R}) \subseteq \mathfrak{F}\mathfrak{I}(T\mathfrak{R})$. If $\mathfrak{F}\mathfrak{I}(\mathfrak{R})$ is finer than $\mathfrak{F}\mathfrak{I}(T\mathfrak{R})$, then $T\mathfrak{F}\mathfrak{I}(\mathfrak{R}) = T\mathfrak{F}\mathfrak{I}(T\mathfrak{R})$.*

Proof. 1. A homomorphic image of a base is a base again.

2. (a) $\bigwedge_{r\in\mathfrak{R}} r \leqslant Tr$, (b) $\bigwedge_{r\in\mathfrak{R}} Tr \leqslant r$.

3. $\mathfrak{R} \subseteq \mathfrak{F}\mathfrak{I}(\mathfrak{R}) \Rightarrow T\mathfrak{R} \subseteq T\mathfrak{F}\mathfrak{I}(\mathfrak{R})$,

(a) $\bigwedge_{x\in T\mathfrak{F}(\mathfrak{R})} \bigvee_{y\in V} (x = Tx = Ty \wedge \bigvee_{r\in\mathfrak{R}} (r \leqslant y \wedge Tr \leqslant x)) \Rightarrow x \in T\mathfrak{F}(T\mathfrak{R})$,

$\bigwedge_{x\in\mathfrak{F}(T\mathfrak{R})} \left(\bigvee_{r\in\mathfrak{R}} Tr \leqslant x \Rightarrow Tr \leqslant Tx\right) \Rightarrow T\mathfrak{F}(T\mathfrak{R}) \subseteq \mathfrak{F}(T\mathfrak{R})$,

(b) dual,

if $\mathfrak{F}\mathfrak{I}(\mathfrak{R})$ is finer than $\mathfrak{F}\mathfrak{I}(T\mathfrak{R})$, then $\mathfrak{F}\mathfrak{I}(T\mathfrak{R}) \subseteq \mathfrak{F}\mathfrak{I}(\mathfrak{R})$ and $T\mathfrak{F}\mathfrak{I}(T\mathfrak{R}) \subseteq T\mathfrak{F}\mathfrak{I}(\mathfrak{R})$. Thus $T\mathfrak{F}\mathfrak{I}(T\mathfrak{R}) = T\mathfrak{F}\mathfrak{I}(\mathfrak{R})$.

Definition 1.3. Let $\mathfrak{R}$ be (a) an F- resp. (b) an I-base on $(V, \leqslant)$.

1. $\bigwedge_{a\in V}$ "a is in $(V, \leqslant)$ (a) F-, resp. (b) I-adherent to $\mathfrak{R}$" $:\Leftrightarrow$
   $\bigwedge_{r\in\mathfrak{R}}$ (a) $a \leqslant r$, resp. (b) $r \leqslant a$,

   "F-, resp. I-adherence" of $\mathfrak{R}$ in $(V, \leqslant) :\Leftrightarrow$
   F-, resp. I-$\mathrm{adh}_{(V,\leqslant)}\mathfrak{R} := \{a | a \in V \wedge a$ in $(V, \leqslant)$
   (a) F-, resp. (b) I-adherent to $\mathfrak{R}\}$,
2. $\bigwedge_{c\in V}$ "c is in $(V, \leqslant)$ (a) F-, resp. (b) I-co-adherent to $\mathfrak{R}$" $:\Leftrightarrow$

   (a) F-$\mathrm{adh}_{(V,\leqslant)}\mathfrak{R} \leqslant \{c\}$, resp.

   (b) $\{c\} \leqslant I$-$\mathrm{adh}_{(V,\leqslant)}\mathfrak{R}$,
   "F-, resp. I-co-adherence" of $\mathfrak{R}$ in $(V, \leqslant) :=$
   F-, resp. I-co-$\mathrm{adh}_{(V,\leqslant)}\mathfrak{R} :=$
   $\{c | c \in V \wedge c$ in $(V, \leqslant)$ (a) F-, resp. (b) I-co-adherent to $\mathfrak{R}\}$,
3. for either being all F- or all I-adherences and co-adherences:

$$\mathrm{adh}_{(V,\leqslant)}\,\mathrm{adh}_{(V,\leqslant)}\mathfrak{R} := \text{co-adh}_{(V,\leqslant)}\mathfrak{R},$$
$$\text{co-adh}_{(V,\leqslant)}\,\mathrm{adh}_{(V,\leqslant)}\mathfrak{R} := \mathrm{adh}_{(V,\leqslant)}\mathfrak{R},$$
$$\mathrm{adh}_{(V,\leqslant)}\,\text{co-adh}_{(V,\leqslant)}\mathfrak{R} := \mathrm{adh}_{(V,\leqslant)}\mathfrak{R},$$
$$\text{co-adh}_{(V,\leqslant)}\,\text{co-adh}_{(V,\leqslant)}\mathfrak{R} := \text{co-adh}_{(V,\leqslant)}\mathfrak{R}.$$

If $\mathfrak{R}$ is considered to be an F- respectively an I-base, and if the range of $\mathfrak{R}$ is clear, we omit the prefix F respectively I and the subscript $(V, \leqslant)$.

Definition 1.4. Let $\mathfrak{B}$ be the set of all bases on $(V, \leqslant)$.

$$\bigwedge_{\mathfrak{R}_1, \mathfrak{R}_2 \in \mathfrak{B}} \text{"}\mathfrak{R}_1 \text{ is adherence-equivalent to } \mathfrak{R}_2\text{"} :\Leftrightarrow$$
$$\text{"}\mathfrak{R}_1 \text{ and } \mathfrak{R}_2 \text{ are stable to each other"} :\Leftrightarrow$$

$$\mathfrak{R}_1 \underset{\mathrm{adh}}{\sim} \mathfrak{R}_2 :\Leftrightarrow \begin{cases} \mathrm{adh}\,\mathfrak{R}_1 = \mathrm{adh}\,\mathfrak{R}_2 \text{ if } \mathfrak{R}_1 \text{ and } \mathfrak{R}_2 \text{ are both } F\text{-bases,} \\ \qquad \text{or both } I\text{-bases,} \\ \mathrm{adh}\,\mathfrak{R}_1 = \text{co-adh}\,\mathfrak{R}_2 \text{ if } \mathfrak{R}_1 \text{ is an } F\text{- and } \mathfrak{R}_2 \text{ is an } I\text{-base,} \\ \qquad \text{or vice versa.} \end{cases}$$

Theorem 1.3. *Stability or adherence-equivalence is an equivalence relation on* $\mathfrak{B}$.

The proof is evident.

Theorem 1.4. *Given* $(V, \leqslant, T)$ *and* $\mathfrak{R} \in \mathfrak{B}$.

1. $T\mathrm{adh}\,\mathfrak{R} \subseteq T\mathrm{adh}\,T\mathfrak{R} = \mathrm{adh}\,T\mathfrak{R} \cap TV = \mathrm{adh}_{(TV,\leqslant)}T\mathfrak{R} \subseteq \mathrm{adh}\,T\mathfrak{R}$.

2. $\mathfrak{R}' \in \mathfrak{B} \wedge \mathfrak{R}' \underset{\mathrm{adh}}{\sim} \mathfrak{R} \Rightarrow$

(*i*) $\mathfrak{R}' \subseteq \text{co-adh}\,\mathfrak{R} \wedge T\mathfrak{R}' \subseteq \text{co-adh}_{(TV,\leqslant)}T\mathfrak{R}$
if $\mathfrak{R}'$, $\mathfrak{R}$ *are two F-bases or two I-bases,*

(*ii*) $\mathfrak{R}' \subseteq \mathrm{adh}\,\mathfrak{R} \wedge T\mathfrak{R}' \subseteq \mathrm{adh}_{(TV,\leqslant)}T\mathfrak{R}$
if $\mathfrak{R}'$ *is an I-base and* $\mathfrak{R}$ *is an F-base.*

3. $\mathrm{adh}\,\mathfrak{R} = \mathrm{adh}\,\mathfrak{F}\mathfrak{I}(\mathfrak{R})$.

4. $\mathfrak{R}' \in \mathfrak{B} \wedge (\mathfrak{R}$ *finer than* $\mathfrak{R}') \Rightarrow \mathrm{adh}\,\mathfrak{R} \subseteq \mathrm{adh}\,\mathfrak{R}'$.

Proof.

1. (a) $\mathrm{adh}\,\mathfrak{R} \leqslant \mathfrak{R} \Rightarrow T\mathrm{adh}\,\mathfrak{R} \Rightarrow T\mathfrak{R} \Rightarrow T\mathrm{adh}\,\mathfrak{R} \subseteq \mathrm{adh}\,T\mathfrak{R}$,
similarly $T\mathrm{adh}\,T\mathfrak{R} \subseteq \mathrm{adh}\,T\mathfrak{R}$, thus $T\mathrm{adh}\,\mathfrak{R} \subseteq T\mathrm{adh}\,T\mathfrak{R} \subseteq \mathrm{adh}\,T\mathfrak{R}$.
Obviously
$T\mathrm{adh}\,T\mathfrak{R} = \mathrm{adh}\,T\mathfrak{R} \cap TV = \mathrm{adh}_{(TV,\leqslant)}T\mathfrak{R}$.

(b) dual.

2. Follows from Definitions 1.3 and 1.4.

3. $\mathfrak{R} \subseteq \mathfrak{F}\mathfrak{I}(\mathfrak{R}) \wedge \bigwedge_{x \in \mathfrak{F}\mathfrak{I}(\mathfrak{R})} \bigvee_{r \in \mathfrak{R}}$

(a) $r \leqslant x$, resp. (b) $x \leqslant r \Rightarrow \mathrm{adh}\,\mathfrak{R} = \mathrm{adh}\,\mathfrak{F}\mathfrak{I}(\mathfrak{R})$.

4. Follows from Definition 1.3 and from the definition of what it means that $\mathfrak{R}$ is finer than $\mathfrak{R}'$.

Definition 1.5. Given $(V, \leqslant, T)$ and $\mathfrak{R} \in \mathfrak{B}$. (a) $\sqcap T\mathfrak{R}$, resp. (b) $\sqcup T\mathfrak{R}$ are assumed to exist in $(V, \leqslant)$. Then

$$\text{“}\mathfrak{R} \text{ converges to } x \text{ in } (V, \leqslant) \text{ with respect to } T\text{”} :\Leftrightarrow \mathfrak{R} \underset{T}{\rightarrow} x :\Leftrightarrow$$
$$x = T\text{-}\lim_{(TV, \leqslant)} \mathfrak{R} :\Leftrightarrow$$
$$\text{(a) } x = \sqcap T\mathfrak{R}, \quad \text{resp.} \quad \text{(b) } x = \sqcup T\mathfrak{R}.$$

In particular: If $T = \Delta$, then

$$T\mathfrak{R} = \mathfrak{R}, \text{ and } \mathfrak{R} \rightarrow x :\Leftrightarrow$$
$$\mathfrak{R} \underset{\Delta}{\rightarrow} x, \lim_{(V, \leqslant)} \mathfrak{R} := \Delta - \lim_{(V, \leqslant)} \mathfrak{R}.$$

If it is clear that the limits are taken in $(V, \leqslant)$ we omit the subscript $(V, \leqslant)$.

If $\mathfrak{R}$ is any convergent base in $(V, \leqslant, T)$, then in case of rounded computations we assign to it the base $T\mathfrak{R}$ and the questions arising are: How are $\lim \mathfrak{R}$, $T \lim \mathfrak{R}$, $\lim T\mathfrak{R}$ (if existing), and $T \lim T\mathfrak{R}$ related to each other, and under which assumptions may $\lim T\mathfrak{R}$ be taken in $(TV, \leqslant)$ instead of $(V, \leqslant)$? An answer to these questions is given by

Theorem 1.5.

1. $\mathfrak{R} \underset{T}{\rightarrow} x \Rightarrow$

(*a*) $x = \sqcap T\mathfrak{R} = \sqcup \operatorname{adh} T\mathfrak{R} = \sqcap \operatorname{co-adh} T\mathfrak{R}$,

(*b*) $x = \sqcup T\mathfrak{R} = \sqcap \operatorname{adh} T\mathfrak{R} = \sqcup \operatorname{co-adh} T\mathfrak{R}$.

2. $\mathfrak{R} \underset{T}{\rightarrow} x \Leftrightarrow \mathfrak{F}\mathfrak{I}(T\mathfrak{R}) \rightarrow x \Leftrightarrow T\mathfrak{R} \rightarrow x$.

3. $\mathfrak{R} \underset{T}{\rightarrow} x \Rightarrow$

(*i*) $\{x\} = \operatorname{adh} T\mathfrak{R} \cap \operatorname{co-adh} T\mathfrak{R}$;

(*ii*) $\operatorname{adh} T\mathfrak{R}$ *is* (*a*) *an ideal, resp.* (*b*) *a filter*;
co-adh T *is* (*a*) *a filter, resp.* (*b*) *an ideal*;

(*iii*) $\operatorname{adh} T\mathfrak{R}$ *is the union of*
(*a*) *all I-bases* $\mathfrak{S} \in \mathfrak{B}$, *resp.*
(*b*) *all F-bases* $\mathfrak{S} \in \mathfrak{B}$ *with* $\mathfrak{S} \underset{\text{adh}}{\sim} T\mathfrak{R}$,

co-adh $T\mathfrak{R}$ *is the union of*
(*a*) *all F-bases* $\mathfrak{S} \in \mathfrak{B}$, *resp.*
(*b*) *all I-bases* $\mathfrak{S} \in \mathfrak{B}$ *with* $\mathfrak{S} \underset{\text{adh}}{\sim} T\mathfrak{R}$.

4. If the following meets and joins exist, then

(*a*) $T \lim \mathfrak{R} = \sqcup T \operatorname{adh} \mathfrak{R} \leqslant \sqcup T \operatorname{adh} T\mathfrak{R} =$
$T \sqcup \operatorname{adh} T\mathfrak{R} = T \lim T\mathfrak{R} \leqslant \lim T\mathfrak{R}$,

(*b*) $T\lim \mathfrak{R} = \sqcap T\operatorname{adh} \mathfrak{R} \geqslant \sqcap T\operatorname{adh} T\mathfrak{R} =$
$T \sqcap \operatorname{adh} T\mathfrak{R} = T\lim T\mathfrak{R} \geqslant \lim T\mathfrak{R}$.

5. (*a*) $T\lim T\mathfrak{R} \geqslant \lim T\mathfrak{R} \Rightarrow$
$T\lim T\mathfrak{R} = \lim T\mathfrak{R}$,

(*b*) $T\lim T\mathfrak{R} \leqslant \lim T\mathfrak{R} \Rightarrow$
$T\lim T\mathfrak{R} = \lim T\mathfrak{R}$.

6. $\lim \mathfrak{R} \in \overset{-1}{T} T\mathfrak{R} \Rightarrow T\lim \mathfrak{R} = \lim T\mathfrak{R}$.

7. $x = \lim T\mathfrak{R} = T\lim T\mathfrak{R} \Rightarrow$
$x = \lim_{(TV, \leqslant)} T\mathfrak{R}$.

Proof.

1. Follows from the definition of the meet $\sqcap$ and the join $\sqcup$, and the Definitions 1.3, 1.5.
2. Follows from Theorem 1.4 (3) and Definition 1.5.
3. (i) is a consequence of Definitions 1.3, 1.5.

   (ii) (a) $\bigwedge_{r,s \in \operatorname{adh} T\mathfrak{R}} \{r, s\} \leqslant \{x\}$,

   $\bigwedge_{r,s \in \text{co-adh}\, T\mathfrak{R}} \{x\} \leqslant \{r, s\}$,

   (b) dual.

   (iii) can be concluded from

   $\operatorname{adh} T\mathfrak{R} \to \lim T\mathfrak{R} \wedge \text{co-adh}\, T\mathfrak{R} \to \lim T\mathfrak{R}$

   together with Theorem 1.4 (2) and (i).

4. (a) $x := \lim \mathfrak{R} = \sqcap \mathfrak{R} = \sqcup \operatorname{adh} \mathfrak{R}$,
   $\lim T\mathfrak{R} = \sqcap T\mathfrak{R} = \sqcup \operatorname{adh} T\mathfrak{R}$,

   $(\operatorname{adh} \mathfrak{R} \leqslant \{x\} \Rightarrow T\operatorname{adh} \mathfrak{R} \leqslant \{Tx\}) \wedge$
   $(Tx \in T\operatorname{adh} \mathfrak{R}) \Rightarrow T \sqcup \operatorname{adh} \mathfrak{R} = \sqcup T\operatorname{adh} \mathfrak{R}$,
   and analogously $T \sqcup \operatorname{adh} T\mathfrak{R} = \sqcup T\operatorname{adh} T\mathfrak{R}$,

   according to Theorem 1.4 (1) we have

   $T\operatorname{adh} \mathfrak{R} \subseteq T\operatorname{adh} T\mathfrak{R} \subseteq \operatorname{adh} T\mathfrak{R} \Rightarrow$
   $\sqcup T\operatorname{adh} \mathfrak{R} \leqslant \sqcup T\operatorname{adh} T\mathfrak{R} \leqslant \sqcup \operatorname{adh} T\mathfrak{R}$,

   (b) dual.

5. (a) $T\lim T\mathfrak{R} \geqslant \lim T\mathfrak{R} \wedge T\lim T\mathfrak{R} \leqslant \lim T\mathfrak{R} \Rightarrow$
   $T\lim T\mathfrak{R} = \lim T\mathfrak{R}$,

   (b) dual.

6. $\lim \mathfrak{R} \in \overset{-1}{T} T\mathfrak{R} \Rightarrow T\lim \mathfrak{R} \in T\mathfrak{R} \Rightarrow$

   $T\lim \mathfrak{R} = \lim T\mathfrak{R}$.

7. is a consequence of Theorem 1.4 (1).

Theorem 1.6. *Let* $\mathfrak{I} \in \mathfrak{B}$ *be an I-base and* $\mathfrak{F} \in \mathfrak{B}$ *be an F-base, with* $\mathfrak{I} \to x$, $\mathfrak{F} \to y$, *and* $x \leqslant y$. *Then* $\langle x, y \rangle = I\text{-adh}\,\mathfrak{I} \cap F\text{-adh}\,\mathfrak{F}$.

Proof. Follows from Definition 1.3.

Theorem 1.6 applies to monotone inclusions of an element in the interval $\langle x, y \rangle$. In particular, if $\mathfrak{I} = \{x\}$, respectively $\mathfrak{F} = \{y\}$, then x is bounded from above by the elements of F-co-adh $\mathfrak{F}$, respectively y is bounded from below by the elements of I-co-adh $\mathfrak{I}$.

Finally it should be pointed out that the existence of limits in some parts of the Theorems 1.5, 1.6 is no severe restriction. By the well-known completion theorem any non-empty ordered set may be isomorphically embedded into a complete lattice.

2. Metrization of $\mathfrak{B}$ and the Contraction Principle

In many applications the problem is not only to approximate an element x of $(V, \leqslant)$ by the elements of a base $\mathfrak{R} \in \mathfrak{B}$ which converges towards x, but to approximate $\mathfrak{R}$ itself by other bases $\mathfrak{R}_k \in \mathfrak{B}$. This, for example, always happens in case of iteration procedures and also in case of rounded or disturbed bases. In order to measure the "difference" between any two bases $\mathfrak{R}_1 \in \mathfrak{B}$ and $\mathfrak{R}_2 \in \mathfrak{B}$ we introduce a generalized distance function on $\mathfrak{B} \times \mathfrak{B}$.

Definition 2.1. Let $d\colon \mathfrak{B} \times \mathfrak{B} \to (\mathfrak{P}(V), \subseteq)$ such that

$$\bigwedge_{\mathfrak{R}_1, \mathfrak{R}_2 \in \mathfrak{B}} d(\mathfrak{R}_1, \mathfrak{R}_2) := \begin{cases} A_1 \backslash A_2 \cup A_2 \backslash A_1 \text{ where} \\ A_k := \text{adh}\,\mathfrak{R}_k \text{ if } \mathfrak{R}_k \text{ is an } F\text{-base}, \\ A_k := \text{co-adh}\,\mathfrak{R}_k \text{ if } \mathfrak{R}_k \text{ is an } I\text{-base}, \\ \text{for } k \in \{1, 2\}. \end{cases}$$

Remarks. 1. The value of $d(\mathfrak{R}_1, \mathfrak{R}_2)$ is a "symmetrical difference".

2. As range of d any other complete boolean lattice which is isomorphic to $(\mathfrak{P}(V), \subseteq)$ can be used.

Theorem 2.1. $(\mathfrak{B}, d)$ *is a non-separated uniform space with the generalized distance d. For all bases* $\mathfrak{R}_1, \mathfrak{R}_2, \mathfrak{R} \in \mathfrak{B}$ *holds*

(*1*) $d(\mathfrak{R}_1, \mathfrak{R}_2) = d(\mathfrak{R}_2, \mathfrak{R}_1)$,

(*2*) $d(\mathfrak{R}_1, \mathfrak{R}_2) \supseteq \emptyset$,
$d(\mathfrak{R}_1, \mathfrak{R}_2) = \emptyset \Leftrightarrow \mathfrak{R}_1 \underset{\text{adh}}{\sim} \mathfrak{R}_2$,

(*3*) $d(\mathfrak{R}_1, \mathfrak{R}_2) \subseteq d(\mathfrak{R}_1, \mathfrak{R}) \cup d(\mathfrak{R}, \mathfrak{R}_2)$.

Proof. (1) and (2) are obvious, (3) is known. The symmetrical difference of two sets equals zero if and only if the two sets are equal which by Definition 1.4 results in adherence equivalence. To prove that d generates a uniform structure on $\mathfrak{B}$ one has to show that the inverse images of d form a fundamental system of neighbourhoods. For this we refer to [2].

Theorem 2.2. *Let the relation* $\prec$ *on* $\mathfrak{B}$ *be defined by*

$$\bigwedge_{\mathfrak{R}_1, \mathfrak{R}_2 \in \mathfrak{B}} \mathfrak{R}_1 \prec \mathfrak{R}_2 :\Leftrightarrow \begin{cases} \operatorname{adh} \mathfrak{R}_1 \subset \operatorname{adh} \mathfrak{R}_2 \\ \textit{if } \mathfrak{R}_1, \mathfrak{R}_2 \textit{ are two F- or two I-bases.} \end{cases}$$

Further, let $\mathfrak{L} := \{\mathfrak{R}_i\}_{i \in I}$ *be a non-empty set of bases* $\mathfrak{R}_i \in \mathfrak{B}$, *and let*

$$\mathfrak{L}_0 := \{\mathfrak{R} | \mathfrak{R} \in \mathfrak{L} \wedge \neg \bigvee_{\mathfrak{R}' \in \mathfrak{L}} \mathfrak{R}' \prec \mathfrak{R}\}.$$

If the following assumptions hold

(1) $\bigwedge_{\mathfrak{R} \in \mathfrak{L} \setminus \mathfrak{L}_0} \bigvee_{\tilde{\mathfrak{R}} \in \mathfrak{L}_0} \tilde{\mathfrak{R}} \prec \mathfrak{R}$,

(2) the space $(\mathfrak{L}, d)$ *is complete,*

(3) $\bigwedge_{\mathfrak{R}_1, \mathfrak{R}_2 \in \mathfrak{L}} d(\mathfrak{R}_1, \mathfrak{R}_2) \neq \emptyset \Rightarrow \bigvee_{\mathfrak{R} \in \mathfrak{L}} \mathfrak{R} \prec \mathfrak{R}_1 \vee \mathfrak{R} \prec \mathfrak{R}_2$,

then

1. $\mathfrak{L}_0 \neq \emptyset$, $\bigwedge_{\tilde{\mathfrak{R}}_1, \tilde{\mathfrak{R}}_2 \in \mathfrak{L}_0} d(\tilde{\mathfrak{R}}_1, \tilde{\mathfrak{R}}_2) = \emptyset$,

2. $\bigwedge_{\mathfrak{R}_0 \in \mathfrak{L} \setminus \mathfrak{L}_0} \bigvee_{\mathfrak{A} \subseteq \mathfrak{L}} \mathfrak{A}$ *a* $\prec$*-chain* $\wedge\ \mathfrak{A} \prec \{\mathfrak{R}_0\} \wedge \bigwedge_{\tilde{\mathfrak{R}} \in \mathfrak{L}_0} \mathfrak{A} \xrightarrow[d]{} \tilde{\mathfrak{R}}$,

3. $\bigwedge_{\mathfrak{A} \subseteq \mathfrak{L}} \mathfrak{A}$ *a* $\prec$*-chain* $\wedge \neg \bigvee_{\mathfrak{R} \in \mathfrak{L} \setminus \mathfrak{L}_0} \{\mathfrak{R}\} \prec \mathfrak{A} \Rightarrow \bigwedge_{\tilde{\mathfrak{R}} \in \mathfrak{L}_0} \mathfrak{A} \xrightarrow[d]{} \tilde{\mathfrak{R}}$.

Proof. 1. $\mathfrak{L}_0 \neq \emptyset$ by (1). Let be $\tilde{\mathfrak{R}}_1, \tilde{\mathfrak{R}}_2 \in \mathfrak{L}_0$. Then (3) contradicts $d(\tilde{\mathfrak{R}}_1, \tilde{\mathfrak{R}}_2) \neq \emptyset$.

2. Let be $\mathfrak{R}_0 \in \mathfrak{L} \setminus \mathfrak{L}_0$. By (1) there exists $\tilde{\mathfrak{R}} \in \mathfrak{L}_0$ with $\tilde{\mathfrak{R}} \prec \mathfrak{R}_0$. We have $d(\tilde{\mathfrak{R}}, \mathfrak{R}_0) \neq \emptyset$. By (3) there is an $\mathfrak{R}_1 \in \mathfrak{L}$ with $\mathfrak{R}_1 \prec \mathfrak{R}_0$, and by (1) and 1. $d(\tilde{\mathfrak{R}}, \mathfrak{R}_1) \subset d(\tilde{\mathfrak{R}}, \mathfrak{R}_0)$. If $d(\tilde{\mathfrak{R}}, \mathfrak{R}_1) \neq \emptyset$, then $\tilde{\mathfrak{R}} \prec \mathfrak{R}_1$ and the above reasoning can be applied again. Hence there exist strictly monotonic chains (named $\prec$-chain and d-chain)

$$\cdots \mathfrak{R}_1 \prec \mathfrak{R}_0,$$

$$\cdots d(\tilde{\mathfrak{R}}, \mathfrak{R}_1) \subset d(\tilde{\mathfrak{R}}, \mathfrak{R}_0).$$

Because of the completeness of $(\mathfrak{P}(V), \subseteq)$ the d-chain converges to a value d_0. As a result of the triangular inequality the $\prec$-chain is a Cauchy-chain and by (2) converges to an $\hat{\mathfrak{R}} \in \mathfrak{L}$. If $d(\tilde{\mathfrak{R}}, \hat{\mathfrak{R}}) \neq \emptyset$ then $\hat{\mathfrak{R}} \prec \cdots \mathfrak{R}_1 \prec \mathfrak{R}_0$ and the continuation of the chains is possible.

3. If $\mathfrak{A}$ is a $\prec$-chain according to part 3 of the theorem, then

$$\bigwedge_{\tilde{\mathfrak{R}} \in \mathfrak{L}_0} \left(\bigwedge_{\mathfrak{R}_1, \mathfrak{R}_2 \in \mathfrak{A}} \mathfrak{R}_1 \prec \mathfrak{R}_2 \Rightarrow d(\tilde{\mathfrak{R}}, \mathfrak{R}_1) \subset d(\tilde{\mathfrak{R}}, \mathfrak{R}_2) \right) \wedge d(\tilde{\mathfrak{R}}, \mathfrak{A}) \to \emptyset.$$

Remarks. 1. Theorem 2.2 includes the case of monotonic inclusions $\mathfrak{R}_1 \prec \tilde{\mathfrak{R}} \prec \mathfrak{R}_2$, $\tilde{\mathfrak{R}} \in \mathfrak{L}_0$.

2. For $\mathfrak{R}_1, \mathfrak{R}_2 \in \mathfrak{L}$ and $d(\mathfrak{R}_1, \mathfrak{R}_2) \neq \emptyset$ there exist $\mathfrak{R}'_1, \mathfrak{R}'_2 \in \mathfrak{L}$ such that the following "contraction property" holds:

(i) $\mathfrak{R}'_1 \prec \mathfrak{R}_1$, $\mathfrak{R}'_2 \prec \mathfrak{R}_2$ and $d(\mathfrak{R}'_1, \mathfrak{R}'_2) \subset d(\mathfrak{R}_1, \mathfrak{R}_2)$ if $\mathfrak{R}_1, \mathfrak{R}_2 \in \mathfrak{L} \setminus \mathfrak{L}_0$, and

(ii) $\mathfrak{R}'_1 \prec \mathfrak{R}_1$ and $d(\mathfrak{R}'_1, \mathfrak{R}_2) \subset d(\mathfrak{R}_1, \mathfrak{R}_2)$ if $\mathfrak{R}_1 \in \mathfrak{L} \setminus \mathfrak{L}_0$ and $\mathfrak{R}_2 \in \mathfrak{L}_0$.

3. Ordered Sets with Two Roundings

Let be given a non-empty ordered set $(V, \leqslant, T)$ with a rounding T, a base $\mathfrak{R}$ on $(V, \leqslant)$, and an endomorphism $F: (V, \leqslant) \to (V, \leqslant)$.

Definition 3.1.

$$\text{“}F \text{ is } T\text{-continuous with respect to } \mathfrak{R}\text{”} :\Leftrightarrow$$

$$F \| \mathfrak{R}\ T\text{-continuous} :\Leftrightarrow \operatorname{adh} FT\mathfrak{R} \subseteq \operatorname{adh} TF\mathfrak{R}.$$

Now we consider a second rounding S on $(V, \leqslant)$ and investigate the relations between T and S.

Theorem 3.1.

$$S\|\mathfrak{R}T\textit{-continuous} \Leftrightarrow \mathfrak{F}\mathfrak{I}(TS\mathfrak{R}) \subseteq \mathfrak{F}\mathfrak{I}(ST\mathfrak{R})$$

$$\Leftrightarrow \text{co-adh}(TS\mathfrak{R}) \subseteq \text{co-adh}(ST\mathfrak{R}),$$

i.e. $\mathfrak{F}\mathfrak{I}(ST\mathfrak{R})$ *is finer than* $\mathfrak{F}\mathfrak{I}(TS\mathfrak{R})$.

This simply follows from Definition 1.3 and Theorem 1.4 (3) and (4).

Theorem 3.2. *Let be*

(1) $\mathfrak{R} \subseteq TV \wedge ((a)\ S\mathfrak{R} \subseteq V_{T\uparrow}$, *resp.* $(b)\ S\mathfrak{R} \subseteq V_{T\downarrow})$,

(2) *(a)* $\bigwedge_{r \in \mathfrak{R}} STr \leqslant Sr \wedge S\mathfrak{R} \subseteq V_{T\uparrow}$, *resp.*

(b) $\bigwedge_{r \in \mathfrak{R}} STr \geqslant Sr \wedge S\mathfrak{R} \subseteq V_{T\downarrow}$.

Then (1) $\vee$ (2) $\Rightarrow S\|\mathfrak{R}$ *T-continuous.*

Proof.

(1) (a) $\bigwedge_{r \in \mathfrak{R}} Sr = STr \leqslant TSr$, (b) dual,

(2) (a) $\bigwedge_{r \in \mathfrak{R}} STr \leqslant Sr \leqslant TSr$, (b) dual.

Theorem 3.3.

1. Let (a) $\mathfrak{R} \subseteq V_{T\uparrow}$, *resp. (b)* $\mathfrak{R} \subseteq V_{T\downarrow}$. *Then*

1.1. $\operatorname{adh} S\mathfrak{R} \subseteq \operatorname{adh} ST\mathfrak{R}$,
1.2. $\operatorname{adh} TS\mathfrak{R} \subseteq \operatorname{adh} TST\mathfrak{R}$.

2. Let (a) $S\mathfrak{R} \subseteq V_{T\uparrow}$, *resp. (b)* $S\mathfrak{R} \subseteq V_{T\downarrow}$. *Then*

2.1. $\operatorname{adh} S\mathfrak{R} \subseteq \operatorname{adh} TS\mathfrak{R}$,
2.2. $\operatorname{adh} S\mathfrak{R} \subseteq \operatorname{adh} STS\mathfrak{R}$,
2.3. $\operatorname{adh} TS\mathfrak{R} \subseteq \operatorname{adh} TSTS\mathfrak{R}$,
2.4. $T\mathfrak{R}$ *finer* $S\mathfrak{R} \Rightarrow S\|\mathfrak{R}$ *T-continuous,*
2.5. $\mathfrak{R} \subseteq TV \Rightarrow S\|\mathfrak{R}$ *T-continuous.*

3. *Let be $S\|\mathfrak{R}$ T-continuous. Then*

   3.1. adh $ST\mathfrak{R} \subseteq$ adh $TS\mathfrak{R}$,
   3.2. adh $TST\mathfrak{R} \subseteq$ adh $TS\mathfrak{R}$.

4. *Let be $S\|T\mathfrak{R}$ T-continuous. Then*

   4.1. adh $ST\mathfrak{R} \subseteq$ adh $TST\mathfrak{R}$,
   4.2. adh $ST\mathfrak{R} \subseteq$ adh $STST\mathfrak{R}$.

5. *Let be $S\|S\mathfrak{R}$ T-continuous. Then*

   5.1. adh $STS\mathfrak{R} \subseteq$ adh $TS\mathfrak{R}$,
   5.2. adh $TSTS\mathfrak{R} \subseteq$ adh $TS\mathfrak{R}$.

6. *From 3., 4., 5. it follows*

   adh $ST\mathfrak{R} \subseteq$ adh $TST\mathfrak{R} \subseteq$ adh $TS\mathfrak{R}$.

   If in addition 2. holds, then

   adh $TSTS\mathfrak{R}$ = adh $TS\mathfrak{R}$.

   If 1. through 5. hold, then

   adh $S\mathfrak{R} \subseteq$ adh $ST\mathfrak{R} \subseteq$ adh $TST\mathfrak{R}$ =
   = adh $TS\mathfrak{R}$ = adh $TSTS\mathfrak{R}$.

7. *For $TS|\mathfrak{R} = ST|\mathfrak{R}$ we have*

   adh $S\mathfrak{R} \subseteq$ adh $ST\mathfrak{R}$ = adh $TST\mathfrak{R}$ = adh $TS\mathfrak{R}$ =
   adh $STS\mathfrak{R}$ = adh $STST\mathfrak{R}$ = adh $TSTS\mathfrak{R}$.

   For $S|\mathfrak{R} = TS|\mathfrak{R} = ST|\mathfrak{R}$ we have in addition

   adh $S\mathfrak{R}$ = adh $ST\mathfrak{R}$.

Proof. 1. to 5. can be obtained by interpretation of the assumptions made, application of T respectively S on the interpreting relations and combining the results. 6. and 7. are consequences of 1. to 5.

A survey of the relations of Theorem 3.3 is given in Fig. 1.

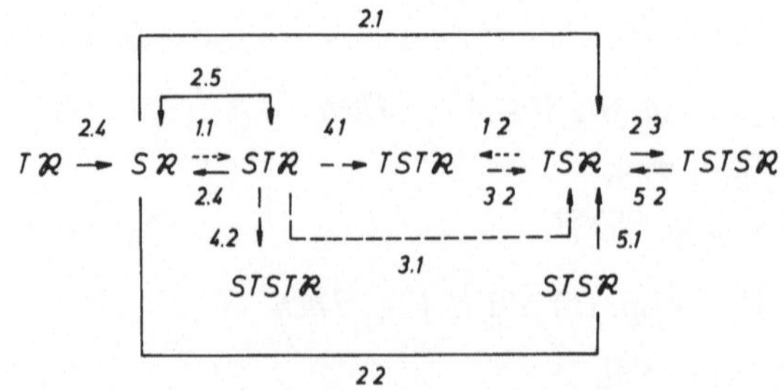

Fig. 1

According to Theorem 3.3 (7), particularly simple facts are obtainable if one chooses $S = TS$, respectively $S = TS = ST$ on $(V, \leqslant)$.

Theorem 3.4.

1. If $S = TS$ on $(V, \leqslant)$, then

1.1. $S = TS = ST$ on $(TV, \leqslant)$, i.e. $\bigwedge_{T\mathfrak{R} \in \mathfrak{B}} S \| T\mathfrak{R}$ T-continuous,

1.2. if (a) $\mathfrak{R} \subseteq V_{S\uparrow}$, resp. (b) $\mathfrak{R} \subseteq V_{S\downarrow}$, then $T\mathfrak{R}$ is finer $S\mathfrak{R}$.

2. If $S = TS = ST$ on $(V, \leqslant)$ then $\bigwedge_{\mathfrak{R} \in \mathfrak{B}} S \| \mathfrak{R}$ T-continuous.

Proof.

1.1. $\bigwedge_{v \in TV} Sv = TSv = STv$,

1.2. (a) $\bigwedge_{r \in \mathfrak{R}} r \leqslant Sr \Rightarrow Tr \leqslant TSr = Sr$, (b) dual.

2. follows from Definition 3.1.

An example of $S = TS = ST$ is illustrated in Fig. 2.

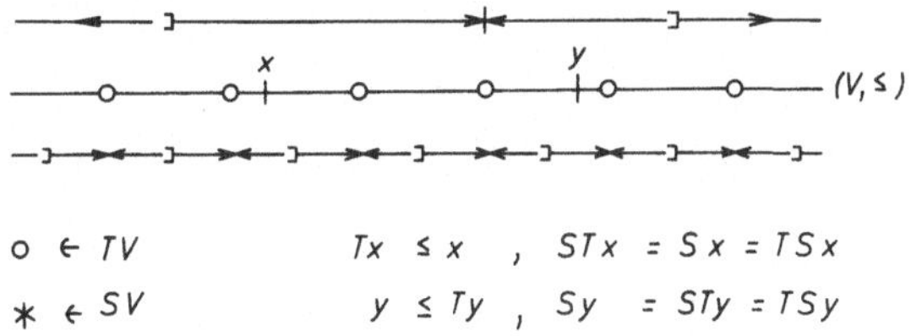

Fig. 2

4. Approximations to a Mapping

Let there be given two non-empty ordered sets, each with a rounding on it:

$$(X, \leqslant_X, T_X), (Y, \leqslant_Y, T_Y).$$

Let $\mathfrak{B}_X$ respectively $\mathfrak{B}_Y$ denote the set of bases on $(X, \leqslant_X)$ respectively on $(Y, \leqslant_Y)$, and let be $\mathfrak{R} \in \mathfrak{B}_X$.

At first, let us intuitively explain the problem we are concerned with in this section:

Assume, there are two non-empty appropriate sets $X_0, X_1 \subseteq X$ and two homomorphisms $f_0\colon X_0 \to Y, f\colon X_1 \to Y$. The mapping problem, to be approximated, is:

(P): given $x_0 \in X_0$,
seeked $y_0 := f_0(x_0)$.

Let be $\mathfrak{R} \subseteq X_1$. Then (P) is approximated by the problems:

(P_i): given $x_i \in \mathfrak{R}$,
seeked $y_i := f(x_i)$,

for $i \in I$. The problems (P_i) may be in some way "easier" to solve than (P).

The questions arising are:

Does the base $(y_i)_{i\in I}$ converge? Does the base $(T_Y y_i)$ converge?

If $(y_i)_{i\in I} \to y_1$ and/or $(T_Y y_i)_{i\in I} \to y_2$, how are y_0, $T_Y y_0$, y_1, $T_Y y_1$, y_2, $T_Y y_2$ related?

Are there "neighbouring" bases $(\tilde{y}_j)_{j\in J}$ to $(y_i)_{i\in I}$, respectively to $(T_Y y_i)_{i\in I}$ leading to the same relation among the limits?

Are there "neighbouring" bases $(\tilde{x}_j)_{j\in J}$ to $\mathfrak{R}$ leading to the same relation among the limits?

May $T_X\mathfrak{R}$ be used instead of $\mathfrak{R}$?

In order to tackle these problems we need, besides of the notions "convergence" and "stability", already introduced in Sect. 1, the notion of "consistency":

Definition 4.1. Let f_0: adh $\mathfrak{R} \to Y$ and f: $\mathfrak{R} \to Y$ be homomorphisms.

(1) "f is consistent, respectively fully consistent with f_0 with respect to $\mathfrak{R}$":$\Leftrightarrow (f_0,f)\|\mathfrak{R}$ is c., resp. is f.c. :$\Leftrightarrow$ co-adh $f_0(\text{adh}\,\mathfrak{R}) \subseteq$ resp. $=$ adh $f(\mathfrak{R})$,

(2) "f is T_Y-consistent, respectively T_Y-fully consistent with f_0 with respect to $\mathfrak{R}$":$\Leftrightarrow (f_0,f)\|\mathfrak{R}$ is T_Y-c., resp. T_Y-f.c. :$\Leftrightarrow$ co-adh $T_Y f_0(\text{adh}\,\mathfrak{R}) \subseteq$ resp. $=$ co-adh $T_Y\,\text{adh}\,f(\mathfrak{R})$.

In the following we shorten the notations by the conventions already made.

Theorem 4.1.

1. Let $(f_0,f)\|\mathfrak{R}$ *be* c., *resp.* f.c. $\Rightarrow$
$(f_0,f)\|\mathfrak{R}$ *is* T-c., *resp.* T-f.c.

2. Let $(f_0,f)\|\mathfrak{R}$ *be* c., *resp.* T-c. $\Rightarrow$
F: $\mathfrak{R} \cup \text{adh}\,\mathfrak{R} \to Y$, *defined by*
$F|\mathfrak{R} = (f \textit{ resp. } Tf) \wedge$
$F|\text{adh}\,\mathfrak{R} = (f_0 \textit{ resp. } Tf_0)$
is a homomorphism.

Let G: $\mathfrak{R} \cup \text{adh}\,\mathfrak{R} \to Y$ *be a homomorphism, then* $(G|\text{adh}\,\mathfrak{R}, G|\mathfrak{R})\|\mathfrak{R}$ *is* c. *and is* T-c.

3. Let $(f_0,f)\|\mathfrak{R}$ *be* f.c., *resp.* T-f.c. $\wedge\ \mathfrak{R} \to x_0 \Rightarrow$
$f_0(\text{adh}\,\mathfrak{R}) \to f_0(x_0) \wedge f(\mathfrak{R}) \to f_0(x_0)$, *resp.*
$Tf_0(\text{adh}\,\mathfrak{R}) \to Tf_0(x_0) \wedge Tf(\mathfrak{R}) \to Tf_0(x_0)$.

4. Let the assumption of 3. hold.

$(\mathfrak{T} \in \mathfrak{B}_Y \wedge \mathfrak{T} \underset{\text{adh}}{\sim} f(\mathfrak{R})$, *resp.*

$\mathfrak{T}' \in \mathfrak{B}_Y \wedge \mathfrak{T}' \underset{\text{adh}}{\sim} Tf(\mathfrak{R})) \Rightarrow$

$\mathfrak{T} \to f_0(x_0)$, *resp.* $\mathfrak{T}' \to Tf_0(x_0)$.

Let be $\mathfrak{S} \in \mathfrak{B}_X \wedge \mathfrak{R} \underset{\text{adh}}{\sim} \mathfrak{S} \wedge f$: $\mathfrak{R} \cup \mathfrak{S} \to Y$ *a homomorphism* $\wedge\ (f_0,f)\|\mathfrak{S}$ f.c., *resp.* T-f.c. $\Rightarrow f(\mathfrak{S}) \to f_0(x_0)$ *resp.* $Tf(\mathfrak{S}) \to Tf_0(x_0)$.

Proof.

1. Part 1 follows by observing

$$\text{co-adh} f_0(\text{adh}\,\mathfrak{R}) \subseteq \text{adh} f(\mathfrak{R}) \Rightarrow$$
$$f_0(\text{adh}\,\mathfrak{R}) \subseteq \text{adh} f(\mathfrak{R})$$

and Definition 4.1.

2. (i): $F(\text{adh}\,\mathfrak{R}) \subseteq \text{adh}\,F(\mathfrak{R})$, and
$TF(\text{adh}\,\mathfrak{R}) \subseteq \text{co-adh}\,T\,\text{adh}\,F(\mathfrak{R}) \subseteq \text{adh}\,TF(\mathfrak{R})$
prove F to be a homomorphism,

(ii): $G(\text{adh}\,\mathfrak{R}) \subseteq \text{adh}\,G(\mathfrak{R}) \wedge$
$TG(\text{adh}\,\mathfrak{R}) \subseteq T\,\text{adh}\,G(\mathfrak{R}) \Rightarrow$
$\text{co-adh}\,G(\text{adh}\,\mathfrak{R}) \subseteq \text{adh}\,G(\mathfrak{R}) \wedge$
$\text{co-adh}\,TG(\text{adh}\,\mathfrak{R}) \subseteq \text{co-adh}\,T\,\text{adh}\,G(\mathfrak{R})$.

3. $x_0 \in \text{adh}\,\mathfrak{R}$, and f_0 is a homomorphism, thus $\lim \text{co-adh} f_0(\text{adh}\,\mathfrak{R}) = f_0(x_0)$. Consistency implies $f_0(x_0) \in \text{adh} f(\mathfrak{R})$, in case of full consistency $f_0(x_0) = \lim \text{adh} f(\mathfrak{R}) = \lim f(\mathfrak{R})$. Analogously $Tf_0(x_0) = \lim Tf(\mathfrak{R})$.

4. (i): A base stable with respect to a convergent base converges to the same limit,

(ii): $\text{co-adh} f_0(\text{adh}\,\mathfrak{R}) = \text{co-adh} f_0(\text{adh}\,\mathfrak{S}) = \text{adh} f(\mathfrak{S})$,
thus $f_0(x_0) = \lim \text{adh} f(\mathfrak{S})$, analogously $Tf_0(x_0) = \lim Tf(\mathfrak{S})$.

Remarks. 1. Theorem 4.1 (3), (4) states: If for some base $\mathfrak{R} \in \mathfrak{B}_X$ holds: $\mathfrak{R} \to x_0$, f is fully consistent with f_0 with respect to $\mathfrak{R}$, respectively T-fully consistent, then any base $\mathfrak{T} \in \mathfrak{B}_Y$, respectively $\mathfrak{T}' \in \mathfrak{B}_Y$ which is stable with respect to $f(\mathfrak{R})$, respectively to $Tf(\mathfrak{R})$, converges to $f_0(x_0)$, respectively to $Tf_0(x_0)$.

2. The totality of bases stable with respect to $f(\mathfrak{R})$, respectively $Tf(\mathfrak{R})$, are those converging to $f_0(x_0)$, respectively $Tf_0(x_0)$.

3. Implicitely, we already used consistency in Theorem 2.2 in assumption (2).

4. Another abstract theory of convergence, consistency, and stability was given by Stummel [5]. It differs to ours by using Moore-Smith sequences instead of F- and I-bases in an ordered set, by different notions of convergence and consistency, and by not taking into account roundings ($T = \Delta$).

Now we change the meaning of f_0 and f in the following

Theorem 4.2. *Let*

$$f_0\colon \text{adh}\,T\mathfrak{R} \to Y \textit{ and}$$

$$f\colon T\mathfrak{R} \to Y$$

be homomorphisms.

1. $(f_0,f)\|T\mathfrak{R}$ *is* c. $\Rightarrow$
$f_0(T\,\mathrm{adh}\,\mathfrak{R}) \subseteq f_0(T\,\mathrm{adh}\,T\mathfrak{R}) = f_0(\mathrm{adh}_{(TX,\leqslant)}T\mathfrak{R}) \subseteq$
$f_0(\mathrm{adh}\,T\mathfrak{R}) \subseteq \mathrm{adh}\,f(T\mathfrak{R})$.

2. $(f_0,f)\|T\mathfrak{R}$ *is* f.c. $\Rightarrow$
$Tf_0(T\,\mathrm{adh}\,\mathfrak{R}) \subseteq Tf_0(T\,\mathrm{adh}\,T\mathfrak{R}) = Tf_0(\mathrm{adh}_{(TX,\leqslant)}T\mathfrak{R})$
$\subseteq Tf_0(\mathrm{adh}\,T\mathfrak{R}) \subseteq T\,\mathrm{adh}\,f(T\mathfrak{R}) \subseteq T\,\mathrm{adh}\,Tf(T\mathfrak{R}) =$
$= \mathrm{adh}_{(TY,\leqslant)}Tf(T\mathfrak{R}) \subseteq \mathrm{adh}\,Tf(T\mathfrak{R})$.

3. Let be $\mathfrak{S}, \mathfrak{S}' \in \mathfrak{B}_X$ *and* $\mathfrak{T}, \mathfrak{T}' \in \mathfrak{B}_Y$ *such that*

$$\mathfrak{S} \underset{\mathrm{adh}}{\sim} \mathfrak{R},\ \mathfrak{S}' \underset{\mathrm{adh}}{\sim} T\mathfrak{R},$$
$$\mathfrak{T} \underset{\mathrm{adh}}{\sim} f(T\mathfrak{R}),\ \mathfrak{T}' \underset{\mathrm{adh}}{\sim} Tf(T\mathfrak{R}).$$

Then we may substitute adh $\mathfrak{R}$ *by* adh $\mathfrak{S}$, adh $T\mathfrak{R}$ *by* adh $\mathfrak{S}'$, adh $f(T\mathfrak{R})$ *by* adh $\mathfrak{T}$, adh $Tf(T\mathfrak{R})$ *by* adh $\mathfrak{T}'$.

4. If the following limits exist, then

4.1. (*a*) $f_0(T\lim\mathfrak{R}) \leqslant f_0(T\lim T\mathfrak{R}) = f_0(\lim_{(TX,\leqslant)}T\mathfrak{R})$
$\leqslant f_0(\lim T\mathfrak{R}) \leqslant \lim f(T\mathfrak{R})$,

(*b*) *dual to* (*a*),

4.2. (*a*) $Tf_0(T\lim\mathfrak{R}) \leqslant Tf_0(T\lim T\mathfrak{R}) =$
$Tf_0(\lim_{(TX,\leqslant)}T\mathfrak{R}) \leqslant Tf_0(\lim T\mathfrak{R})$
$\leqslant T\lim f(T\mathfrak{R}) \leqslant T\lim Tf(T\mathfrak{R}) =$
$\lim_{(TY,\leqslant)}Tf(T\mathfrak{R}) \leqslant \lim Tf(T\mathfrak{R})$,

(*b*) *dual to* (*a*).

Proof. Part 1 and part 2 can be obtained by applying f_0 respectively Tf_0 to the inclusion relations of Theorem 1.4 (1) and by using the consistency property. Part 3 is obvious. Part 4 is a consequence of part 1 and 2.

Remarks. 1. If $\mathrm{adh}\,f(T\mathfrak{R}) = \text{co-adh}\,T\,\mathrm{adh}\,f(T\mathfrak{R})$ holds (for instance if $T_X = T_{X\uparrow}$ and $T_Y = T_{Y\uparrow}$ are classical topologies and $\mathfrak{R}$ is an F-base), then part 1 of Theorem 4.2 may be coupled with part 2 by $\mathrm{adh}\,f(T\mathfrak{R}) = \text{co-adh}\,T\,\mathrm{adh}\,f(T\mathfrak{R}) \subseteq \mathrm{adh}\,Tf(T\mathfrak{R})$.

2. Theorem 4.2 (1), (2) implies

$(f_0T, fT)\|\mathfrak{R}$ f.c. $\Rightarrow (f_0, f)\|T\mathfrak{R}$ f.c.,
$(Tf_0T, TfT)\|\mathfrak{R}$ f.c. $\Rightarrow (Tf_0, Tf)\|T\mathfrak{R}$ f.c.

References

[1] Albrecht, R.: Grundlagen einer Theorie gerundeter algebraischer Verknüpfungen in topologischen Vereinen. Computing, Suppl. 1, pp. 1–14. Wien-New York: Springer 1977.
[2] Albrecht, R., Karrer, G.: Fixpunktsätze in uniformen Räumen. Math. Zeitschr. **74**, 387–391 (1960).
[3] Kulisch, U.: Grundlagen des Numerischen Rechnens. (Reihe Informatik, Bd. 19.) Zürich: Bibliograph. Institut 1976.

[4] Nöbeling, G.: Grundlagen der analytischen Topologie. Berlin-Göttingen-Heidelberg: Springer 1954.

[5] Stummel, F.: Discrete convergence of mappings. Topics in numerical analysis. Proceedings of the Royal Irish Academy Conference on Numerical Analysis 1972, pp. 285 – 310. London: Academic Press 1973.

Prof. Dr. R. Albrecht
Institut für Informatik
und Numerische Mathematik
Universität Innsbruck
Innrain 52
A-6020 Innsbruck
Austria

Computing, Suppl. 2, 33–49 (1980)

Interval Analysis in the Extended Interval Space $\mathbb{I}\mathbb{R}$

E. Kaucher, Karlsruhe

Abstract

This paper shows, how the extended Interval Space $\mathbb{I}\mathbb{R}$ can be used to write formulas, theorems, and proofs in a closed form, i.e. without using the left and right interval bounds. So a basic generalization and moreover a simplification and improvement of the theorems and proofs is achieved.

0. Introduction

The extension to generalized intervals (with negative width) retaining all important properties of interval analysis like isotoneness etc. is leading to a more closed space in algebraic as well as in lattice-theoretic sense. These advantages enable us to write formulas and proofs in a closed form without using the left and right interval bounds. So the theorems and their proofs can be shortened in many cases, the statements are more general and extended to generalized intervals. Furthermore the Interval Analysis resembles with classical analysis because the ideas of norm, metric etc. can be handled more easily. Some recently appeared papers show that this extended interval analysis facilitates or makes possible formulation and solving problems as described in [4], [8], [13], [14]. Moreover the extended interval space allows to "underestimate" interval expressions, that means a rounding "to the inner", and to transform this into the usual interval analysis. This problem occurs in safety problems, where a minimum set for the solutions instead of an inclusion is asked for.

1. The Extended Interval Space

In the following we regard only the interval analysis $\mathbb{I}\mathbb{R}$ over the field of real numbers. We can do this w.l.o.g., because the formulas in $\mathbb{I}\mathbb{R}^n$, $\mathbb{I}\mathbb{C}$ etc. are of analogous form as showed in [5] and [6].

The algebraic structure $(I\mathbb{R}, +, *, \subseteq)$ is a regular commutative semigroup with respect to addition. It can be embedded in an isotone group $(\mathbb{I}\mathbb{R}, +, \subseteq)$ as shown in [4], [5] and [6]. Moreover $(I\mathbb{R}, *, \subseteq)$ satisfies all assumptions requested in [6], so that $(I\mathbb{R}, +, *, \subseteq)$ can be embedded in the high algebraic structure $(\mathbb{I}\mathbb{R}, +, *, \subseteq)$ to be introduced now. Furthermore, $(I\mathbb{R}, \subseteq)$ can be extended to $(\mathbb{I}\mathbb{R}, \subseteq)$ so that $(\mathbb{I}\mathbb{R}, \bar{\cap}, \underline{\cup}, \subseteq)$ turns out to be a complete lattice. For special details see also [4], [5], [6], [7] and [9]. To shorten this paper some proofs and properties are neglected which are summarized in [7] and [9].

In the following $A, B, C, \ldots, Z \in \mathbb{I}\mathbb{R}$ are elements of the extended interval space and $a = [a, a] \in \mathbb{I}\mathbb{R}$ is the with a identified interval. With $A = [a, b]$ the left and right bounds of A are denoted by $\lambda(A) := a$ and $\rho(A) := b$ whereas the midpoint and radius of A are denoted by $\mu(A) := \frac{a+b}{2}$ and $\delta(A) := \frac{b-a}{2}$. The latter yields a so-called midpoint-radius designation of $A = (\mu(A), \delta(A))$. Furthermore, we define:

$$\mathcal{S}^* = \mathcal{S} \cup -\mathcal{S} \quad \mathcal{S} := \{A \in \mathbb{IR} \mid 0 \leqslant \lambda A \wedge 0 \leqslant \rho A\} \quad -\mathcal{S} := \{-A \mid A \in \mathcal{S}\}$$

$$\mathcal{T}^* = \mathcal{T} \cup \bar{\mathcal{T}} \quad \mathcal{T} := \{A \in \mathbb{IR} \mid \lambda A \leqslant 0 \leqslant \rho A\} \quad \bar{\mathcal{T}} := \{\bar{A} \mid A \in \mathcal{T}\}$$

$$\mathcal{L} = \mathcal{L}_1 \cup \mathcal{L}_2 \quad \mathcal{L}_1 := \{A \in \mathbb{IR} \mid \lambda A = 0\} \quad \mathcal{L}_2 := \{A \in \mathbb{IR} \mid \rho A = 0\}$$

(I0) $A = B :\Leftrightarrow \lambda A = \lambda B \wedge \rho A = \rho B$
$A \subseteq B :\Leftrightarrow \lambda B \leqslant \lambda A \wedge \rho A \leqslant \rho B$
$A + B := [\lambda A + \lambda B, \rho A + \rho B] = (\mu A + \mu B, \delta A + \delta B)$
$A - B := A + (-B)$ with $-B := [-\rho B, -\rho B] = (-\mu B, \delta B)$
$A \cdot B := [\lambda A \cdot \lambda B, \rho A \cdot \rho B] = (\mu A \mu B + \delta A \delta B, \mu A \delta B + \mu B \delta A)$
(the so-called hyperbolic product "$\cdot$").
$A * B :=$ Table 1
$A / B := A * 1 / B$ for $B \in \mathcal{S}^* \backslash \mathcal{L}$ and $1 / B := [1/\rho B, 1/\lambda B] = B \frac{1}{\mu(B)^2 - \delta(B)^2}$
$A \bar{\cap} B = \inf(A, B) = [\lambda A \sqcup \lambda B, \rho A \sqcap \rho B]$
$A \underline{\cup} B = \sup(A, B) = [\lambda A \sqcap \lambda B, \rho A \sqcup \rho B]$
$\bar{A} := [\rho A, \lambda A] = (\mu(A), -\delta(A))$ (conjugation)
$\mathcal{J} := [-1, 1]$, $\bar{\mathcal{J}} := [1, -1]$
$\mathcal{Q}(A) := \{T \in \mathbb{IR} \mid \bar{A} \subseteq T \subseteq A \vee A \subseteq T \subseteq \bar{A}\}$.

Table 1

$*$	$B \in \mathcal{S}$	$B \in \mathcal{T}$	$B \in -\mathcal{S}$	$B \in \bar{\mathcal{T}}$
$A \in \mathcal{S}$	$A \cdot B$	$\rho(A) \cdot B$	$\bar{A} \cdot B$	$\lambda(A) \cdot B$
$A \in \mathcal{T}$	$A \cdot \rho(B)$	$[\lambda(A)\rho(B) \sqcap \rho(A)\lambda(B), \lambda(A)\lambda(B) \sqcup \rho(A)\rho(B)]$	$\bar{A} \cdot \lambda(B)$	0
$A \in -\mathcal{S}$	$A \cdot \bar{B}$	$\lambda(A) \cdot \bar{B}$	$\bar{A} \cdot \bar{B}$	$\rho(A) \cdot \bar{B}$
$A \in \bar{\mathcal{T}}$	$A \cdot \lambda(B)$	0	$\bar{A} \cdot \rho(B)$	$[\lambda(A)\lambda(B) \sqcup \rho(A)\rho(B), \lambda(A)\rho(B) \sqcap \rho(A)\lambda(B)]$

So we get the following properties in $(\mathbb{IR}, +, *, /, \cdot, \bar{\cap}, \underline{\cup}, \subseteq)$:

Theorem 1.1.

(I1) $A + B = B + A$

(I2) $(A + B) + C = A + (B + C)$

(I3) $0 + A = A$

(I4) $A + (-\bar{A}) = 0$

(I5) $A \subseteq B \Leftrightarrow A + C \subseteq B + C$

(I6) $\left.\begin{array}{l}(A \mathbin{\underline{\cup}} B) + C = (A + C) \mathbin{\underline{\cup}} (B + C)\\ (A + B) \mathbin{\underline{\cup}} C = (A \mathbin{\underline{\cup}} (C - \bar{B})) + B\end{array}\right\}$ *(analogously for* $\bar{\cap}$*)*

(I7) $\overline{A + B} = \bar{A} + \bar{B},\ A + \bar{A} \in \mathbb{R}$

(I8) $A * B = B * A$

(I9) $(A * B) * C = A * (B * C)$

(I10) $1 * A = A,\ 0 * A = 0$

(I11) $A * 1/\bar{A} = 1 \Leftrightarrow A \in \mathscr{S}^* \backslash \mathscr{L}$

(I12) $A \subseteq B \Rightarrow A * C \subseteq B * C$

(I13) (i) $\left.\begin{array}{l}\displaystyle\bigwedge_{\substack{A,B \in \mathscr{S}^*\\ A * B \in \mathscr{S}}} (A \mathbin{\underline{\cup}} B) * C = A * C \mathbin{\underline{\cup}} B * C\\ \qquad\quad (A * B) \mathbin{\underline{\cup}} C = (A \mathbin{\underline{\cup}} (C/\bar{B})) * B\end{array}\right\}$ *(analogously for* $\bar{\cap}$*)*

in general: $(A \mathbin{\underline{\cup}} B) * C \subseteq A * C \mathbin{\underline{\cup}} B * C,\ (A \mathbin{\bar{\cap}} B) * C \supseteq A * C \mathbin{\bar{\cap}} A * C$

(ii) $\displaystyle\bigwedge_{A,B \in \mathbb{IR}} A \subseteq B \Rightarrow \bigwedge_{u,v \in \mathbb{R}} \bigvee_{x \in \mathbb{R}} A * u \mathbin{\bar{\cap}} B * v \subseteq A * x \subseteq B * v$

(I14) $\overline{A * B} = \bar{A} * \bar{B},\ A * \bar{A} \in \mathbb{R}$

(I15) $A * B \in \mathscr{T} \Leftrightarrow A \in \mathscr{T} \vee B \in \mathscr{T}$
$A * B \in \bar{\mathscr{T}} \Leftrightarrow A \in \bar{\mathscr{T}} \vee B \in \bar{\mathscr{T}}$

(I16) $\displaystyle\bigwedge_{\substack{A,B \in \mathbb{IR}\\ a \in \mathbb{R}}} a * (A + B) = a * A + a * B$

(I17) $\displaystyle\bigwedge_{\substack{A \in \mathbb{IR}\\ a,b \in \mathbb{R}\\ a \cdot b \geqslant 0}} A * (a + b) = A * a + A * b$

(I18) (i) $\displaystyle\bigwedge_{A,B,C \in \mathbb{IR}} A * (B + C) \subseteq A * B + A * C$

(ii) $\displaystyle\bigwedge_{A,B,C \in \overline{\mathbb{IR}}} A * (B + C) \supseteq A * B + A * C$

(iii) $\displaystyle\bigwedge_{A,C \in \mathbb{IR}} \bigwedge_{B + \bar{C} \in \mathbb{IR}} A * (B + \bar{C}) \supseteq A * B + \bar{A} * \bar{C}$

(I19) $\displaystyle\bigwedge_{A,B \in S^* \backslash \mathscr{L}} 1/(A * B) = 1/A * 1/B$

(I20) $\displaystyle\bigwedge_{A,B \in \mathscr{S}^* \backslash \mathscr{L}} 1/(A/B) = B/A$

(I21) $\displaystyle\bigwedge_{A \in \mathscr{S}^* \backslash \mathscr{L}} 1/A = A/(A * \bar{A})$

(I22) $\displaystyle\bigwedge_{\substack{A,B \in \mathscr{S}^* \backslash \mathscr{L}\\ A * B \in \mathscr{S}}} \begin{array}{l}1/(A \mathbin{\underline{\cup}} B) = 1/A \mathbin{\underline{\cup}} 1/B\\ 1/(A \mathbin{\bar{\cap}} B) = 1/A \mathbin{\bar{\cap}} 1/B\end{array}$

(I23) $A \cdot B = B \cdot A$

(I24) $(A \cdot B) \cdot C = A \cdot (B \cdot C)$

(I25) $1 \cdot A = A$

(I26) $\bigwedge_{A \notin \mathscr{L}} A \cdot 1/\bar{A} = 1$

(I27) $\overline{A \cdot B} = \bar{A} \cdot \bar{B},\ A \cdot \bar{A} \in \mathbb{R}$

(I28) $(A + B) \cdot C = A \cdot C + B \cdot C$

(I29) $A \cdot B = 0 \Leftrightarrow (A = a \cdot \bar{B} \in \mathscr{L} \wedge a \in \mathbb{R})$

(I30) $A \subseteq B \Leftrightarrow \bar{A} \supseteq \bar{B}$

(I31) $\overline{A \bar{\cap} B} = \bar{A} \underline{\cup} \bar{B}$

(I32) $(\mathbb{IR}, \bar{\cap}, \underline{\cup}, \subseteq)$ *conditionally complete lattice*

(I33) $(\mathbb{IR}, +, \subseteq)$ *isotone group*

(I34) $(\mathbb{IR}, *, \subseteq)$ *isotone semigroup, but* $(\mathbb{IR} \backslash \mathscr{T}^*, *, \subseteq)$ *isotone group*

(I35) $\mathbb{IR} = \mathrm{I}\mathbb{R} \cup \overline{\mathrm{I}\mathbb{R}}, \qquad \mathbb{R} = \mathrm{I}\mathbb{R} \cap \overline{\mathrm{I}\mathbb{R}},$ *with* $\overline{\mathrm{I}\mathbb{R}} = \{\bar{A} \mid A \in \mathrm{I}\mathbb{R}\}$

(I36) $A \subseteq B \subseteq C \Rightarrow \mathscr{Q}(B) \subseteq \mathscr{Q}(A) \cup \mathscr{Q}(C)$
especially: $B \in \mathscr{Q}(A) \Rightarrow \mathscr{Q}(B) \subseteq \mathscr{Q}(A)$.

With arithmetical expressions we can construct the so-called interval functions. A further method to get interval functions is the direct extension of continuous real functions from $\mathbb{R}$ to $\mathbb{IR}$. For this purpose we give the following definitions:

Definition 1.2. Let $\tilde{f} \in C^0_{\mathscr{D}}$ with $\mathscr{D} = Y \times Y \times \cdots \times Y \times A_1 \times A_2 \times \cdots \times A_m$ be a continuous function, then $\tilde{f}$ can be extended on $\mathscr{D}^* \in \mathbb{IR}$ with $\mathscr{D}^* = \mathscr{Q}(Y) \times \cdots \times \mathscr{Q}(Y) \times \mathscr{Q}(A_1) \times \cdots \times \mathscr{Q}(A_m)$ and $(X, \mathscr{B}) \in \mathscr{D}^*$:

$$f(X, \mathscr{B}) = \tilde{f}(X, \ldots, X, B_1, \ldots, B_m) = \mathop{\mathscr{W}}_{\substack{x \in \mathscr{Q}(X) \times \cdots \times \mathscr{Q}(X) \\ a \in \mathscr{Q}(B_1) \times \cdots \times \mathscr{Q}(B_m)}} \{\tilde{f}(x, a)\} \tag{1.1}$$

For arbitrary $Z \in \mathbb{IR}$ the operator $\mathscr{W}$ is defined as:

$$\mathop{\mathscr{W}}_{z \in \mathscr{Q}(Z)} :\equiv \begin{cases} \bigcup\limits_{z \in Z} & \text{for} \quad Z \in \mathbb{IR} \\ \bigcap\limits_{z \in \bar{Z}} & \text{for} \quad Z \in \overline{\mathbb{IR}} \end{cases} \tag{1.2}$$

This operator is dependent on the order of evaluation, i.e.

$$\mathop{\mathscr{W}}_{(u,v) \in \mathscr{Q}(U) \times \mathscr{Q}(V)} :\equiv \mathop{\mathscr{W}}_{u \in \mathscr{Q}(U)} \cdot \mathop{\mathscr{W}}_{v \in \mathscr{Q}(V)}. \tag{1.3}$$

2. Topological Properties of $\mathbb{IR}$, Norm, Metric, Sequence, Width

$(\mathbb{IR}, +, *, /)$ is a normed space in the following sense:

Definition 2.1.

$$\bigwedge_{A \in \mathbb{IR}} |A| := \inf\{t \in \mathbb{R}^+ \mid t * \bar{\mathscr{J}} \subseteq A \subseteq t * \mathscr{J}\} =$$

$$= \inf\{t \in \mathbb{R}^+ \mid A \in t * \mathcal{Q}(\mathcal{J})\}. \tag{2.1}$$

This is a norm with the following properties:

Lemma 2.2.

(N1) $|A| \geqslant 0,\ |A| = 0 \Leftrightarrow A = 0$

(N2) $|A + B| \leqslant |A| + |B|$

(N3) $|a * A| = |a|\,|A|$

(N4) $|A * B| \leqslant |A|\,|B|$

(N5) $|1/A| = \dfrac{|A|}{A * \bar{A}}$ *for* $A \in \mathcal{S}^* \backslash \mathcal{L}$

(N6) $|A + B| = |A| + |B|$ *if* $A = -A \wedge B = -B \wedge A, B \in \mathcal{T}$ *or* $A, B \in \bar{\mathcal{T}}$ *or if* $A * B \in \mathcal{S}^* \backslash \mathcal{L}$, *resp.*

(N7) $|A * B| = |A|\,|B|$ *for* $A, B \in \mathrm{I}\mathbb{R}$ *or* $A, B \in \overline{\mathrm{I}\mathbb{R}}$

(N8) $A * B = |A| * B$ *for* $B = -B \wedge A, B \in \mathrm{I}\mathbb{R}$ *or* $A, B \in \overline{\mathrm{I}\mathbb{R}}$

(N9) $|A \cdot B| \leqslant |A|\,|B|$

(N10) $A \subseteq B \subseteq C \Rightarrow |B| \leqslant |A| \sqcup |C|$, *especially*: $a \subseteq A \Rightarrow |a| \leqslant |A|$

(N11) $|A \underline{\cup} B| \leqslant |A| \sqcup |B|, |A \bar{\cap} B| \leqslant |A| \sqcup |B|$, *but*
$|A \bar{\cap} B| \leqslant |A| \sqcap |B| \wedge |A \underline{\cup} B| = |A| \sqcup |B|$, *for* $A, B \in \mathrm{I}\mathbb{R}$ *and*
$|A \underline{\cup} B| \leqslant |A| \sqcap |B| \wedge |A \bar{\cap} B| = |A| \sqcup |B|$, *for* $A, B \in \overline{\mathrm{I}\mathbb{R}}$

(N12) $|\pm A| = |A|, |\bar{A}| = |A|$

(N13) $|A| \subseteq A * \operatorname{sign}(A)$ *for* $A \in \mathrm{I}\mathbb{R}$; $|A| \supseteq A * \operatorname{sign}(A)$ *for* $A \in \overline{\mathrm{I}\mathbb{R}}$

$$\textit{with} \quad \operatorname{sign}(A) = \begin{cases} +1 & A \in \mathcal{S} \\ \mathcal{J} & A \in \mathcal{T} \\ \bar{\mathcal{J}} & A \in \bar{\mathcal{T}} \\ -1 & A \in -\mathcal{S} \end{cases}$$

For the proof we use the following properties:

Lemma 2.3.

$$|W|\bar{\mathcal{J}} \subseteq W \subseteq |W|\mathcal{J} \tag{2.2}$$

$$t\bar{\mathcal{J}} \subseteq W \subseteq t\mathcal{J} \Leftrightarrow \bigwedge_{\substack{w \in \mathcal{Q}(W) \\ w \in \mathbb{R}}} t\bar{\mathcal{J}} \subseteq w \subseteq t\mathcal{J} \tag{2.3}$$

$$|W| = \sup_{w \in \mathcal{Q}(W)} |w| = |\lambda W| \sqcup |\rho W| \tag{2.4}$$

Proof. See [9].

Proof (to Lemma 2.2).

(N1) $|A| \geqslant 0$ by definition
$A = 0 \Leftrightarrow 0 \subseteq A \subseteq 0 \Leftrightarrow t = 0 \wedge t\bar{\mathcal{J}} \subseteq A \subseteq t\mathcal{J} \Leftrightarrow |A| = t = 0$

(N2) $|A|\bar{\mathscr{I}} \subseteq A \subseteq |A|\mathscr{I},\ |B|\bar{\mathscr{I}} \subseteq B \subseteq |B|\mathscr{I} \overset{(I5)(I17)}{\Rightarrow} (|A| + |B|)\bar{\mathscr{I}} \subseteq A + B \subseteq$

$$(|A| + |B|)\mathscr{I} \Rightarrow |A + B| \leqslant |A| + |B|$$

(N3) $|A|\bar{\mathscr{I}} \subseteq A \subseteq |A|\mathscr{I} \overset{(I12)}{\Leftrightarrow}$

$$a * |A| * \bar{\mathscr{I}} \subseteq a * A \subseteq a * |A| * \mathscr{I} \Leftrightarrow$$
$$|a|\,|A| * \bar{\mathscr{I}} \subseteq a * A \subseteq |a| * |A| * \mathscr{I} \Leftrightarrow |a * A| \leqslant |a|\,|A|$$

on the other hand we have

$$|a * A|\bar{\mathscr{I}} \subseteq a * A \subseteq |a * A|\mathscr{I} \overset{(I12)}{\Leftrightarrow}$$

$$\frac{|a * A|}{a}\bar{\mathscr{I}} \subseteq A \subseteq \frac{|a * A|}{a}\mathscr{I} \Leftrightarrow \qquad \text{for } \pm\mathscr{I} = \mathscr{I}$$

$$\frac{|a * A|}{|a|}\bar{\mathscr{I}} \subseteq A \subseteq \frac{|a * A|}{|a|}\mathscr{I} \Leftrightarrow |A| \leqslant \frac{|a * A|}{|a|} \Leftrightarrow |a|\,|A| \leqslant |a * A|$$

(N4) $|A|\bar{\mathscr{I}} \subseteq A \subseteq |A|\mathscr{I} \wedge |B|\bar{\mathscr{I}} \subseteq B \subseteq |B|\mathscr{I} \overset{(I12)}{\Rightarrow}$

$$|A|\,|B| * \bar{\mathscr{I}} \subseteq A * B \subseteq |A|\,|B|\mathscr{I} \Rightarrow |A * B| \leqslant |A|\,|B|$$

Equality does not hold in generality, as shown by the following example for $\mathrm{I}\mathbb{R} \times \overline{\mathrm{I}\mathbb{R}}$:

$$0 = |[0,1] * [1,0]| \neq |[0,1]|\,|[1,0]| = 1 \text{ cf. (N7).}$$

(N5) From (I21) we get at once

$$|1/A| = \left|\frac{A}{A * \bar{A}}\right| = \frac{|A|}{A * \bar{A}} \quad \text{for} \quad A * \bar{A} \in \mathbb{R}^+$$

(N6) See [9].

(N7) See [9].

(N8) For $A, B \in \mathrm{I}\mathbb{R}$ we have with $B = B * \mathscr{I} = |B|\mathscr{I}$

$A \subseteq |A|\mathscr{I} \Rightarrow A * B \subseteq |A|\,|B|\mathscr{I} * \mathscr{I} = |A|\,|B| * \mathscr{I} = |A|B$ and with (N13)
$|A| \subseteq A * \mathrm{sign}(A) \subseteq A * \mathscr{I} \Rightarrow |A|B \subseteq A * B * \mathscr{I} = A * B$, i.e.

$$A * B = |A|B \text{ (N12)}$$

For $A, B \in \overline{\mathrm{I}\mathbb{R}}$ we have $\bar{A} * \bar{B} = |\bar{A}| * \bar{B} = |A| * \bar{B}$, hence

$$A * B = \overline{|A| * \bar{B}} = |A| * B$$

(N9) $|A \cdot B| = |\lambda A \lambda B| \sqcup |\rho A \rho B| = |\lambda A|\,|\lambda B| \sqcup |\rho A|\,|\rho B| \leqslant$
$\leqslant (|\lambda A| \sqcup |\rho A|)(|\lambda B| \sqcup |\rho B|) = |A|\,|B|$

(N10) With $|A|\bar{\mathscr{I}} \subseteq A \subseteq |A|\mathscr{I}$ and $|C|\bar{\mathscr{I}} \subseteq C \subseteq |C|\mathscr{I}$
follows $|A|\bar{\mathscr{I}} \subseteq A \subseteq B \subseteq C \subseteq |C|\mathscr{I}$ and therefore $|B| \leqslant |A| \sqcup |C|$.
If $a \subseteq A$ then $\bar{A} \subseteq a \subseteq A$ and with (N12) $|a| \leqslant |A| \sqcup |\bar{A}| = |A|$

(N11) With $|A|\bar{\mathcal{J}} \subseteq A \subseteq |A|\mathcal{J}$ and $|B|\bar{\mathcal{J}} \subseteq B \subseteq |B|\mathcal{J}$
we get $|A|\bar{\mathcal{J}} \subseteq |B|\bar{\mathcal{J}} \subseteq A \subseteq B \subseteq |A|\mathcal{J} \subseteq |B|\mathcal{J} \Leftrightarrow$

$|A| * \mathcal{J} \underline{\cup} |B| * \mathcal{J} \overset{(I13)}{=} (|A| \underline{\cup} |B|) * \mathcal{J}$ and in the same way

$|A| * \bar{\mathcal{J}} \underline{\cup} |B| * \bar{\mathcal{J}} = (|A| \underline{\cup} |B|) * \bar{\mathcal{J}}$, so that with (N10) holds:

$$|A \underline{\cup} B| \underset{(N4)}{\leqslant} |(|A| \underline{\cup} |B|) * \mathcal{J}| \sqcup |(|A| \underline{\cup} |B|) * \bar{\mathcal{J}}|$$
$$\leqslant ||A| \underline{\cup} |B|| = |A| \sqcup |B|$$

The other properties are proved in [9] analogously.

(N12) $\pm T = T$ and $t\bar{\mathcal{J}} \subseteq A \subseteq t\mathcal{J} \overset{(I30)}{\Leftrightarrow} t\bar{\mathcal{J}} \subseteq \bar{A} \subseteq t\mathcal{J}$ proves the assertion

(N13) see [9].

The norm defined in Definition 2.1 is inducing a metric in $\mathbb{IR}$ in the following well-known way:

Definition 2.4.

$$\bigwedge_{A, B \in \mathbb{IR}} q(A, B) := |A - \bar{B}| \tag{2.5}$$

This is a metric with the following properties:

Lemma 2.5.

(M1) $q(A, B) \geqslant 0,\ q(A, B) = 0 \Leftrightarrow A = B,\ q(A, B) = q(B, A)$

(M2) $q(A, B) \leqslant q(A, C) + q(C, B)$

(M3) $q(A + B, A + C) = q(B, C)$

(M4) $q(A + B, C + D) \leqslant q(A, C) + q(B, D)$

(M5) $q(a * B, a * C) = |a| q(B, C),\ a \in \mathbb{R}$
$q(A * b, A * c) = |A|\, q(b, c)$ *for* $b \cdot c \geqslant 0$

(M6) $q(A * B, A * C) \leqslant |A|\, q(B, C)$

(M7) $q(A/B, A/C) \leqslant \dfrac{|A|}{|B * C|} q(B, C)$

(M8) $A \subseteq B \subseteq C \Rightarrow q(B, C) \leqslant q(A, C) \wedge q(A, B) \leqslant q(A, C) \Leftrightarrow$
$q(B, C) \sqcup q(A, B) \leqslant q(A, C)$

(M9) $A_1 \subseteq X \subseteq A_2 \wedge B_1 \subseteq Y \subseteq B_2 \Rightarrow q(X, Y) \leqslant q(A_1, B_2) \sqcup q(A_2, B_1)$
For $X \in \mathcal{Q}(A)$ *and* $Y \in \mathcal{Q}(B)$ *we have*: $q(X, Y) \leqslant q(A, B)$

(M10) $q(A, B) = q(\bar{A}, \bar{B})$

(M11) $|A| - |B| \leqslant q(A, B)$

(M12) $$q(A, B) \leqslant x \Leftrightarrow \bigvee_{C \in \mathbb{IR}} \begin{cases} \bigwedge_{a \subseteq A + C} \bigvee_{b \subseteq B + C} q(a, b) \leqslant x \wedge \\ \bigwedge_{b \subseteq B + C} \bigvee_{a \subseteq A + C} q(a, b) \leqslant x \end{cases}$$

(M13) *The set* $\mathscr{R}(A,B) := \{Z | q(A,B) = q(A,Z) + q(Z,B)\}$ *is non-empty for arbitrary* $A, B \in \mathbb{IR}$ *and is representing a rectangle in the* μ-δ-*plane with A and B as diagonal corners.*

(M14) $Z \in \mathscr{R}(A,0) \Rightarrow q(A,Z) = |A| - |Z|$

Proof.

(M1) follows from (N1) and (N12)

(M2) $|A - \bar{B}| \overset{(I4)}{=} |A - \bar{C} + C - \bar{B}| \overset{(N2)}{\leqslant} |A - C| + |C - \bar{B}|$

(M3) $|A + B - \overline{A + C}| = |A - \bar{A} + B - \bar{C}| \overset{(I4)}{=} |B - \bar{C}|$

(M4) $|A + B - \overline{C + D}| = |A - \bar{C} + B - \bar{D}| \overset{(N2)}{\leqslant} |A - \bar{C}| + |B - \bar{D}|$

(M5) $|a * B - \overline{a * C}| = |a * B - a * \bar{C}| \overset{(I16)}{=} |a(B - \bar{C})| \overset{(N3)}{=} |a|\,|B - C|$

For the second formula we can assume $c \leqslant b$ w.l.o.g., so that with $b \cdot c \geqslant 0$ follows $0 \leqslant \frac{c}{b} \leqslant 1$. Hence with the just proved assertion we have

$$|A * b - \overline{A * c}| = |b| \left| A * 1 - \overline{A * \frac{c}{b}} \right| = |b| \left| A * \left(\frac{c}{b} + 1 - \frac{c}{b}\right) - \overline{A * \frac{c}{b}} \right| =$$

$$\overset{(I17)}{=} |b| \left| A * \frac{c}{b} + A * \left(1 - \frac{c}{b}\right) - \overline{A * \frac{c}{b}} \right| =$$

$$\overset{(I4)}{=} |b| \left| A * \left(1 - \frac{c}{b}\right) \right| \overset{(N3)}{=} |A|\,|b| \left| 1 - \frac{c}{b} \right| = |A|\,|b - c|$$

(M6) with $f(x,a) = a \cdot x$ and the interval function $f(X,A) = A * X$ we derive from Theorem 2.6

$$|ax - ay| \leqslant \sup_{a \in \mathscr{Q}(A)} |a|\,|x - y| = |A|\,|x - y|$$

and so

$$q(A * X, A * Y) \leqslant |A|\, q(X, Y).$$

(M7) With (M6) we have $q(A/B, A/C) \leqslant |A|\, q(1/B, 1/C)$. Furthermore

$$|1/B - \overline{1/C}| = |\bar{C}/(B * C) - \overline{\bar{B}/(B * C)}|$$

$$\overset{(M6)}{\leqslant} \frac{1}{|B * C|} |\bar{C} - B| = \frac{1}{|B * C|} q(B, C)$$

(M8) From $A \subseteq B \subseteq C \Leftrightarrow A - \bar{C} \subseteq B - \bar{C} \subseteq 0 \wedge 0 \subseteq B - \bar{A} \subseteq C - \bar{A}$ we have with (N10)

$$|B - \bar{C}| \leqslant |A - \bar{C}| \wedge |A - \bar{B}| \leqslant |A - \bar{C}|.$$

(M9) $A_1 - \bar{B}_2 \subseteq X - \bar{Y} \subseteq A_2 - \bar{B}_1 \overset{(N10)}{\Rightarrow} |X - \bar{Y}| \leqslant |A_1 - \bar{B}_2| \sqcup |A_2 - \bar{B}_1|$

(M10) follows from (N12)

(M11) see [9]

(M12) "$\Rightarrow$" proves by $q(a,b) \overset{(M9)}{\leqslant} q(A+C, B+C) \overset{(M3)}{=} q(A,B) \leqslant x$

"$\Leftarrow$": $A + C \in \mathrm{I}\mathbb{R}$ and $B + C \in \mathrm{I}\mathbb{R}$ always holds and therefore with $|a-b| \leqslant x \Leftrightarrow a - b \subseteq \mathscr{J}x \wedge b - a \subseteq \mathscr{J}x$ we have

$$\bigwedge_{a \subseteq A+C} \bigvee_{b \subseteq B+C} a \subseteq \mathscr{J}x + b \wedge \bigwedge_{b \subseteq B+C} \bigvee_{a \subseteq A+C} b \subseteq \mathscr{J}x + a \Rightarrow$$

$$\bigwedge_{a \subseteq A+C} a \subseteq \mathscr{J}x + B + C \wedge \bigwedge_{b \subseteq B+C} b \subseteq \mathscr{J}x + A + C.$$

Therefore we have $A + C \subseteq \mathscr{J}x + B + C$ and $B + C \subseteq \mathscr{J}x + A + C$ such that

$A - \bar{B} \subseteq \mathscr{J}x$ and $B - \bar{A} \subseteq \mathscr{J}x$. Hence

$A - \bar{B} \subseteq \mathscr{J}x$ and $A - \bar{B} \supseteq \bar{\mathscr{J}}x$; therefore $\bar{\mathscr{J}}x \subseteq A - \bar{B} \subseteq x\mathscr{J}$

and with (N10) finally $|A - \bar{B}| \leqslant x$.

(M14) We have $|A| = q(A,0) = q(A,Z) + q(Z,0) \Rightarrow q(A,Z) = |A| - |Z|$

The following two theorems state important estimations for the interval analysis theoretically as well as in practical applications. They show, roughly spoken, that the topological properties of real functions hold in their interval extension, too.

Theorem 2.6. *Let $f(x,a)\colon D_1 \times \cdots \times D_n \times A_1 \times \cdots \times A_m \to \mathbb{R}$ with $D_i \in \mathbb{IR}$, $A_j \in \mathbb{IR}$ for $1 \leqslant i \leqslant n$, $1 \leqslant j \leqslant m$ be a continuous function in the sense of Lipschitz, which is (see Definition 1.2) extended on the interval domain $\mathscr{D} = \mathscr{Q}(D_1) \times \cdots \times \mathscr{Q}(D_n)$ with $\mathscr{X} = (X_1, \ldots, X_n)$ and $\mathscr{A} = (A_1, \ldots, A_m)$ (parameter domain) as in (1.1):*

$$f(\mathscr{X}, \mathscr{A}) = \mathop{\mathscr{W}}_{\substack{x \in \mathscr{Q}(X_1) \times \cdots \times \mathscr{Q}(X_n) \\ a \in \mathscr{Q}(A_1) \times \cdots \times \mathscr{Q}(A_m) = \mathscr{Q}(\mathscr{A})}} \{f(x,a)\}$$

The continuousity in the sense of Lipschitz shall hold for each variable x_i, $1 \leqslant i \leqslant n$ in the form

$$|f(x_1, \ldots, x_{i-1}, s, x_{i+1}, \ldots, x_n, \mathscr{A}) - f(x_1, \ldots, x_{i-1}, t, x_{i+1}, \ldots, x_n, \mathscr{A})| \leqslant l_i|s-t| \quad (*)$$

for $x_j \in D_j$, $1 \leqslant j \leqslant n$, $j \neq i$ and $a \in \mathscr{Q}(\mathscr{A})$. Then we have for all $\mathscr{X}, \mathscr{Y} \in \mathscr{D}$:

$$q(f(\mathscr{X}, \mathscr{A}), f(\mathscr{Y}, \mathscr{A})) \leqslant \sum_{i=1}^{n} l_i q(X_i, Y_i) \tag{2.6}$$

i.e. the interval extension is continuous in the sense of Lipschitz, too.

Proof. With (∗) we have immediately from the triangle inequation

$$q(f(x,a),f(y,a)) = |f(x,a) - f(y,a)| \overset{(*)}{\leqslant} \sum_{i=n}^{1} l_i|x_i - y_i| \overset{(M9)}{\leqslant} \sum_{i=1}^{n} l_i q(X_i, Y_i) \qquad (**)$$

Let now $C \in \mathbb{IR}$ such that

(i) for every $u = f(x, a) \subseteq f(\mathscr{X}, \mathscr{A}) + C$ with $x \in \mathscr{Q}(\mathscr{X}) \wedge a \in \mathscr{Q}(\mathscr{A})$ there always exists a $v = f(y, a) \subseteq f(\mathscr{Y}, \mathscr{A}) + C$ with $y \in \mathscr{Q}(\mathscr{Y})$. Then

$$q(u, v) \overset{(**)}{\leqslant} \sum_{i=1}^{n} l_i q(X_i, Y_i),$$

(ii) and for every $v = f(y, b) \subseteq f(\mathscr{Y}, \mathscr{A}) + C$ with $y \in \mathscr{Q}(\mathscr{Y}) \wedge b \in \mathscr{Q}(\mathscr{A})$ there always exists a $u = f(x, b) \subseteq f(\mathscr{X}, \mathscr{A}) + C$ with $x \in \mathscr{Q}(\mathscr{X})$. Then

$$q(u, v) \overset{(**)}{\leqslant} \sum_{i=1}^{n} l_i q(X_i, Y_i).$$

So we get with (M12)

$$q(f(\mathscr{X}, \mathscr{A}), f(\mathscr{Y}, \mathscr{A})) \overset{(M3)}{=} q(f(\mathscr{X}, \mathscr{A}) + C, f(\mathscr{Y}, \mathscr{A}) + C)) \overset{(M12)}{\leqslant}$$

$$\overset{(i)(ii)}{\leqslant} \sum_{i=1}^{n} l_i q(X_i, Y_i).$$

Theorem 2.7. *If in Theorem* 2.6 $D_1 = \cdots = D_n = D$ *and holds*

$$f(X, \mathscr{A}) = \underset{\substack{x \in \mathscr{D} \\ a \in \mathscr{Q}(\mathscr{A})}}{\mathscr{W}} \{\mathfrak{f}(x, a)\} = \mathfrak{f}(X, \ldots, X, \mathscr{A})$$

then for every $X, Y \in \mathscr{D}$

$$q(f(X, \mathscr{A}), f(Y, \mathscr{A})) \leqslant \sum_{i=1}^{n} l_i q(X, Y) = l q(X, Y).$$

After introducing a metric $\mathbb{IR}$ becomes a metric and therefore a topological space. Therefore we can use Cauchy sequences, convergence and continuousity as usually.

Lemma 2.8.

(2.7) *Every sequence in* $\mathbb{IR}$ *has at least one limes point in* $\mathbb{IR}$.

(2.8) $\{A_k\}_{k \geqslant 0}$ *is a Cauchy sequence if* $\{\lambda A_k\}_{k \geqslant 0}$ *and* $\{\rho A_k\}_{k \geqslant 0}$ *are both Cauchy sequences, i.e.* $\lim_{k \to \infty} A_k = A \Leftrightarrow \lim_{k \to \infty} \lambda A_k = \lambda A \; \lim_{k \to \infty} \rho A_k = \rho A$.

(2.9) *The operations* $+, -, *, /, \bar{\cap}, \underline{\cup}$ *and conjugation are continuous operators. Moreover* $f(\mathscr{X}, A)$ *is a continuous function if* $f(x, a)$ *like in* (1.1) *is one.*

(2.10) *Every sequence* $\{A_k\}_{k \geqslant 0}$ *with* $A_0 \supseteq A_1 \supseteq \cdots \supseteq B$ *is a Cauchy sequence and convergent against the interval* $A = \bigcap A_k \supseteq B$.

Proof.

(2.7) Follows from the homeomorphism of $(\mathbb{IR}, |\cdot|)$ and $(\mathbb{R}^2, \|\cdot\|)$ with respect to the maximum norm and (2.4).

(2.8) $|A_k - \bar{A}| \overset{(2.4)}{=} |\lambda A_k - \lambda A| \sqcup |\rho A_k - \rho A|$ proves the assertion.

(2.9) see [9].

(2.10) The sequence $\{A_i - \bar{B}\}_{i \geq 0}$ is lower bounded by 0 and therefore convergent and so a Cauchy sequence with limes A. For $\overline{\bigcap}$ is isotone we have for every n

$\overline{\bigcap_{i \geq 0}^{n}} A_i \supseteq A_n \supseteq A \supseteq B$ and hence the assertion for $n \to \infty$.

For the quality of numerical algorithms in interval analysis the width is an important criterion.

Definition 2.9. The functional

(2.11) $\bigwedge_{A \in \mathbb{IR}} d(A) := |A - A| = q(A, \bar{A})$

is called the width (diameter) of the interval A.

For d the following properties hold:

Lemma 2.10.

(2.12) $d(A) = |\rho A - \lambda A|, \quad d(\bar{A}) = d(A)$

(2.13) $d(A) = 0 \Leftrightarrow A = a \in \mathbb{R}$

(2.14) $A \subseteq B \subseteq C \Rightarrow d(B) \leq d(A) \sqcup d(C)$

(2.15) $d(A + B) = d(A) + d(B)$

(2.16) $d(A * B) \leq d(A)|B| + |A|\, d(B)$

(2.17) (i) $d(A * B) \geq |A|\, d(B) \sqcup d(A)\, |B|$ *for* $A, B \in \mathrm{I}\mathbb{R}$ *or* $A, B \in \overline{\mathrm{I}\mathbb{R}}$

(ii) $d(A * B) \geq d(A)\, d(B)$ *for* $A, B \in \mathscr{T}^*$

(2.18) (i) $d(a * B) = |a|\, d(B)$ *for* $a \in \mathbb{R}$

(ii) $d(A * B) = |B|\, d(A)$ *for* $A, B \in \mathrm{I}\mathbb{R}$, $A \in \mathscr{T}$, $B \in \mathscr{S}^*$ *or* $A, B \in \overline{\mathrm{I}\mathbb{R}}$, $A \in \bar{\mathscr{T}}$, $B \in \mathscr{S}^*$

(2.19) $d(A^n) \leq n|A|^{n-1} d(A)$ *if* $A^n := \underbrace{A * A * \cdots * A}_{n \text{ factors}}$

(2.20) $d((A - B)^n) \leq 2\, d(A)^n$ *for* $B \in \mathscr{Q}(A)$, $n \geq 1$
$\leq 2^{1-n} d(A)^n$ *for* $\mu(A) \subseteq B$, $n \geq 1$

(2.21) $|A| \leq d(A) \leq 2|A|$ *for* $A \in \mathscr{T}^*$

(2.22) (i) $d(1/A) = \dfrac{d(A)}{A * \bar{A}}$ *for* $A \notin \mathscr{T}^*$

(ii) $A \subseteq X/Y \subseteq B \Rightarrow d(X/Y) \leqslant \left(\frac{d(X)}{|X|} + \frac{d(Y)}{|Y|}\right)(|A| \sqcup |B|)$

for $Y \notin \mathcal{T}^*$ *and* $|X * Y| = |X|\,|Y|$ (*cf.* N7, N8).

(2.23) $\bar{B} \subseteq \bar{A} \subseteq A \subseteq B \Rightarrow q(A, B) \leqslant q(\bar{A}, B) \leqslant q(B, \bar{B}) = d(B)$

$$\bar{B} \subseteq \bar{A} \subseteq A \subseteq B \Rightarrow \frac{d(B) - d(A)}{2} \leqslant q(A, B) \leqslant d(B) - d(A)$$

(2.24) $d(A \mathbin{\bar{\sqcap}} B) \leqslant d(A) \sqcap d(B)$ *for* $A, B \in I\mathbb{R}$

Proof.

(2.12) $A - A = [\lambda A - \rho A, \rho A - \lambda A] \overset{(2.4)}{\Rightarrow} |A - A| = |\rho A - \lambda A|$

$$d(\bar{A}) = |\bar{A} - \bar{A}| \overset{(N12)}{=} |A - A| = d(A)$$

(2.13) $d(A) = 0 \overset{(2.12)}{\Rightarrow} \rho A = \lambda A \Rightarrow A \in \mathbb{R}$

(2.14) $A \subseteq B \subseteq C \Rightarrow A - A \overset{(I5)}{\subseteq} B - A \subseteq B - B \subseteq C - B \subseteq C - C \overset{(N10)}{\Rightarrow}$

$$d(B) = |B - B| \leqslant |C - C| \sqcup |A - A| = d(C) \sqcup d(A)$$

(2.15) $|(A + B) - (A + B)| = |A - A + B - B| \overset{(N2)}{\leqslant}$

$$|A - A| + |B - B| = d(A) + d(B)$$

(2.16) $d(A * B) = q(A * B, \overline{A * B}) \overset{(M2)}{\leqslant} q(A * B, \bar{A} * B) + q(\bar{A} * B, \overline{A * B}) \leqslant$

$$\overset{(M6)}{\leqslant} |B| q(A, \bar{A}) + |A| q(B, \bar{B}) = |B|\, d(A) + |A|\, d(B)$$

(2.17) (i) For $A, B \in I\mathbb{R}$ we have

$$\left.\begin{array}{l} 0 \subseteq A * (B - B) \overset{(I18,i)}{\subseteq} A * B - A * B \\ \text{as well as} \\ 0 \subseteq B * (A - A) \overset{(I18,i)}{\subseteq} A * B - A * B \end{array}\right\} \overset{(N4)(N10)}{\Rightarrow}$$

$$|A|\, d(B) = |A|\,|B - B| \overset{(N7)}{=} |A * (B - B)| \overset{(N10)}{\leqslant} |A * B - A * B| = d(A * B)$$

as well as

$$|B|\, d(A) = |B|\,|A - A| \overset{(N7)}{=} |B * (A - A)| \overset{(N10)}{\leqslant} |B * A - B * A| = d(A * B),$$

and therefore $|A|\, d(B) \sqcup |B|\, d(A) \leqslant d(A * B)$.

For $A, B \in \overline{I\mathbb{R}}$ the assertion follows with
$d(A * B) = d(\overline{A * B}) = d(\bar{A} * \bar{B})$

(ii) proves with (i) and (2.21)

(2.18) (i) $d(a*B) \overset{(I16)}{=} |a*B - a*B| = |a*(B-B)| =$

$$\overset{(N3)}{=} |a|\,|B-B| = |a|\,d(B)$$

(ii) $d(A*B) \overset{(N8)}{=} d(A*|B|) \overset{(i)}{=} |B|\,d(A)$

(2.19) True for $n = 0$, so take induction over n:

$$d(A^{n+1}) = d(A^n * A) \overset{(2.16)}{\leqslant} |A|\,d(A^n) + |A|^n d(A) \overset{(N4)}{\leqslant}$$

$$\leqslant |A| n |A|^{n-1} d(A) + |A|^n d(A) = (n+1)|A|^n d(A)$$

(2.20) see [9]

(2.21) see [9]

(2.22) (i) $d(1/A) \overset{(I11)}{=} d(1/A * A/\bar{A}) = d(A * 1/(A*\bar{A})) \overset{(2.18)}{=} \dfrac{d(A)}{A*\bar{A}}$

(ii) $A \subseteq X/Y \subseteq B \Rightarrow Y*\bar{Y}*A \subseteq Y*X \subseteq Y*\bar{Y}*B \Rightarrow$

$$\Rightarrow |Y*X| \overset{(N7)}{=} |Y|\,|X| \overset{(N10)}{\leqslant} Y*\bar{Y}*(|A| \sqcup |B|) \text{ and so}$$

$$|X| \leqslant \frac{Y*\bar{Y}*(|A|\sqcup|B|)}{|Y|} \text{ as well as } |Y| \leqslant \frac{Y*\bar{Y}(|A|\sqcup|B|)}{|X|} \qquad (*)$$

On the other hand

$$d\left(\frac{X}{Y}\right) = \frac{d(X*Y)}{Y*\bar{Y}} \overset{(2.16)}{\leqslant} \frac{d(X)|Y|}{Y*\bar{Y}} + \frac{d(Y)|X|}{Y*\bar{Y}} \overset{(*)}{\leqslant} \left(\frac{d(X)}{|X|} + \frac{d(Y)}{|Y|}\right)(|A|\sqcup|B|).$$

(2.23) $\bar{B} \subseteq \bar{A} \subseteq A \subseteq B \Rightarrow q(A,B) \leqslant q(\bar{A},B) \leqslant q(\bar{B},B) = d(B)$

on the other hand

$$d(B) = q(B,\bar{B}) \leqslant q(B,A) + q(A,\bar{A}) + q(A,\bar{B}) \overset{(M13)(M10)}{=} 2q(A,B) + d(A),$$

i.e. $\dfrac{d(B)-d(A)}{2} \leqslant q(A,B)$.

(2.24) see [9].

To conclude this chapter we give some important estimations for interval functions, at first an analogous estimation to (2.6) for the width.

Theorem 2.11. *Under the assumption of Theorem* 2.6 *we have for all* $\mathscr{X} \in \mathscr{D}$

$$d(f(\mathscr{X})) \leqslant \sum_{i=1}^{n} l_i\,d(X_i) \qquad (2.25)$$

where $\mathscr{A} = a \in \mathbb{R}^m$ *are no intervals.*

Proof.

$$d(f(\mathscr{X})) = q(f(\mathscr{X}),\overline{f(\mathscr{X})}) = q(f(\mathscr{X}),f(\bar{\mathscr{X}})) \overset{(2.6)}{\leqslant} \sum_{i=1}^{n} l_i q(X_i,\bar{X}_i) \overset{(2.11)}{=} \sum_{i=1}^{n} l_i d(X_i).$$

Furthermore we give estimations for the distance of interval functions with properties depending on the width of the arguments. Let be in the following $\mathscr{D} \subset \mathbb{IR}$.

Theorem 2.12. *Let $g(X,\mathscr{A})$ be the interval extension of $g(x,a)$ like in Theorem 2.6 and let $f(X,\mathscr{A})$ be an arbitrary interval function satisfying*

$$\bigwedge_{X\in\mathscr{D}} \bigvee_{T\in\mathscr{Q}(X)} g(T,\mathscr{A}) \subseteq f(X,\mathscr{A}) \subseteq g(X,\mathscr{A}) \tag{2.26}$$

Then for all $X \in \mathscr{D}$

$$q(f(X,\mathscr{A}),g(X,\mathscr{A})) \leqslant \sum_{i=1}^{n} l_i d(X) = l\,d(X) \tag{2.27}$$

If in (2.26) $\mu(X) \subseteq T$ *or* $f(X,\mathscr{A}) = \underset{\substack{x\in\mathscr{Q}(X)\\ a\in\mathscr{Q}(A)}}{\mathscr{W}} \{g(x,a)\}$

then we have

$$q(f(X,\mathscr{A}),g(X,\mathscr{A})) \leqslant l/2\, d(X) \tag{2.28}$$

Proof.

(2.27): From (2.26) we get

$$q(f(X,\mathscr{A}),g(X,\mathscr{A})) \leqslant q(g(T,\mathscr{A}),g(X,\mathscr{A})) \leqslant$$

$$\overset{(2.6)}{\leqslant} \sum_{i=1}^{n} l_i q(T,X) \leqslant \tag{$*$}$$

(respecting $\bar{X} \subseteq T \subseteq X$) $\quad \overset{(M8)}{\leqslant} \sum_{i=1}^{n} l_i q(\bar{X},X) = l\,d(X)$

(2.28): From $\mu(X) \subseteq T \subseteq X$ we derive

$$q(T,X) \leqslant q(\mu(X),X) = q(X/2+\bar{X}/2,X) \overset{(M3)}{=} q(\bar{X}/2,X/2) = d(X)/2,$$

so $(*)$ is estimated by $l/2\,d(X)$.

If $f(X,\mathscr{A}) = \underset{\substack{x\in\mathscr{Q}(X)\\ a\in\mathscr{Q}(\mathscr{A})}}{\mathscr{W}} \{g(x,a)\}$ then trivially

$g(\mu(X),\mathscr{A}) \subseteq f(X,\mathscr{A}) \subseteq g(X,\mathscr{A})$ holds.

Theorem 2.13. *Let $h(X,\mathscr{A})$ with $X\in\mathscr{D}$ be the interval extension of $h(x,a)$ for $a\in\mathscr{Q}(\mathscr{A})$ and let h be continuous in the sense of Lipschitz with constant l. Let for $G\in\mathbb{IR}$*

$$g(X,\mathscr{A}) = G + h(X,\mathscr{A}) * (X - Z)$$

be an interval function in (generalized) centralized form [3] *with arbitrary $Z\in\mathscr{Q}(X)$*

and let $f(X, \mathscr{A})$ be an arbitrary interval function with the following property:

$$\bigwedge_{X \in \mathscr{D}} \bigwedge_{T \in \mathscr{Q}(X)} G + h(T, \mathscr{A}) * (X - \mathscr{A}) \subseteq f(X, \mathscr{A}) \subseteq g(X, \mathscr{A}) \tag{2.29}$$

then

$$\text{for all } X \in \mathscr{D}\colon q(f(X, \mathscr{A}), g(X, \mathscr{A})) \leqslant l \cdot d(X)^2 \tag{2.30}$$

holds. Moreover for $\mu(X) \subseteq T$ or $\mu(X) = Z$ in (2.29) we have

$$q(f(X, \mathscr{A}), g(X, \mathscr{A})) \leqslant (l/2) \cdot d(X)^2. \tag{2.31}$$

If $\mu(X) \subseteq T$ and $\mu(X) = Z$ then

$$q(f(X, \mathscr{A}), g(X, \mathscr{A})) \leqslant (l/4) \cdot d(X)^2. \tag{2.32}$$

Proof.

(2.30): From (2.29) we get

$$\begin{aligned} q(f(X, \mathscr{A}), g(X, \mathscr{A})) &\overset{(M8)}{\leqslant} q(G + h(T, \mathscr{A}) * (X - Z), G + h(X, \mathscr{A}) * (X - Z)) \\ &\overset{(M3)}{=} q(h(T, \mathscr{A}) * (X - Z), h(X, \mathscr{A}) * (X - Z)) \leqslant \\ &\overset{(M6)}{\leqslant} |X - Z|\, q(h(T, \mathscr{A}), h(X, \mathscr{A})) \leqslant |X - Z| \cdot l \cdot q(T, X) \leqslant \\ &\overset{(N10, M8)}{\leqslant} |X - X| \cdot l \cdot q(\bar{X}, X) \overset{(2.11)}{\leqslant} d(X) \cdot l \cdot d(X) = l \cdot d(X)^2 \end{aligned} \tag{$*$}$$

with $\bar{X} - \bar{X} \subseteq X - Z \subseteq X - X$ and $\bar{X} \subseteq T \subseteq X$.

(2.31): From $\mu(X) = (X + \bar{X})/2 \subseteq T \subseteq X$ we get with (M8)

$$q(T, X) \leqslant q(X/2 + \bar{X}/2, X) \overset{(M3)}{=} q(\bar{X}/2) = d(X)/2$$

or in the case $\mu(X) = Z$ and

$$|X - Z| = |X - X/2 - \bar{X}/2| = |X/2 - X/2| = d(X)/2$$

with line $(*)$ the assertion.

(2.32): With both estimations in (2.31) we achieve the factor $l/4$.

Lemma 2.14. *If $\mathscr{A} = a \in \mathbb{R}^m$, $Z \in \mathscr{Q}(X)$ and $f(X) = \mathscr{W}_{x \subseteq X}\{g(Z) + h(x) * (x - Z)\}$, then with $G = g(Z)$ we have $f(X) \subseteq g(X)$ and from*

$$\mathscr{W}_{x \subseteq X}\{g(Z) + h(x) * (x - Z)\} = g(Z) + h(w) * (X - Z)$$

with $|h(w)| = \min_{x \subseteq X} |h(x)|$, $T = w \in \mathbb{R}$ and $0 \subseteq X - Z$ on the other hand

$$G + h(w) * (X - Z) \subseteq f(X)$$

so that (2.29) holds. Therefore we have

$$q(\mathscr{W}_{x \subseteq X}\{g(x)\}, g(Z) + h(X) * (X - Z)) \leqslant l\, d(X)^2. \tag{2.33}$$

Remarks. Both the Theorems 2.12 and 2.13 can be generalized (to several variables) as well as in the properties (2.26) and (2.29). For instance in (2.26) suffices the weaker assumption

$$\bigwedge_{X \in \mathscr{D}} \bigvee_{T \subseteq X \subseteq U} g(T, \mathscr{A}) \subseteq f(X, \mathscr{A}) \subseteq g(U, \mathscr{A}). \tag{2.26'}$$

Finally we can state, that if an interval function f can be included in this way by the interval function g, then f and g have the same topological properties.

If g is continuous, then so is f, if g is continuous in the sense of Lipschitz, then so is f and this of the same degree. This theorem is a very important instrument in the interval analysis.

Theorem 2.15. *Let* $g \in C^1_{\mathscr{D}}, g'$ *continuous in the sense of Lipschitz with constant* l, $g(X) = g(Z) + g'(X) * (X - Z)$ *the interval extension of* g *with* $Z \in \mathscr{Q}(X)$ *and* $g'(X)$ *the interval extension of* g' *for* $X \in \mathscr{D}$. *If then for an arbitrary function* $f(X)$

$$\bigwedge_{X \in \mathscr{D}} \bigvee_{T \in \mathscr{Q}(X)} g(Z) + g'(T)(X - Z) \subseteq f(X) \subseteq g(X) \tag{2.34}$$

holds, then

$$q(f(X), g(X)) \leqslant l\, d(X)^2 \tag{2.35}$$

For $z = \mu(X)$ *or* $\mu(X) \subseteq T$ *we have*

$$q(f(X), g(X)) \leqslant l/2\, d(X)^2 \tag{2.36}$$

If $z = \mu(X)$ *and* $\mu(X) \subseteq T$ *then*

$$q(f(X), g(X)) \leqslant l/4\, d(X)^2. \tag{2.37}$$

The following theorem is a generalization of Theorem 2.15 and introduces the application of Taylor-expansions.

Theorem 2.16. *Let* $L(X, \mathscr{A})$ *be as in Theorem* 2.13 *and for arbitrary* $Z \in \mathscr{Q}(X)$ *and* $G_i \in \mathbb{IR}$, $0 \leqslant i \leqslant n - 1$

$$g(X, \mathscr{A}) = \sum_{i=0}^{n-1} G_i * (X - Z)^i + h(X, \mathscr{A}) * (X - Z)^n$$

be an interval function. Moreover for the interval function $f(X, \mathscr{A})$

$$\bigwedge_{X \in \mathscr{D}} \bigvee_{T \in \mathscr{Q}(X)} \sum_{i=0}^{n-1} G_i * (X - Z)^i + h(T, \mathscr{A}) * (X - Z)^n \subseteq f(X, \mathscr{A}) \subseteq g(X, \mathscr{A}) \tag{2.38}$$

let be satisfied. Then for all $X \in \mathscr{D}$

$$q(f(X, \mathscr{A}), g(X, \mathscr{A})) \leqslant l \cdot d(X)^{n+1}$$

holds. If $Z = \mu(X)$ *or* $\mu(X) \subseteq T$, *then the constant can be halved as in Theorem* 2.15.

Proof. Like for Theorem 2.13, where $|(X - Z)^n| \leqslant |X - Z|^n \leqslant d(X)^n$ and $q(T, X) = \frac{1}{2} d(X)$ have now to be estimated, resp.

In many cases we have with $z \in \mathbb{R}$, $z \in X$:

$$G_i = f^{(i)}(z)/i! \text{ and } h(X, \mathscr{A}) = \frac{f^{(n)}(X, \mathscr{A})}{n!}$$

where $z = \mu(X)$ is to be preferred in general.

3. Conclusion

The methods used here improve the estimations of the Theorems 3, 4, and 6 in [1] with a factor 2 and, moreover the assertions were widely generalized and reduced to few, clear lines. So the Interval Analysis now becomes a calculus which is comparable to those of classical Analysis: The handling of norm and metric are very similar to norm and metric in linear spaces. As a by-product of the extension to generalized intervals the Interval Analysis turns out to be more independent of set theory because many results are derivable without using set theory.

Some further assertions are summarized in [9]. Nevertheless a lot of properties have to be investigated in the future to complete the theory.

References

[1] Alefeld, G., Herzberger, J.: Einführung in die Intervallrechnung. (Reihe Informatik.) BI 1974.

[2] Albrecht, R.: Grundlagen einer Theorie gerundeter algebraischer Verknüpfungen in topologischen Vereinen, Computing, Suppl. 1, pp. 1–14. Wien-New York: Springer 1977.

[3] Hansen, E.: Topics in Interval Analysis. Oxford University Press 1969.

[4] Kaucher, E.: Über metrische und algebraische Eigenschaften einiger beim numerischen Rechnen auftretender Räume. Dissertation, Universität Karlsruhe, 1973.

[5] Kaucher, E.: Allgemeine Einbettungssätze algebraischer Strukturen unter Erhaltung von verträglichen Ordnungs- und Verbandsstrukturen mit Anwendung in der Intervallrechnung. ZAMM **56**, T296 (1976).

[6] Kaucher, E.: Algebraische Erweiterung der Intervallrechnung unter Erhaltung der Ordnungs- und Verbandsstrukturen. Computing, Suppl. 1, pp. 68–69. Wien-New York: Springer 1977.

[7] Kaucher, E.: Über Eigenschaften und Anwendungsmöglichkeiten der erweiterten Intervallrechnung und des hyperbolischen Fastkörpers über *R*. Computing, Suppl. 1, pp. 82–94. Wien-New York: Springer 1977.

[8] Kaucher, E.: Über eine Überlaufarithmetik auf Rechenanlagen und deren Anwendungsmöglichkeiten. ZAMM **57**, T286 (1977).

[9] Kaucher, E.: Formelsammlung zur erweiterten Intervallrechnung. Interner Bericht des Institutes für Angewandte Mathematik, Universität Karlsruhe, 1979.

[10] Kulisch, U.: Grundlagen des Numerischen Rechnens. (Reihe Informatik, Bd. 19.) Zürich: Bibliograph. Institut 1976.

[11] Moore, R. E.: Interval Analysis. Prentice-Hall 1966.

[12] Nickel, K.: Interval Mathematics, in: Lecture Notes in Computer Science. Berlin-Heidelberg-New York: Springer 1975.

[13] Wolff von Gudenberg, J.: Determination of minimum sets of the set of zeros of a function. Computing **24** (1980).

[14] Hansen, E.: A globally convergent interval method for computing and bounding real roots. BIT **18**, 415 (1978).

Dr. E. Kaucher
Institut für Angewandte Mathematik
Universität Karlsruhe
Kaiserstrasse 12
D-7500 Karlsruhe
Federal Republic of Germany

Computing, Suppl. 2, 51 – 67 (1980)

Arithmetic Operations in Interval Spaces

U. W. Kulisch* and **W. L. Miranker,** Yorktown Heights, New York

Abstract

If M is an ordered algebraic structure every operation $*$ can be extended to intervals over M by the following two steps: 1. Extend the operation to the powerset $\mathbf{P}M$ of M and 2. Round this result into the set of intervals $\mathbf{I}M$ by taking the least interval which includes the result in $\mathbf{P}M$. This process leads to the best possible approximation of the operation in $\mathbf{P}M$ by that in $\mathbf{I}M$. Because of the nature of the definition of the powerset operation, however, this process does not lead to executable formulas if in actual applications M is replaced by the set of vectors or matrices over the real numbers, the complex numbers or the set of vectors or matrices over the complex numbers. We show that for all of these sets different definitions of the operations can be given which via isomorphism lead to executable formulas for the optimal arithmetic even on subsystems of the various sets mentioned above. Interval arithmetic in product spaces was originally defined by these executable formulas. In such an approach, however, the connection with the powerset operations and several other interpretations, which are of fundamental interest in interval analysis, are missing.

1. Introduction

We begin with a general consideration of computer arithmetic and develop different methods of defining the operations in product sets. These methods are then used in the following sections to obtain computer executable formulas for the arithmetic operations in interval spaces.

Numerical algorithms are usually defined in the space $\mathbf{R}$ of real numbers and the vectors $V\mathbf{R}$ or matrices $M\mathbf{R}$ over the real numbers. The algorithms often apply to complex valued quantities also, so that the complex spaces $\mathbf{C}$, $V\mathbf{C}$ and $M\mathbf{C}$ are also employed. All of these spaces are ordered with respect to the order relation, $\leqslant$. (In all product spaces the order relation is defined componentwise.) Recently numerical analysts have begun to define and study algorithms for intervals defined over these spaces. If we denote the set of intervals over an ordered set $\{M, \leqslant\}$ by $\mathbf{I}M$, we obtain the spaces $\mathbf{IR}$, $\mathbf{I}V\mathbf{R}$, $\mathbf{I}M\mathbf{R}$, and $\mathbf{IC}$, $\mathbf{I}V\mathbf{C}$, $\mathbf{I}M\mathbf{C}$.

In Fig. 1 to which we will repeatedly refer, we present a tableau of spaces and operations. For example, in the second column of this figure we list the various spaces in which arithmetic is performed and which we have introduced in the previous paragraph.

For arithmetic purposes a real number is usually represented by an infinite b-adic expansion, with operations performed on these expansions being defined as the

* On leave from the University of Karlsruhe.

	I	II	III	IV	V
1		$\mathbf{R}$	$\supset D$	$\supset S$	$+ - \cdot /$
					$\times$
2		$V\mathbf{R}$	$\supset VD$	$\supset VS$	$+ -$
					$\times$
3		$M\mathbf{R}$	$\supset MD$	$\supset MS$	$+ - \cdot$
4	$\mathbf{PR}$	$\supset \mathbf{IR}$	$\supset \mathbf{I}D$	$\supset \mathbf{I}S$	$+ - \cdot /$
					$\times$
5	$\mathbf{P}V\mathbf{R}$	$\supset \mathbf{I}V\mathbf{R}$	$\supset \mathbf{I}VD$	$\supset \mathbf{I}VS$	$+ -$
					$\times$
6	$\mathbf{P}M\mathbf{R}$	$\supset \mathbf{I}M\mathbf{R}$	$\supset \mathbf{I}MD$	$\supset \mathbf{I}MS$	$+ - \cdot$
7		$\mathbf{C}$	$\supset \mathbf{C}D$	$\supset \mathbf{C}S$	$+ - \cdot /$
					$\times$
8		$V\mathbf{C}$	$\supset V\mathbf{C}D$	$\supset V\mathbf{C}S$	$+ -$
					$\times$
9		$M\mathbf{C}$	$\supset M\mathbf{C}D$	$\supset M\mathbf{C}S$	$+ - \cdot$
10	$\mathbf{PC}$	$\supset \mathbf{IC}$	$\supset \mathbf{IC}D$	$\supset \mathbf{IC}S$	$+ - \cdot /$
					$\times$
11	$\mathbf{P}V\mathbf{C}$	$\supset \mathbf{I}V\mathbf{C}$	$\supset \mathbf{I}V\mathbf{C}D$	$\supset \mathbf{I}V\mathbf{C}S$	$+ -$
					$\times$
12	$\mathbf{P}M\mathbf{C}$	$\supset \mathbf{I}M\mathbf{C}$	$\supset \mathbf{I}M\mathbf{C}D$	$\supset \mathbf{I}M\mathbf{C}S$	$+ - \cdot$

Fig. 1. Table of the spaces occurring in numerical computations

limit of the sequence of results obtained by operating on finite portions of the expansions. In principle, a computer could approximate this limiting process but the apparent inefficiency of such an approach rules out its implementation even on the fastest computers. In fact, for arithmetic purposes the real numbers are approximated by a subset S of $\mathbf{R}$ in which all operations are simple and rapidly performable. The most common choice for this subset S is the so-called floating-point system with a fixed number of digits in the mantissa. If a prescribed accuracy for the computation cannot be achieved by operating within S, we use a larger subset D of $\mathbf{R}$ with the property $\mathbf{R} \supset D \supset S$. For arithmetic purposes we define vectors, matrices, intervals and so forth as well as the corresponding complexifications over S and D. So doing, we obtain the spaces VS, MS, $\mathbf{I}S$, $\mathbf{I}VS$, $\mathbf{I}MS$, $\mathbf{C}S$, $V\mathbf{C}S$, $M\mathbf{C}S$, $\mathbf{IC}S$, $\mathbf{I}V\mathbf{C}S$, $\mathbf{I}M\mathbf{C}S$ and the corresponding spaces over D. These two collections of spaces are listed in the third and fourth columns of Fig. 1. For example, $\mathbf{C}S$ is the set of all pairs of elements of S, $V\mathbf{C}S$ the set of all n-tuples of such pairs, $\mathbf{IC}S$ the set of all intervals over the ordered set $\{\mathbf{C}S, \leqslant\}$ and so forth.

S and D are usually chosen as the sets of floating point numbers of single and double length, respectively. However in Fig. 1, S and D are generic symbols for a whole system of subsets of $\mathbf{R}$ with arithmetic properties which we will subsequently define.

Having defined the sets listed in the third and fourth columns of Fig. 1, we turn to the question of defining operations within these sets. These operations are supposed to approximate operations which are defined on the corresponding sets listed in the second column of Fig. 1. Fig. 1 is composed of four blocks of rows, and in every set in each of the first rows of each block, we are to define an addition, a subtraction, a multiplication and a division. For the sets in the last row of each such block, for instance, we need to define an addition, a subtraction and a multiplication, etc. These required operations are listed in the fifth column of Fig. 1. Furthermore the rows in Fig. 1 are not mutually independent, arithmetically. By this we mean, for instance, that a vector can be multiplied by a scalar as well as by a matrix; an interval vector can be multiplied by an interval as well as by an interval matrix. These latter multiplication types are indicated in Fig. 1 by means of a $\times$-sign between rows in the fifth column therein.

Let S and D be floating-point numbers of single and double mantissa length. By computer arithmetic, we understand the totality of operations which have to be defined in all of the sets listed in the third and fourth column in Fig. 1. In a good programming system these operations should be available as operators for all admissible combinations of data types.

It turns out that there are in principle two different basic methods for defining computer arithmetic. In the sense of Fig. 1, these methods may be called the vertical and the horizontal method. By way of illustration of these two possibilities, we consider in Fig. 2, a simple detail of Fig. 1. In Fig. 2 we display the sets $\mathbf{R}$, D and S as well as the spaces of matrices $M\mathbf{R}$, MD and MS. In D and S, an arithmetic can only be defined by the horizontal method which we will describe below.

$$\text{a)}\quad \begin{array}{ccccc} \mathbf{R} & \rightarrow & D & \rightarrow & S \\ \downarrow & & \downarrow & & \downarrow \\ M\mathbf{R} & & MD & & MS \end{array}$$

$$\text{b)}\quad \begin{array}{ccccc} \mathbf{R} & \rightarrow & D & \rightarrow & S \\ \downarrow & & & & \\ M\mathbf{R} & \rightarrow & MD & \rightarrow & MS \end{array}$$

Fig. 2

Let us assume that an arithmetic in D and in S is defined. By the vertical definition of the arithmetic in MD and MS, we mean that the operations in MD and MS are defined by the operations in D and S and the usual formulas for the addition and multiplication of real matrices (see Fig. 2a). On most computers this is precisely the method of definition of addition and multiplication for matrices.

The horizontal method of defining the arithmetic operations in the sets of the third and fourth columns of Fig. 1 is a process which passes from left to right by using a certain set of formulas. In the case of the first row in Fig. 2b, these formulas employ a mapping called a rounding, $\square: \mathbf{R} \rightarrow S$ and are given as follows:

(R1) $\bigwedge_{a \in S} \square a = a$

(R2) $\bigwedge_{a,b \in \mathbf{R}} (a \leqslant b \Rightarrow \square a \leqslant \square b)$

(R4) $\bigwedge_{a \in \mathbf{R}} \square(-a) = -\square a$

(RG) $\bigwedge_{a,b \in S} a \boxed{*} b := \square(a * b)$ for all $* \in \{+, -, \cdot, /\}$.

In case of the second row of Fig. 2b, we employ the same set of formulas, where $\square$: $M\mathbf{R} \to MS$ denotes a mapping from $M\mathbf{R}$ into MS:

(R1) $\bigwedge_{A \in MS} \square A = A$

(R2) $\bigwedge_{A, B \in M\mathbf{R}} (A \leqslant B \Rightarrow \square A \leqslant \square B)$

(R4) $\bigwedge_{A \in M\mathbf{R}} \square(-A) = -\square A$

(RG) $\bigwedge_{A, B \in MS} A \boxed{*} B := \square(A * B)$ for all $* \in \{+, -, \cdot\}$.

These formulas show that the mapping $\square$ is close to being a homomorphism. (R1) is a quite natural property, which every rounding should have. (R2), (R4) and (RG) can be shown to be necessary conditions for a homomorphism [8], [9]. However, it can be shown that a proper homomorphism cannot be established. Because of this, we call a mapping which fulfills (R1), (R2), (R4) and (RG) – for all inner and outer operations – a semimorphism. In the cases of rows of Fig. 1 which contain spaces of intervals, we assume that a semimorphism has the additional property

(R3) $\bigwedge_{A \in \mathbf{IR}} A \subseteq \square A$.

The horizontal method of defining computer arithmetic is one which employs the semimorphism, $\square$. Because of the properties (R1), (R2) and (RG), this definition furnishes the arithmetic in the subset which is best possible in a sense to be made precise.

It can be shown that the horizontal method or the definition of arithmetic operations by means of the semimorphism can be applied in all rows of Fig. 1. It seems to be the most natural such definition for all rows of Fig. 1 which include interval sets. We now focus our attention on these rows.

2. Interval Arithmetic Defined by Semimorphisms

Intervals are defined over ordered sets. From Fig. 1 we see that the latter are the sets $\mathbf{R}$, $V\mathbf{R}$, $M\mathbf{R}$ and $\mathbf{C}$, $V\mathbf{C}$ and $M\mathbf{C}$. The arithmetic operations in these sets are well defined. Now let M denote one of these sets, and let $*$ be one of the operations defined in M. We extend the operation $*$ to the elements of the powerset $\mathbf{P}M$ of M, which is the set of all subsets of M, by means of the following definition:

$$\bigwedge_{A, B \in \mathbf{P}M} A * B := \{a * b \mid a \in A \wedge b \in B\}. \tag{1}$$

We do this for all operations $*$ defined in M.

The set of intervals $\mathbf{I}M$ over the ordered set $\{M, \leqslant\}$ is a subset of the powerset $\mathbf{P}M$. In general $\mathbf{I}M$ is not closed under the powerset operations. To arrange that an operation on intervals leads to an *interval* which includes the result of the corresponding powerset operation on these intervals, we define the interval operation by means of a semimorphism by using the monotone upwardly directed rounding $\square$: $\mathbf{P}M \to \mathbf{I}M$. This process is described by the following formulas:

(R1) $\bigwedge_{A \in \mathbf{I}M} \square A = A$

(R2) $\bigwedge_{A, B \in \mathbf{P}M} A \subseteq B \Rightarrow \square A \subseteq \square B$

(R3) $\bigwedge_{A \in \mathbf{P}M} A \subseteq \square A$

(R4) $\bigwedge_{A \in \mathbf{P}M} \square(-A) = -\square A$

(RG) $\bigwedge_{A, B \in \mathbf{I}M} A \boxast B := \square(A * B)$ for all $*$ defined in $\mathbf{P}M$.

Let $U(A)$ denote the subset of $\mathbf{P}M$ all of whose elements contain A itself. Then because of the well-known formula (see [8]) for the monotone upwardly directed rounding $\square$:

(R) $\bigwedge_{A \in \mathbf{P}M} \square A = \inf(U(A) \cap \mathbf{I}M)$,

the above definition (RG) gives the best possible approximation in $\mathbf{I}M$ of the results of powerset operations when the latter are applied to intervals.

Applying this process, arithmetic can be defined in all interval sets which occur in the second column in Fig. 1.

In the derivation of the arithmetic in $\mathbf{I}M$ given here, M is used as a generic symbol for the sets $\mathbf{R}$, $V\mathbf{R}$, $M\mathbf{R}$, $\mathbf{C}$, $V\mathbf{C}$, and $M\mathbf{C}$. The operations defined for intervals in $\mathbf{I}M$, therefore, are not executable on a computer, because the operations in $\mathbf{R}$ are not executable on a computer. Thus we are obliged to approximate M by a subset $N \subseteq M$ which is representable on a computer. The subset $\mathbf{I}N \subseteq \mathbf{I}M$ is not closed under the operations $\boxast$ defined in the set $\mathbf{I}M$. We require that an operation on intervals in $\mathbf{I}N$ leads to an interval in $\mathbf{I}N$ (since the latter is representable on the computer). This requires an approximation of the arithmetic defined in $\mathbf{I}M$ by that in $\mathbf{I}N$. For reasons inherent to interval arithmetic, the result of an operation in $\mathbf{I}N$ has to include the result of the corresponding operation in $\mathbf{I}M$. We also require that the deviation of the results of both operations be as small as possible. This means that we have to define the operations in $\mathbf{I}N$ by means of semimorphism, using the monotone upwardly directed rounding $\diamond$: $\mathbf{I}M \to \mathbf{I}N$. This process is described by the following formulas:

(R1) $\bigwedge_{A \in \mathbf{I}N} \diamond A = A$

(R2) $\bigwedge_{A, B \in \mathbf{I}M} (A \subseteq B \Rightarrow \diamond A \subseteq \diamond B)$

(R3) $\bigwedge_{A \in \mathbf{I}M} A \subseteq \diamond A$

(R4) $\bigwedge_{A \in \mathbf{I}M} \diamond(-A) = -\diamond A$

(RG) $\bigwedge_{A, B \in IN} A \diamondsuit\!\!\!\!\ast\; B := \diamond (A \boxast B)$ for all $\boxast$ in $\mathbf{I}M$.

Because of the property (see [8], [9])

(R) $\bigwedge_{A \in IM} \diamond A = \inf(U(A) \cap \mathbf{I}N)$,

the definition (RG) gives the best possible approximation in $\mathbf{I}N$ of the results of operations $\boxast$ in $\mathbf{I}M$ when the latter are applied to elements of $\mathbf{I}N$.

This process defines the arithmetic in all subsets of the interval sets $\mathbf{IR}$, $\mathbf{I}V\mathbf{R}$, $\mathbf{I}M\mathbf{R}$, $\mathbf{IC}$, $\mathbf{I}V\mathbf{C}$, and $\mathbf{I}M\mathbf{C}$ which are displayed in Fig. 1.

It is well known that this process can be used to develop explicit formulas for the computation of the result of all interval operations in the subsets $\mathbf{I}D$ and $\mathbf{I}S$ of $\mathbf{IR}$ (see [8], [9]). This result can be summarized by the following formula. For all $A = [a_1, a_2]$, $B = [b_1, b_2] \in \mathbf{I}S$, we obtain

$$A \diamondsuit\!\!\!\!\ast\; B := \diamond (A \boxast B) = \left[\min_{i,j=1,2} (a_i \,\triangledown\!\!\!\!\ast\; b_j),\ \max_{i,j=1,2} (a_i \,\triangle\!\!\!\!\ast\; b_j) \right]. \tag{2}$$

Here the operations $\triangledown\!\!\!\!\ast$ and $\triangle\!\!\!\!\ast$, $\ast \in \{+, -, \cdot, /\}$, are defined by the monotone downwardly resp. upwardly directed rounding $\nabla: \mathbf{R} \to S$ resp. $\Delta: \mathbf{R} \to S$ by:

$$\bigwedge_{a, b \in S} a \,\triangledown\!\!\!\!\ast\; b := \nabla (a * b), \qquad a \,\triangle\!\!\!\!\ast\; b := \Delta (a * b), * \in \{+, -, \cdot, /\}, \tag{3}$$

(cf. [7], [8], [9]).

We observed above that the intervals $\mathbf{I}M$, M standing for $\mathbf{R}$, $V\mathbf{R}$, $M\mathbf{R}$, $\mathbf{C}$, $V\mathbf{C}$, $M\mathbf{C}$, are not representable on a computer. Therefore we approximated them by intervals of $\mathbf{I}N$, N standing for S, VS, MS, $\mathbf{C}S$, $V\mathbf{C}S$, $M\mathbf{C}S$, for instance. Now we recognize that apart from the row $\mathbf{IR} \to \mathbf{I}D \to \mathbf{I}S$, the operations defined in $\mathbf{I}N$ are not executable on a computer even though the operands and results are representable on a computer. Indeed according to definition (1), these operations are ultimately based on operations in the powerset which are not executable in a computer. Therefore we seek to express the operations $\diamondsuit\!\!\!\!\ast$ in $\mathbf{I}N$ in terms of executable formulas. We do this by means of isomorphisms. (We may except the operations in $\mathbf{I}N$ which correspond to the row $\mathbf{IR} \to \mathbf{I}D \to \mathbf{I}S$ from this development since these operations are executable, as we have just observed (see (2), (3)), by employing the roundings Δ and ∇.)

3. Interval Matrices

To obtain explicit formulas for the operations in $\mathbf{I}M\mathbf{R}$, we consider the set $M\mathbf{IR}$. The elements of $M\mathbf{IR}$ are matrices the components of which are intervals over $\mathbf{R}$. Using the operations $\boxast$, $\ast \in \{+, -, \cdot\}$, and the order relation $\leqslant$ defined for intervals in $\mathbf{IR}$, we define operations $\circledast$, $\ast \in \{+, -, \cdot\}$, and an order relation $\leqslant$ in $M\mathbf{IR}$ by means of the vertical method:

$$\bigwedge_{\mathbf{A} = (A_{ij}), \mathbf{B} = (B_{ij}) \in M\mathbf{IR}} \left(\mathbf{A} \oplus \mathbf{B} := (A_{ij} \boxplus B_{ij}), \right.$$

$$\mathbf{A} \ominus \mathbf{B} := (A_{ij} \boxminus B_{ij}),$$

$$\mathbf{A} \odot \mathbf{B} := \left(\boxed{\Sigma}_{\nu=1}^{n} A_{i\nu} \boxdot B_{\nu j} \right).$$

$$(A_{ij}) \leqslant (B_{ij}) :\Leftrightarrow \bigwedge_{i=1(1)n} \bigwedge_{j=1(1)n} A_{ij} \leqslant B_{ij}.$$

Here $\boxed{\Sigma}$ denotes the repeated summation in **IR**.

Now we define a mapping

$$\chi : M\mathbf{IR} \to \mathbf{I}M\mathbf{R},$$

which for matrices $\mathbf{A} = (A_{ij}) \in M\mathbf{IR}$ with $A_{ij} = [a_{ij}^{(1)}, a_{ij}^{(2)}] \in \mathbf{IR}$, $i, j = 1(1)n$, has the property

$$\chi\mathbf{A} = \chi(A_{ij}) = \chi([a_{ij}^{(1)}, a_{ij}^{(2)}]) := [(a_{ij}^{(1)}), (a_{ij}^{(2)})].$$

Obviously χ is a one-to-one mapping of $M\mathbf{IR}$ onto $\mathbf{I}M\mathbf{R}$. It is shown in [7] (see also [8], [9]) that the mapping χ establishes an algebraic as well as an order isomorphism between the ordered algebraic structures $\{M\mathbf{IR}, \oplus, \odot, \leqslant\}$ and $\{\mathbf{I}M\mathbf{R}, \boxplus, \boxdot, \leqslant\}$. This isomorphism relates the nonperformable operations $\boxed{*}$ in $\mathbf{I}M\mathbf{R}$ to performable operations $\circledast$ in $M\mathbf{IR}$. The isomorphism expresses the fact that the vertical definition of the operations in $M\mathbf{IR}$ by means of those of **IR** and the horizontal definition of the operations in $\mathbf{I}M\mathbf{R}$ by the semimorphism $\square : \mathbf{P}M\mathbf{R} \to \mathbf{I}M\mathbf{R}$ lead to the same result. Fig. 3 illustrates this connection.

$$\begin{array}{ccccc}
\mathbf{R} & & \longrightarrow & & M\mathbf{R} \\
\downarrow & & & & \downarrow \\
\mathbf{PR} & & & & \mathbf{P}M\mathbf{R} \\
\downarrow & & & & \downarrow \\
\mathbf{IR} & \to & M\mathbf{IR} & \underset{\chi}{\leftrightarrow} & \mathbf{I}M\mathbf{R}
\end{array}$$

Fig. 3. Illustration of the isomorphism χ

The algebraic isomorphism $\chi : M\mathbf{IR} \to \mathbf{I}M\mathbf{R}$ is formally expressed by the equality

$$\bigwedge_{\mathbf{A}, \mathbf{B} \in M\mathbf{IR}} \chi\mathbf{A} \boxed{*} \chi\mathbf{B} = \chi(\mathbf{A} \circledast \mathbf{B}), \; * \in \{+, \cdot\}.$$

Because of the isomorphism χ, corresponding elements of $M\mathbf{IR}$ and $\mathbf{I}M\mathbf{R}$ can be identified with each other. This allows us to define an inclusion relation even for elements $\mathbf{A} = (A_{ij})$, $\mathbf{B} = (B_{ij}) \in M\mathbf{IR}$ by:

$$\mathbf{A} \subseteq \mathbf{B} :\Leftrightarrow \bigwedge_{i,j=1(1)n} A_{ij} \subseteq B_{ij}. \tag{4}$$

We already observed above that the operations $\diamond\!\!\!\!\ast$ in $\mathbf{I}M S$, which are defined by the semimorphism $\Diamond : \mathbf{I}M\mathbf{R} \to \mathbf{I}M S$, are not executable on a computer because of their ultimate dependence on powerset operations. We are now going to express them in terms of executable formulas. Using the monotone upwardly directed rounding $\Diamond : \mathbf{IR} \to \mathbf{I}S$, a rounding $\Diamond : M\mathbf{IR} \to M\mathbf{I}S$ and operations in $M\mathbf{I}S$ can be defined by

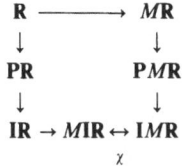

$$\bigwedge_{\mathbf{A}=(A_{ij})\in M\mathbf{IR}} \Diamond \mathbf{A} := (\Diamond A_{ij})$$

$$\bigwedge_{\mathbf{A},\mathbf{B}\in M\mathbf{I}S} \mathbf{A} \,\hat{\circledast}\, \mathbf{B} := \Diamond (\mathbf{A} \circledast \mathbf{B}), * \in \{+, -, \cdot\}. \tag{5}$$

We show that the mapping χ defined above also establishes an isomorphism between the ordered algebraic structures $\{M\mathbf{I}S, \hat{+}, \hat{\cdot}, \leqslant\}$ and $\{\mathbf{I}MS, \hat{+}, \hat{\cdot}, \leqslant\}$. In order to do this we denote the matrix $\mathbf{A} \circledast \mathbf{B} \in M\mathbf{IR}$ by

$$(C_{ij}) := \mathbf{A} \circledast \mathbf{B} \text{ with } C_{ij} = [c_{ij}^1, c_{ij}^2] \text{ and } c_{ij}^1, c_{ij}^2 \in \mathbf{R}.$$

For the expression $\chi(\mathbf{A} \,\hat{\circledast}\, \mathbf{B})$ in which we are interested, we obtain

$$\chi(\mathbf{A} \,\hat{\circledast}\, \mathbf{B}) = \chi(\Diamond\, C_{ij}) = \chi(\inf(U(C_{ij}) \cap \mathbf{I}S)).$$

If we exchange the infima with the matrix parenthesis, we obtain

$$\chi(\mathbf{A} \,\hat{\circledast}\, \mathbf{B}) = \chi\{\inf(U(C_{ij}) \cap \mathbf{I}S)\}.$$

Since the inclusion and intersection (see (4)) in $M\mathbf{IR}$ are defined componentwise, we may exchange the matrix parenthesis with the upper bounds and obtain

$$\chi(\mathbf{A} \,\hat{\circledast}\, B) = \chi\{\inf U\{([c_{ij}^1, c_{ij}^2]) \cap M\mathbf{I}S\}\}.$$

If we now apply the mapping χ, we obtain stepwise

$$\begin{aligned}\chi(\mathbf{A} \,\hat{\circledast}\, \mathbf{B}) &= \inf U\{[(c_{ij}^1), (c_{ij}^2)] \cap \mathbf{I}MS\}\\ &= \inf U\{\chi(\mathbf{A} \circledast \mathbf{B}) \cap \mathbf{I}MS\}\\ &= \inf U\{(\chi\mathbf{A} \boxast \chi\mathbf{B}) \cap \mathbf{I}MS\}\\ &= \Diamond\, (\chi\mathbf{A} \boxast \chi\mathbf{B}).\end{aligned}$$

By the definition of the operations $\hat{\circledast}$ in $\mathbf{I}MS$, this finally leads to

$$\chi(\mathbf{A} \,\hat{\circledast}\, \mathbf{B}) = \chi\mathbf{A} \,\hat{\circledast}\, \chi\mathbf{B}.$$

This isomorphism reduces the non-implementable operations in $\mathbf{I}MS$ to the operations in $M\mathbf{I}S$ which have the properties (see (5))

$$\mathbf{A} \,\hat{+}\, \mathbf{B} = (\Diamond\, (A_{ij} \boxplus B_{ij})) = (A_{ij} \,\hat{+}\, B_{ij})$$

$$\mathbf{A} \,\hat{\cdot}\, \mathbf{B} = \left(\Diamond \boxed{\sum_{\nu=1}^{n}} A_{i\nu} \boxdot B_{\nu j}\right).$$

These operations are executable on a computer. The addition can be done componentwise in terms of addition in $\mathbf{I}S$. The multiplication can be executed using the well-known table for the multiplication in $\mathbf{IR}$ (see [8], [9]), and the algorithm for scalar products with optimal accuracy given by G. Bohlender [4], [5].

We illustrate these results with a diagram in Fig. 4 which is related to Fig. 1. Preparatory to this, we note that the passage which we made from $\mathbf{I}M\mathbf{R}$ to $\mathbf{I}MS$ can also be performed from $\mathbf{I}M\mathbf{R}$ to $\mathbf{I}MD$ or from $\mathbf{I}MD$ to $\mathbf{I}MS$. This results in isomorphisms between the sets $\mathbf{I}M\mathbf{R}$, $\mathbf{I}MD$, $\mathbf{I}MS$ and $M\mathbf{IR}$, $M\mathbf{I}D$, $M\mathbf{I}S$ respectively.

Operations in the sets *M*I*D* and *M*I*S* can also be defined by the vertical method using the operations in I*D* resp. I*S*. We show that the addition is then identical to the one already defined in *M*I*S*. Using (R3) and (RG) for $\diamond$, we also show that the multiplication defined by the vertical method delivers upper bounds for the original multiplication. This can be seen by means of the following formulas wherein $\mathbf{A} = (A_{ij})$, $\mathbf{B} = (B_{ij}) \in M\mathbf{I}S$:

$$\mathbf{A} \mathbin{\Diamond\!\!\!\!\!\!+} \mathbf{B} = (\diamond (A_{ij} \boxplus B_{ij})) = (A_{ij} \mathbin{\Diamond\!\!\!\!\!\!+} B_{ij})$$

$$\mathbf{A} \mathbin{\Diamond\!\!\!\!\!\!\cdot\;} \mathbf{B} = \left(\diamond \boxed{\Sigma}_{\nu=1}^{n} A_{i\nu} \boxdot B_{\nu j} \right) \subseteq \left(\mathop{\Diamond\!\!\!\!\!\!+}_{\nu=1}^{n} A_{i\nu} \mathbin{\Diamond\!\!\!\!\!\!\cdot\;} B_{\nu j} \right). \tag{6}$$

The expressions on the right-hand side of these formulas represent the sum resp. product defined in *M*I*S* by the vertical method. In fact on computers, calculations in *M*I*D* and *M*I*S* are, for simplicity, often performed by using the operations defined by the vertical method. The operations defined by semimorphisms, however, provide optimal accuracy.

In Fig. 4 double headed arrows indicate isomorphisms, horizontal arrows denote semimorphisms and vertical arrows indicate a vertical definition of the arithmetic operations. A slanting arrow shows where the powerset operations come from. The sign $\vee$ used in Fig. 4 is intended to illustrate the inequality (6). Sets the operations of which are executable on a computer are underlined in Fig. 4.

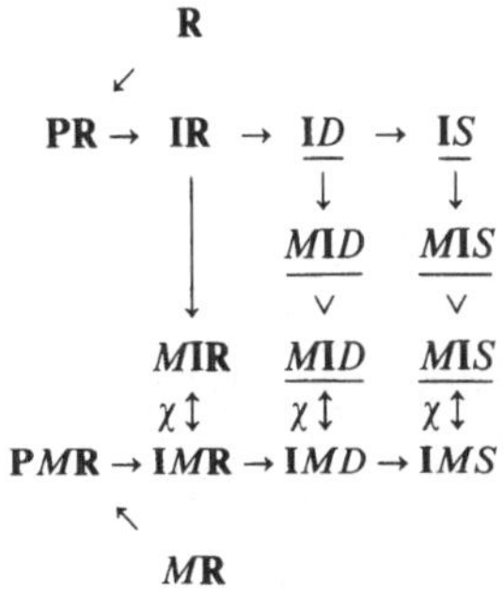

Fig. 4. Interval matrices

4. Interval Vectors

Employing semimorphisms and the horizontal definition of arithmetic in row 5 of Fig. 1, the following interval vectoids[1] are defined: $\{\mathbf{I}V\mathbf{R}, \mathbf{IR}\}$, $\{\mathbf{I}M\mathbf{R}, \mathbf{IR}\}$, $\{\mathbf{I}V\mathbf{R}, \mathbf{I}M\mathbf{R}\}$ as well as $\{\mathbf{I}VS, \mathbf{I}S\}$, $\{\mathbf{I}MS, \mathbf{I}S\}$, $\{\mathbf{I}VS, \mathbf{I}MS\}$.

The principle situation is similar to that which arose in the case of matrix structures. The elements of the first three vectoidal structures cannot be represented on a

[1] We do not in fact require the concept of a vectoid for the purpose of this paper. Foregoing details, we simply mention it here in order to indicate that an outer multiplication is involved (see [8], [9]).

computer while the operations within all six structures cannot in general be executed on a computer since they are based on the corresponding powerset operations.

In order to obtain implementable formulas for the operations we, therefore, consider the set $V\mathbf{IR}$. The elements of this set are n-tuples over $\mathbf{IR}$. Employing the operations $\boxed{*}$ and the order relation $\leqslant$ defined in $\mathbf{IR}$, we now define operations $\circledast$ and an order relation $\leqslant$ in $V\mathbf{IR}$ and $M\mathbf{IR}$ by the vertical method:

$$\bigwedge_{\mathbf{a} = (A_i),\, \mathbf{b} = (B_i) \in V\mathbf{IR}} \mathbf{a} \oplus \mathbf{b} := (A_i \boxplus B_i)$$

$$\bigwedge_{A \in \mathbf{IR}} \bigwedge_{\mathbf{a} = (A_i) \in V\mathbf{IR}} A \odot \mathbf{a} := (A \boxdot A_i)$$

$$\bigwedge_{A \in \mathbf{IR}} \bigwedge_{\mathbf{A} = (A_{ij}) \in M\mathbf{IR}} A \odot \mathbf{A} := (A \boxdot A_{ij})$$

$$\bigwedge_{\mathbf{A} = (A_{ij}) \in M\mathbf{IR}} \bigwedge_{\mathbf{a} = (A_i) \in V\mathbf{IR}} \mathbf{A} \odot \mathbf{a} := \left(\boxed{\Sigma}_{\nu=1}^{n} A_{i\nu} \boxdot A_\nu \right)$$

$$(A_i) \leqslant (B_i) :\Leftrightarrow \bigwedge_{i=1(1)n} A_i \leqslant B_i.$$

Here $\boxed{\Sigma}$ denotes the repeated summation in $\mathbf{IR}$.

We also adopt the definition of the inner operations $\oplus$, $\odot$ and the order relation $\leqslant$ in $M\mathbf{IR}$ from the previous section.

Now we define mappings,

$$\psi\colon\ V\mathbf{IR} \to \mathbf{I}V\mathbf{R}$$

$$\chi\colon\ M\mathbf{IR} \to \mathbf{I}M\mathbf{R},$$

which for vectors and matrices

$$\mathbf{a} = (A_i) \in V\mathbf{IR} \quad \text{with} \quad A_i = [a_i^{(1)}, a_i^{(2)}] \in \mathbf{IR}, \qquad i = 1(1)n,$$

$$\mathbf{A} = (A_{ij}) \in M\mathbf{IR} \quad \text{with} \quad A_{ij} = [a_{ij}^{(1)}, a_{ij}^{(2)}] \in \mathbf{IR}, \qquad i,j = 1(1)n,$$

have the properties

$$\psi\mathbf{a} = \psi(A_i) = \psi([a_i^{(1)}, a_i^{(2)}]) := [(a_i^{(1)}), (a_i^{(2)})]$$

and

$$\chi\mathbf{A} = \chi(A_{ij}) = \chi([a_{ij}^{(1)}, a_{ij}^{(2)}]) := [(a_{ij}^{(1)}), (a_{ij}^{(2)})].$$

We have already noted that the mapping χ establishes an isomorphism between the structures $M\mathbf{IR}$ and $\mathbf{I}M\mathbf{R}$. In complete analogy to the proof of the isomorphism, it is shown in [10] (see also [8], [9]) that the mappings ψ and χ establish algebraic and order isomorphisms between the following pairs of structures:

$$\{V\mathbf{IR}, \mathbf{IR}, \leqslant\} \leftrightarrow \{\mathbf{I}V\mathbf{R}, \mathbf{IR}, \leqslant\}$$

$$\{M\mathbf{IR}, \mathbf{IR}, \leqslant\} \leftrightarrow \{\mathbf{I}M\mathbf{R}, \mathbf{IR}, \leqslant\}$$

$$\{V\mathbf{IR}, M\mathbf{IR}, \leqslant\} \leftrightarrow \{\mathbf{I}V\mathbf{R}, \mathbf{I}M\mathbf{R}, \leqslant\}.$$

The algebraic isomorphisms are expressed by the following relations.

$$\psi(A \odot \mathbf{a}) = A \boxdot \psi\mathbf{a}, \qquad \psi(\mathbf{a} \oplus \mathbf{b}) = \psi\mathbf{a} \boxplus \psi\mathbf{b}$$

$$\chi(A \odot \mathbf{A}) = A \boxdot \chi\mathbf{A}, \qquad \chi(\mathbf{A} \circledast \mathbf{B}) = \chi\mathbf{A} \boxast \chi\mathbf{B}, \ast \in \{+, \cdot\}$$

$$\psi(\mathbf{A} \odot \mathbf{a}) = \chi\mathbf{A} \boxdot \psi\mathbf{a}, \qquad \psi(\mathbf{a} \oplus \mathbf{b}) = \psi\mathbf{a} \boxplus \psi\mathbf{b}.$$

The operations $\boxast$ in $\mathbf{I}V\mathbf{R}$ and $\mathbf{I}M\mathbf{R}$ are not performable because of their definition in terms of powerset operations. The isomorphisms ψ and χ reduce these operations $\boxast$ to the explicit operations $\circledast$ in $V\mathbf{IR}$ resp. $M\mathbf{IR}$.

Because of the isomorphism ψ, corresponding elements in $V\mathbf{IR}$ and $\mathbf{I}V\mathbf{R}$ can be identified with each other. This shows that an inclusion relation for elements $\mathbf{a} = (A_i)$, $\mathbf{b} = (B_i) \in V\mathbf{IR}$ can be defined by:

$$\mathbf{a} \subseteq \mathbf{b} :\Leftrightarrow \bigwedge_{i=1(1)n} A_i \subseteq B_i.$$

We observed above that the (inner and outer) operations $\diamondast$ in $\mathbf{I}VS$, which are defined by the semimorphism $\diamond : \mathbf{I}V\mathbf{R} \to \mathbf{I}VS$, are not executable on the computer. We now express these operations in terms of executable formulas. Using the monotone upwardly directed rounding $\diamond : \mathbf{IR} \to \mathbf{I}S$, a rounding $\diamond : V\mathbf{IR} \to V\mathbf{I}S$ resp. $\diamond : M\mathbf{IR} \to M\mathbf{I}S$ as well as operations in $V\mathbf{I}S$ resp. $M\mathbf{I}S$ can be defined by the following formulas:

$$\bigwedge_{\mathbf{a} = (A_i) \in V\mathbf{I}S} \diamond\, \mathbf{a} := (\diamond\, A_i) \qquad \text{resp.} \qquad \bigwedge_{\mathbf{A} = (A_{ij}) \in M\mathbf{I}S} \diamond\, \mathbf{A} := (\diamond\, \mathbf{A}_{ij}).$$

$$\bigwedge_{\mathbf{a}, \mathbf{b} \in V\mathbf{I}S} \mathbf{a} \diamondplus \mathbf{b} := \diamond\, (\mathbf{a} \oplus \mathbf{b})$$

$$\bigwedge_{A \in \mathbf{I}S} \bigwedge_{\mathbf{a} \in V\mathbf{I}S} A \diamonddot \mathbf{a} := \diamond\, (A \odot \mathbf{a})$$

$$\bigwedge_{A \in \mathbf{I}S} \bigwedge_{\mathbf{A} \in M\mathbf{I}S} A \diamonddot \mathbf{A} := \diamond\, (A \odot \mathbf{A})$$

$$\bigwedge_{\mathbf{A} \in M\mathbf{I}S} \bigwedge_{\mathbf{a} \in V\mathbf{I}S} \mathbf{A} \diamonddot \mathbf{a} := \diamond\, (\mathbf{A} \odot \mathbf{a}).$$

Similar to the matrix case treated above, it can be shown that the mappings ψ and χ establish isomorphisms between the following ordered algebraic structures.

$$\{V\mathbf{I}S, \mathbf{I}S, \leqslant\} \leftrightarrow \{\mathbf{I}VS, \mathbf{I}S, \leqslant\}$$

$$\{M\mathbf{I}S, \mathbf{I}S, \leqslant\} \leftrightarrow \{\mathbf{I}MS, \mathbf{I}S, \leqslant\}$$

$$\{V\mathbf{I}S, M\mathbf{I}S, \leqslant\} \leftrightarrow \{\mathbf{I}VS, \mathbf{I}MS, \leqslant\}.$$

With $\mathbf{a} = (A_i)$ and $\mathbf{A} = (A_{ij})$, we obtain the following relations for the operations defined above.

$$\mathbf{a} \diamondplus \mathbf{b} = \diamond\, (\mathbf{a} \oplus \mathbf{b}) = \diamond\, (A_i \boxplus B_i) = (A_i \diamondplus B_i),$$

$$A \diamonddot \mathbf{a} = \diamond\, (A \odot \mathbf{a}) = \diamond\, (A \boxdot A_i) = (A \diamonddot A_i),$$

$$A \diamonddot \mathbf{A} = \diamond\, (A \odot \mathbf{A}) = \diamond\, (A \boxdot A_{ij}) = (A \diamonddot A_{ij}),$$

$$\mathbf{A} \diamond \mathbf{a} = \diamond (\mathbf{A} \odot \mathbf{a}) = \diamond \left(\mathop{\boxed{\Sigma}}_{v=1}^{n} A_{iv} \boxdot A_v \right) = \left(\diamond \mathop{\boxed{\Sigma}}_{v=1}^{n} A_{iv} \boxdot A_v \right).$$

These formulas show that the operations $\mathbf{a} \oplus \mathbf{b}$, $A \diamond \mathbf{a}$ and $A \diamond \mathbf{A}$ are identical, whether defined by the vertical or the horizontal method. The horizontal and the vertical definition of the outer multiplications in $\{V\mathbf{IS}, M\mathbf{IS}, \leqslant\}$, however, leads to the following inequality

$$\mathbf{A} \diamond \mathbf{a} = \left(\diamond \mathop{\boxed{\Sigma}}_{v=1}^{n} A_{iv} \boxdot A_v \right) \subseteq \left(\mathop{\diamond\!\!\!\!+}_{v=1}^{n} A_{iv} \diamond A_v \right). \tag{7}$$

All operations $\mathbf{a} \oplus \mathbf{b}$, $A \diamond \mathbf{a}$, $A \diamond \mathbf{A}$ and $\mathbf{A} \diamond \mathbf{a}$ defined and described here, furnish optimal accuracy. Moreover they all can be executed on a computer.

In Fig. 5 we display a diagram of the interval vectoids discussed in this section. In Fig. 5 arrows are used in complete analogy with those in the diagram of Fig. 4. Double arrows indicate isomorphisms. Horizontal arrows denote semimorphisms. Vertical arrows indicate a vertical definition of the arithmetic operations. The sign $\vee$ used in Fig. 5 indicates the inequality (7). All structures, the operations of which are executable on a computer, are underlined in Fig. 5.

$$\begin{array}{ccccccc}
 & & \{V\mathbf{R}, \mathbf{R}\} & & & & \\
 & \swarrow & & & & & \\
\{\mathbf{P}V\mathbf{R}, \mathbf{PR}\} & \rightarrow & \{\mathbf{I}V\mathbf{R}, \mathbf{IR}\} & \rightarrow & \{\mathbf{I}VD, \mathbf{I}D\} & \rightarrow & \{\mathbf{I}VS, \mathbf{I}S\} \\
 & & \updownarrow & & \updownarrow & & \updownarrow \\
 & & \{V\mathbf{IR}, \mathbf{IR}\} & \rightarrow & \underline{\{V\mathbf{I}D, \mathbf{I}D\}} & \rightarrow & \underline{\{V\mathbf{I}S, \mathbf{I}S\}} \\
 & & \uparrow & & \uparrow & & \uparrow \\
\mathbf{PR} & \rightarrow & \mathbf{IR} & \rightarrow & \underline{\mathbf{I}D} & \rightarrow & \underline{\mathbf{I}S} \\
 & & \downarrow & & \downarrow & & \downarrow \\
 & & & & \underline{\{V\mathbf{I}D, M\mathbf{I}D\}} & & \underline{\{V\mathbf{I}S, M\mathbf{I}S\}} \\
 & & & & \vee & & \vee \\
 & & \{V\mathbf{IR}, M\mathbf{IR}\} & \rightarrow & \underline{\{V\mathbf{I}D, M\mathbf{I}D\}} & \rightarrow & \underline{\{V\mathbf{I}S, M\mathbf{I}S\}} \\
 & & \updownarrow & & \updownarrow & & \updownarrow \\
\{\mathbf{P}V\mathbf{R}, \mathbf{P}M\mathbf{R}\} & \rightarrow & \{\mathbf{I}V\mathbf{R}, \mathbf{I}M\mathbf{R}\} & \rightarrow & \{\mathbf{I}VD, \mathbf{I}MD\} & \rightarrow & \{\mathbf{I}VS, \mathbf{I}MS\} \\
 & \nwarrow & & & & & \\
 & & \{V\mathbf{R}, M\mathbf{R}\} & & & &
\end{array}$$

Fig. 5. Interval vectors

5. Complex Interval Arithmetic

We have yet to consider the complex interval spaces listed in Fig. 1. We begin these considerations with row 10 of Fig. 1. The horizontal definition of arithmetic by semimorphism leads to operations $\boxed{*}$ in **IC** which are based on the powerset operations. As such these operations are not implementable. In order to obtain implementable formulas for the operations, we consider the set

$$\mathbf{CIR} := \{(X, Y) | X, Y \in \mathbf{IR}\}.$$

For elements $\Phi = (X_1, X_2)$, $\Psi = (Y_1, Y_2) \in \mathbf{CIR}$, we define equality, the order relation $\leqslant$ and operations $\circledast$, $* \in \{+, \cdot, /\}$ by the vertical method:

$$\Phi = \Psi :\Leftrightarrow X_1 = Y_1 \wedge X_2 = Y_2,$$

$$\Phi \leqslant \Psi :\Leftrightarrow X_1 \leqslant Y_1 \wedge X_2 \leqslant Y_2,$$

$$\Phi \oplus \Psi := (X_1 \boxplus Y_1, X_2 \boxplus Y_2),$$

$$\Phi \odot \Psi := (X_1 \boxdot Y_1 \boxminus X_2 \boxdot Y_2, X_1 \boxdot Y_2 \boxplus X_2 \boxdot Y_1),$$

$$\Phi \oslash \Psi := ((X_1 \boxdot Y_1 \boxplus X_2 \boxdot Y_2)/Z, (X_2 \boxdot Y_1 \boxminus X_1 \boxdot Y_2)/Z)$$

with

$$Z := Y_1 \boxdot Y_1 \boxplus Y_2 \boxdot Y_2.$$

In case of division we suppose that $0 \notin Z$.

Complex interval arithmetic is defined in [1] in this manner. (See [2] also.) All operations are expressed explicitly in terms of the operations in **IR**. By employing an isomorphism, it can be shown that this is the best possible complex interval arithmetic. In order to see this we consider the following mapping.

$$\tau : \mathbf{CIR} \to \mathbf{IC}$$

with

$$\tau([\mathbf{x}_1, \mathbf{x}_2], [\mathbf{y}_1, \mathbf{y}_2]) := [(\mathbf{x}_1, \mathbf{y}_1), (\mathbf{x}_2, \mathbf{y}_2)].$$

τ is a one-to-one mapping of **CIR** onto **IC** and an order isomorphism as well. It is shown in [11] (see also [8], [9]) that it is also an isomorphism with respect to addition and multiplication. The following inequality holds for division.

$$\tau\Phi/\tau\Psi \subseteq \square(\tau\Phi/\tau\Psi) = \tau\Phi \boxslash \tau\Psi \subseteq \tau(\Phi \oslash \Psi).$$

The isomorphism τ is illustrated in Fig. 6. Recall that it is an isomorphism with respect to $\leqslant$, addition and multiplication. The algebraic isomorphism is expressed by the formula

$$\tau\Phi \boxast \tau\Psi = \tau(\Phi \circledast \Psi), * \in \{+, \cdot\}.$$

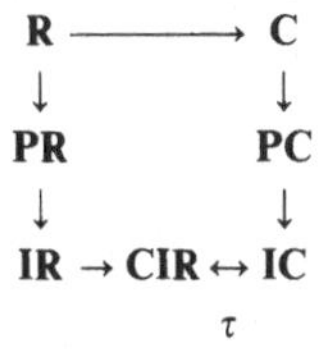

Fig. 6. Illustration of the isomorphism τ: **CIR** ↔ **IC**

Because of this isomorphism, we see that $\Phi \subseteq \Psi :\Leftrightarrow X_1 \subseteq Y_1 \wedge X_2 \subseteq Y_2$ establishes an order relation for elements $\Phi = (X_1, X_2)$, $\Psi = (Y_1, Y_2) \in \mathbf{CI}$.

Now we employ the isomorphism τ to obtain executable formulas for the operations in **IC**S. The operations $\diamond\!\!\!\ast$ in **IC**S are defined by semimorphism with the monotone upwardly directed rounding $\Diamond$: **IC** → **IC**S (R1, 2, 3).

By using the monotone upwardly directed rounding $\Diamond: \mathbf{IR} \to \mathbf{I}S$, a rounding $\Diamond: \mathbf{CIR} \to \mathbf{CI}S$ as well as operations in $\mathbf{CI}S$ can be defined by the following formulas.

$$\bigwedge_{\Phi = (X_1, X_2) \in \mathbf{CIR}} \Diamond \Phi := (\Diamond X_1, \Diamond X_2),$$

$$\bigwedge_{\Phi, \Psi \in \mathbf{CI}S} \Phi \mathbin{\Diamond\!\!\!\!\!\ast} \Psi := \Diamond (\Phi \circledast \Psi), \ast \in \{+, \cdot\}.$$

Now we show in complete analogy to the matrix case which we considered above, that the operations in $\mathbf{IC}S$ and $\mathbf{CI}S$ are isomorphic. Using the notation $(Z_1, Z_2) := \Phi \circledast \Psi$, we obtain the following sequence of relations.

$$\begin{aligned} \tau(\Phi \mathbin{\Diamond\!\!\!\!\!\ast} \Psi) &= \tau(\Diamond (Z_1, Z_2)) = \tau(\Diamond Z_1, \Diamond Z_2) \\ &= \tau(\inf(U(Z_1) \cap \mathbf{I}S), \inf(U(Z_2) \cap \mathbf{I}S)) \\ &= \tau\{\inf(U(Z_1) \cap \mathbf{I}S, U(Z_2) \cap \mathbf{I}S)\} \\ &= \tau\{\inf U\{(Z_1, Z_2) \cap \mathbf{CI}S\}\} \\ &= \inf U\{\tau(\Phi \circledast \Psi) \cap \mathbf{IC}S\} \\ &= \inf U\{(\tau\Phi \boxed{\ast} \tau\Psi) \cap \mathbf{IC}S\} \\ &= \Diamond (\tau\Phi \boxed{\ast} \tau\Psi) = \tau\Phi \mathbin{\Diamond\!\!\!\!\!\ast} \tau\Psi. \end{aligned}$$

This demonstrates the isomorphism of the ordered algebraic structures $\{\mathbf{CI}S, \mathbin{\Diamond\!\!\!\!\!+}, \mathbin{\Diamond\!\!\!\!\!\cdot}, \leqslant, \subseteq\}$ and $\{\mathbf{IC}S, \mathbin{\Diamond\!\!\!\!\!+}, \mathbin{\Diamond\!\!\!\!\!\cdot}, \leqslant, \subseteq\}$. This isomorphism reduces operations in $\mathbf{IC}S$, which are not executable in practice, to operations in $\mathbf{CI}S$ which have the following properties.

$$\Phi \mathbin{\Diamond\!\!\!\!\!+} \Psi = (X_1 \mathbin{\Diamond\!\!\!\!\!+} Y_1, X_2 \mathbin{\Diamond\!\!\!\!\!+} Y_2),$$

$$\Phi \mathbin{\Diamond\!\!\!\!\!\cdot} \Psi = (\Diamond (X_1 \boxdot Y_1 \boxminus X_2 \boxdot Y_2), \Diamond (X_1 \boxdot Y_2 \boxplus X_2 \boxdot Y_1)).$$

These operations are executable on a computer. The addition can be performed in terms of addition in $\mathbf{I}S$. The multiplication can be executed using the formulas for the operations in $\mathbf{I}S$ and the optimally accurate Bohlender algorithm for scalar products [4], [5].

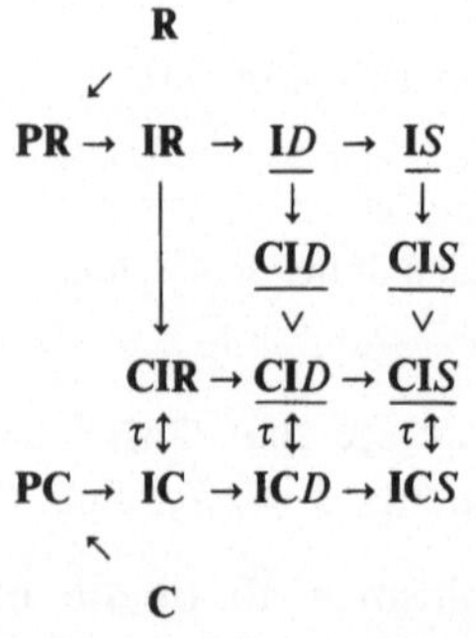

Fig. 7. Complex interval arithmetic

The result of these considerations is illustrated in Fig. 7. The isomorphisms, shown in the figure, however do not include division. The multiplication in **CI***S* can be approximated by means of a vertical definition of a multiplication in **CI***S*. This leads to the following inequality:

$$\Phi \mathbin{\Diamond} \Psi \subseteq (X_1 \mathbin{\Diamond} Y_1 \ominus X_2 \mathbin{\Diamond} Y_2, X_1 \mathbin{\Diamond} Y_2 \oplus X_2 \mathbin{\Diamond} Y_1).$$

6. Complex Interval Matrices and Vectors

In Section 2, operations are defined in **I***M***C** and **I***M***C***S* by employing the semimorphisms, $\square$: **P***M***C** $\to$ **I***M***C** and $\Diamond$: **I***M***C** $\to$ **I***M***C***S*.

Independently of the set **I***M***C**, we now consider the set *M***IC**, i.e., the set of matrices the components of which are intervals over **C**. With the operations defined in $\{$**IC**, $\boxplus, \boxdot, \leqslant, \subseteq\}$, we define operations and an order relation in *M***IC** by means of the vertical method. We denote the operations so defined by $\circledast$, $* \in \{+, \cdot\}$.

It is shown in [11] (see also [8] and [9]), that the two structures $\{$*M***IC**, $\circledast, \odot, \leqslant\}$ and $\{$**I***M***C**, $\boxplus, \boxdot, \leqslant\}$ are isomorphic with respect to the algebraic operations as well as the order relation. Because of the isomorphism τ: **IC** $\to$ **CIR**, the operations in *M***IC** are identical to those in *M***CIR**. The operations in this set are defined and can easily be executed by those in **CIR**. Fig. 8 illustrates these isomorphisms.

C ——→ *M***C**
↓ ↓
PC **P***M***C**
↓ ↓
IC → *M***IC** ↔ **I***M***C**
↘ ↙
*M***CIR**

Fig. 8. Illustration of the isomorphism **I***M***C** ↔ *M***IC**

Now proceeding similarly as in Sections 3 and 5, the analogous isomorphisms for complex interval matrices over *S* and *D* can be derived. Fig. 9 gives the resulting diagram.

For the case of complex interval vectors, we define vector operations, including an outer multiplication with elements of **IC** resp. **IC***S* or of **I***M***C** resp. **I***M***C***S* by using the following semimorphisms.

$\square$: **P***V***C** $\to$ **I***V***C** and $\Diamond$: **I***V***C** $\to$ **I***V***C***S*.

The following isomorphisms can be established.

$\{$**I***V***C**, **IC**, $\leqslant, \subseteq\} \leftrightarrow \{$*V***IC**, **IC**, $\leqslant, \subseteq\} \leftrightarrow \{$*V***CIR**, **CIR**, $\leqslant, \subseteq\}$,

$\{$**I***M***C**, **IC**, $\leqslant, \subseteq\} \leftrightarrow \{$*M***IC**, **IC**, $\leqslant, \subseteq\} \leftrightarrow \{$*M***CIR**, **CIR**, $\leqslant, \subseteq\}$,

$\{$**I***V***C**, **I***M***C**, $\leqslant, \subseteq\} \leftrightarrow \{$*V***IC**, *M***IC**, $\leqslant, \subseteq\} \leftrightarrow \{$*V***CIR**, *M***CIR**, $\leqslant, \subseteq\}$.

This leads to isomorphisms between the corresponding structures over *D* or *S*; for instance

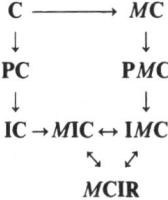

$$\{\mathbf{I}V\mathbf{C}S, \mathbf{IC}S, \leqslant, \subseteq\} \leftrightarrow \{V\mathbf{IC}S, \mathbf{IC}S, \leqslant, \subseteq\} \leftrightarrow \{V\mathbf{CI}S, \mathbf{CI}S, \leqslant, \subseteq\},$$

$$\{\mathbf{I}M\mathbf{C}S, \mathbf{IC}S, \leqslant, \subseteq\} \leftrightarrow \{M\mathbf{IC}S, \mathbf{IC}S, \leqslant, \subseteq\} \leftrightarrow \{M\mathbf{CI}S, \mathbf{CI}S, \leqslant, \subseteq\},$$

$$\{\mathbf{I}V\mathbf{C}S, \mathbf{I}M\mathbf{C}S, \leqslant, \subseteq\} \leftrightarrow \{V\mathbf{IC}S, M\mathbf{IC}S, \leqslant, \subseteq\} \leftrightarrow \{V\mathbf{CI}S, M\mathbf{CI}S, \leqslant, \subseteq\}.$$

Fig. 10 summarizes the results. Consider those structures in Fig. 10 whose names terminate with the letters **CI***D* or **CI***S*. All operations defined by means of semimorphisms within these structures can be executed on computers by the Bohlender algorithm.

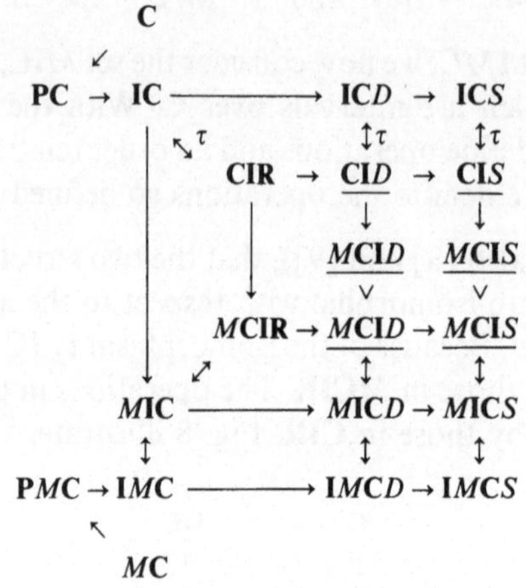

Fig. 9. Complex interval matrices

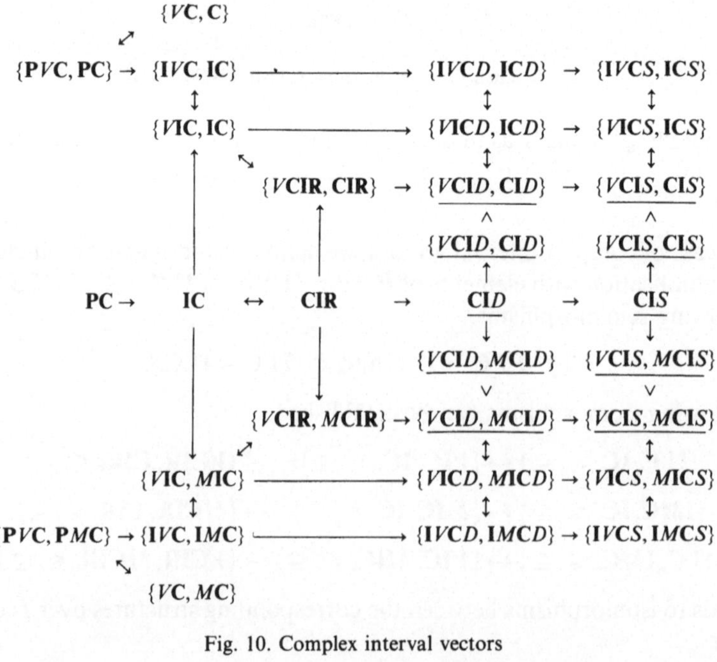

Fig. 10. Complex interval vectors

References

[1] Alefeld, G.: Intervallrechnung über den komplexen Zahlen und einige Anwendungen. Dissertation, Universität Karlsruhe, 1968.

[2] Alefeld, G., Herzberger, J.: Einführung in die Intervallrechnung. Mannheim-Wien-Zürich: Wissenschaftsverlag, Bibliographisches Institut 1974.

[3] Boche, R.: Complex interval arithmetic with some applications. Lockheed Missiles and Space Company, 4-22-66-1, Sunnyvale, California, 1966.

[4] Bohlender, G.: Floating-point computation of functions with maximum accuracy. IEEE Transactions on Computers **C-26** (1977).

[5] Bohlender, G.: Genaue Summation von Gleitkommazahlen. Computing, Suppl. 1, pp. 21–32. Wien-New York: Springer 1977.

[6] Claudio, D. M.: Beiträge zur Struktur der Rechnerarithmetik, Dissertation, Universität Karlsruhe, 1979.

[7] Kulisch, U.: Interval arithmetic over completely ordered ringoids. Mathematics Research Center, The University of Wisconsin, Madison, Wisconsin. TS Report No. 1105, Sept., 1970.

[8] Kulisch, U.: Grundlagen des numerischen Rechnens – Mathematische Begründung der Rechnerarithmetik. Mannheim-Wien-Zürich: Wissenschaftsverlag, Bibliographisches Institut 1976.

[9] Kulisch, U., Miranker, W. L.: Computer arithmetic in theory and practice, R.C. Report 7776, IBM Research Center, Yorktown Heights, N.Y., 1979.

[10] Ullrich, Ch.: Rundungsinvariante Strukturen mit äußeren Verknüpfungen. Dissertation, Universität Karlsruhe, 1972.

[11] Ullrich, Ch.: Über die beim numerischen Rechnen mit komplexen Zahlen und Intervallen vorliegenden mathematischen Strukturen. Computing **14**, 51–65 (1975).

[12] Rokne, J., Lancaster, P.: Complex interval arithmetic. Comm. ACM **14**, 111–112 (1971).

Dr. U. W. Kulisch
Dr. W. L. Miranker
IBM Thomas J. Watson Research Center
Yorktown Heights, NY 10598, U.S.A.

Computing, Suppl. 2, 69–84 (1980)

Some Applications of Extended Interval Arithmetic to Interval Iterations

S. M. Markov, Sofia

Abstract

The calculation of united extensions of real functions of one variable by means of primitive interval operations is considered. It is demonstrated that extended interval arithmetic is a convenient tool for treating this problem. Some direct applications of the results obtained to interval iteration procedures are given.

1. Introduction

Familiar interval mathematics makes use of only two primitive arithmetic operations between intervals: addition $A + B$ and multiplication AB of the intervals A and B. The operations for the familiar subtraction and division are compositions of these two operations. Indeed, the subtraction of $A = [a_1, a_2]$, $B = [b_1, b_2]$ can be written as $A + [-1, -1]B$, and division is $[c, c]AB$, where $c = 1/(b_1 b_2)$.

Therefore we should not consider the last two operations as primitive ones. We shall call them auxiliary subtraction and auxiliary division, respectively, in order to avoid confusion with other operations that are introduced later.

As it is based upon only two primitive operations, familiar interval arithmetic presents a very simple algebraic structure. It is often realized that this algebraic structure is not rich enough to handle various problems arising in interval mathematics. This structure can be substantially enriched if the set of primitive operations is extended by two new operations. This extended interval arithmetic presents an interesting algebraical structure that is especially suitable for handling problems involving interval functions.

In this paper we give a short introduction to extended interval arithmetic and some rules for calculating united extensions of real functions of one variable. We then apply these results to some interval iteration procedures of Newton type.

2. Extended Interval Arithmetic

We shall denote the set of reals by $\mathscr{R}$ and the set of all closed intervals on $\mathscr{R}$ by $\mathscr{I}(\mathscr{R})$. Let a_1 and a_2, $a_1 \leqslant a_2$, be the end-points of $A \in \mathscr{I}(\mathscr{R})$, and write $A = [a_1, a_2]$. By a_c and a_d we mean the end-points of A such that $|a_c| \leqslant |a_d|$; that is, a_c is the end-point of A that is closer to zero than a_d. An interval with end-points α and β (α not necessarily

$\leqslant \beta$) will be denoted by $\alpha \vee \beta$ or $[\alpha \vee \beta]$. For the width of A we write $w(A) = a_2 - a_1$. The absolute value of A is $|A| = \max\{|a_1|, |a_2|\}$. The set of all intervals A, such that $0 \in \bar{A}$, will be denoted by $\mathscr{I}^*(\mathscr{R})$.

Addition of $A, B \in \mathscr{I}(\mathscr{R})$ is defined by

(A) $$A + B = [a_1 + b_1, a_2 + b_2].$$

A convenient formula for multiplication of two intervals $A, B \in \mathscr{I}^*(\mathscr{R})$ is

(M) $$AB = (a_c b_c) \vee (a_d b_d).$$

Auxiliary subtraction $A + (-B)$ will be indicated by $A \ominus B$, and this may be written as

(AS) $$A - B = [a_1 - b_2, a_2 - b_1],$$

for $A, B \in \mathscr{I}(\mathscr{R})$.

Auxiliary division of A and B for $A, B \in \mathscr{I}^*(\mathscr{R})$ can be written as

(AD) $$A \odot B = (a_c/b_d) \vee (a_d/b_c).$$

The operations (A), (M), (AS), and (AD) are well known from familiar interval arithmetic [1, 5, 6, etc.], since we have

$$A + B = \{a + b \colon a \in A, b \in B\}, \qquad A \ominus B = \{a - b \colon a \in A, b \in B\},$$

for $A, B \in \mathscr{I}(\mathscr{R})$ and

$$AB = \{ab \colon a \in A, b \in B\}, \qquad A \odot B = \{a/b \colon a \in A, b \in B\},$$

for $A, B \in \mathscr{I}^*(\mathscr{R})$.

We extend [3] the interval arithmetic by the following two operations: Basic subtraction, given for $A, B \in \mathscr{I}(\mathscr{R})$ by

(S) $$A - B = (a_1 - b_1) \vee (a_2 - b_2),$$

and basic division, defined for $A, B \in \mathscr{I}^*(\mathscr{R})$ by

(D) $$A/B = (a_c/b_c) \vee (a_d/b_d).$$

We shall further consider the operations (A), (M), (S), and (D) as primitive operations. The auxiliary operations (AS) and (AD) are compositions of the primitive operations, since $A \ominus B = A + (-B)$ and $A \odot B = A(1/B)$. Analogously, we may consider an auxiliary addition $A \oplus B = A - (-B)$ and an auxiliary multiplication $A \otimes B = A/(1/B)$.

These operations can be expressed in terms of end-points of the intervals as follows:

(AA) $$A \oplus B = (a_1 + b_2) \vee (a_2 + b_1) \quad \text{for} \quad A, B \in \mathscr{I}(\mathscr{R}),$$

(AM) $$A \otimes B = (a_c b_d) \vee (a_d b_c) \quad \text{for} \quad A, B \in \mathscr{I}^*(\mathscr{R}).$$

Let us give expressions for AB, A/B, $A \otimes B$, and $A \odot B$ for intervals containing zero. In the usual interval arithmetic the operations AB and $A \odot B$ are defined by

$AB = \{ab: a \in A, b \in B\}$ for arbitrary A, B and $A \odot B = \{a/b, a \in A, b \in B\}$ for $0 \in B$ and arbitrary A. By means of the end-points these definitions can be written:

$$\left.\begin{aligned} AB = b_d A &= (a_c b_d) \vee (a_d b_d) \\ A \odot B = (1/b_c)A &= (a_c/b_c) \vee (a_d/b_c) \end{aligned}\right\} \quad \text{for } 0 \in A, 0 \bar{\in} B,$$

and

$$AB = [\min\{a_1 b_2, a_2 b_1\}, \max\{a_1 b_1, a_2 b_2\}] \qquad \text{for } 0 \in A, 0 \in B.$$

Thus we may define $A \otimes B$ and A/B for intervals containing zero by

$$\left.\begin{aligned} A \otimes B = b_c A &= (a_c b_c) \vee (a_d b_c) \\ A/B = (1/b_d)A &= (a_c/b_d) \vee (a_d/b_d) \end{aligned}\right\} \quad \text{for } 0 \in A, 0 \bar{\in} B$$

and

$$A \otimes B = [\max\{a_1 b_2, a_2 b_1\}, \min\{a_1 b_1, a_2 b_2\}] \qquad \text{for } 0 \in A, 0 \in B.$$

Then we have $A \otimes B = A/(1/B)$ and $A \odot B = A(1/B)$ for all $A \in \mathscr{I}(\mathscr{R})$ and $B \in \mathscr{I}^*(\mathscr{R})$.

Table 1 summarizes the definitions of the basic and primitive operations used in extended interval arithmetic.

Table 1

	Basic Operations	Auxiliary Operations
$A, B \in \mathscr{I}(\mathscr{R})$	$A + B = (a_1 + b_1) \vee (a_2 + b_2)$ $A - B = (a_1 - b_1) \vee (a_2 - b_2)$	$A \oplus B = (a_1 + b_2) \vee (a_2 + b_1) = A - (-B)$ $A \ominus B = (a_1 - b_2) \vee (a_2 - b_1) = A + (-B)$
$A, B \in \mathscr{I}^*(\mathscr{R})$	$AB = (a_c b_c) \vee (a_d b_d)$ $A/B = (a_c/b_c) \vee (a_d/b_d)$	$A \otimes B = (a_c b_d) \vee (a_d b_c) = A/(1/B)$ $A \odot B = (a_c/b_d) \vee (a_d/b_c) = A(1/B)$
$0 \in A$ $0 \bar{\in} B$	$AB = b_d A$ $A/B = (1/b_d)A$	$A \otimes B = b_c A = A/(1/B)$ $A \odot B = (1/b_c)A = A(1/B)$
$0 \in A$ $0 \in B$	$AB = [\min\{a_1 b_2, a_2 b_1\},$ $\max\{a_1 b_1, a_2 b_2\}]$	$A \otimes B = [\max\{a_1 b_2, a_2 b_1\},$ $\min\{a_1 b_1, a_2 b_2\}]$

3. Interval Operators and Fixed Points

Let $\mathscr{L}$ be a normed lattice and $\mathscr{I}(\mathscr{L})$ be the corresponding normed interval space over $\mathscr{L}$, see [4]. Denote by $\|\cdot\|$ the norm in $\mathscr{I}(\mathscr{L})$ and consider an interval operator $U: \mathscr{I}(\mathscr{L}) \to \mathscr{I}(\mathscr{L})$. The operator U is a contraction mapping in $\mathscr{I}(\mathscr{L})$ if there is a constant q, $0 < q < 1$, such that the inequality

$$\|U(X) - U(Y)\| \leqslant q\|X - Y\|$$

holds for every $X, Y \in \mathscr{I}(\mathscr{L})$.

Assume further that $\mathscr{L}$ is a Banach lattice and hence is complete. Then we have the following fixed-point theorem (as usual a fixed point of U is an $X^* \in \mathscr{I}(\mathscr{L})$ such that $X^* = U(X^*)$):

Theorem 1. *Let $\mathscr{L}$ be a Banach lattice and U be a contraction mapping in $\mathscr{I}(\mathscr{L})$. Then the operator U possesses a unique fixed point $X^* \in \mathscr{I}(\mathscr{L})$, that is the limit of the sequence of successive approximations*

$$X^{(n+1)} = U(X^{(n)}), \qquad n = 0, 1, 2, \ldots,$$

with $X^{(0)} \in \mathscr{I}(\mathscr{L})$ arbitrary.

Proof. The operation "$-$" in $\mathscr{I}(\mathscr{L})$ enables us to prove Theorem 1 in exactly the same way as in the classical case. Indeed, using the equality $X^{(n+1)} = U(X^{(n)})$ and the properties of the interval norm $\|\cdot\|$, we have for $m > n > 0$:

$$\begin{aligned}\|X^{(n)} - X^{(m)}\| &= \|U(X^{(n-1)}) - U(X^{(m-1)})\| \leqslant q\|X^{(n-1)} - X^{(m-1)}\| \\ &= q\|U(X^{(n-2)}) - U(X^{(m-2)})\| \leqslant q\|X^{(n-2)} - X^{(m-2)}\| \leqslant \cdots \\ &\leqslant q^n\|X^{(0)} - X^{(m-n)}\| \cdots\end{aligned}$$

On the other side, we have

$$\begin{aligned}\|X^{(0)} - X^{(m-n)}\| &\leqslant \|X^{(0)} - X^{(1)}\| + \|X^{(1)} - X^{(2)}\| + \cdots + \|X^{(m-n-1)} - X^{(m-n)}\| \\ &\leqslant (1 + q + q^2 + \cdots + q^{m-n-1})\|X^{(0)} - X^{(1)}\| \\ &\leqslant (1-k)^{-1}\|X^{(0)} - X^{(1)}\|.\end{aligned}$$

Thus

$$\|X^{(n)} - X^{(m)}\| \leqslant q^n(1-q)^{-1}\|X^{(0)} - X^{(1)}\|,$$

and therefore $X^{(n)}$ is an interval Cauchy sequence, that is $\lim_{n\to\infty, n<m} \|X^{(n)} - X^{(m)}\| = 0$. From the completeness of $\mathscr{L}$ it follows then that there exists an X^*, such that

$$\lim_{n\to\infty}\|X^{(n)} - X^*\| = 0.$$

This result and $\|U(X^{(n)}) - U(X^*)\| \leqslant q\|X^{(n)} - X^*\|$ imply $\lim_{n\to\infty}\|U(X^{(n)}) - U(X^*)\| = 0$ as well. Further, from

$$\begin{aligned}\|X^* - U(X^*)\| &\leqslant \|X^* - X^{(n)}\| + \|X^{(n)} - U(X^*)\| \\ &= \|X^* - X^{(n)}\| + \|U(X^{(n-1)}) - U(X^*)\| \to 0\end{aligned}$$

we see that $\|X^* - U(X^*)\| = 0$ and hence $X^* = U(X^*)$. This completes the proof.

We next formulate the fixed-point theorem for the case when U is a contraction mapping only in a neighbourhood of the fixed point.

Theorem 2. *Let $\mathscr{L}$ be a Banach lattice and U be a contraction mapping in a neighbourhood $\mathscr{N} \subset \mathscr{I}(\mathscr{L})$ of its fixed point X^*. Then the iteration process $X^{(n+1)} = U(X^{(n)})$, $n = 0, 1, \ldots$, converges to X^* as $O(q^n)$ for any $X^{(0)} \in \mathscr{N}$.*

Proof. The assumptions on U mean that there exist two positive numbers p and $q < 1$, such that $\|U(X) - U(Y)\| \leqslant q\|X - Y\|$ for every two points $X, Y \in \mathscr{N}(X^*, p) = \{Z : \|Z - X^*\| < p, Z \in \mathscr{I}(\mathscr{L})\}$.

Assume that $X^{(m)} \in \mathscr{N}(X^*, p)$ for some $m = 1, 2, \ldots$. Then

$$\|X^{(m+1)} - X^*\| = \|U(X^{(m)}) - X^*\| \leqslant q\|X^{(m)} - X^*\| < qp < p$$

shows that $X^{(m+1)} \in \mathcal{N}(X^*, p)$ as well. Since $X^{(0)} \in \mathcal{N}(X^*, p)$, we obtain that $X^{(i)} \in \mathcal{N}(X^*, p)$ for all $i = 1, 2, \ldots$. Using the fact that U is a contraction mapping in $\mathcal{N}(X^*, p)$ we can write

$$\|X^{(n)} - X^*\| = \|U(X^{(n-1)}) - U(X^*)\| \leqslant q\|X^{(n-1)} - X^*\| \leqslant \cdots$$
$$\leqslant q^n\|X^{(0)} - X^*\| \leqslant pq^n,$$

and this proves the theorem.

As usual we say that when $\|X^{(n)} - X^*\| = O(q^n)$, the convergence is linear or of order 1. More generally, the convergence is called of order $p > 1$ if $\|X^{(n)} - X^*\| = O(q^{p^n})$. Thus in the above we have linear convergence; later we shall consider interval iteration procedures with convergence of order 2.

4. United Extensions of Real Functions and Their Computation by Means of Interval Arithmetic Operations

Let f be a real function defined on the interval D and $\mathcal{I}(D)$ be the set of all subintervals of D. A mapping $F^*: \mathcal{I}(D) \to \mathcal{I}(\mathcal{R})$ is called an interval extension of f, if the restriction of F^* to D is equal to f, $F^*|D = f$. A special case of interval extension is the united extension, defined by

$$F(X) = \{\vee f(x): x \in X\}, \qquad \text{for} \quad X \in \mathcal{I}(D). \tag{1}$$

Here $\{\vee f(x): x \in X\}$ means the interval of minimal width, containing all $f(x)$ for $x \in X$.

We shall denote the united extension of the functions $f, g, \ldots$ by $F, G, \ldots$, respectively.

If f is continuous, then (1) can be written

$$F(X) = \{f(x): x \in X\} = [\min_{x \in X} f(x), \max_{x \in X} f(x)].$$

We shall now consider the following problem: How do we calculate the united extensions of the functions $f + g$, $f - g$, fg, and f/g if the united extensions of the functions f and g are known? We shall answer this question in case that f and g satisfy certain monotonicity conditions.

Denote by $\mathcal{M}(D)$ the set of all real functions that are monotone on the interval D. Note that the united extension of $f \in \mathcal{M}(D)$ is easily computed by

$$F(X) = [f(x_1) \vee f(x_2)] \qquad \text{for} \quad X = [x_1, x_2] \in \mathcal{I}(D). \tag{2}$$

We shall say that $f, g \in \mathcal{M}(D)$ satisfy the monotonicity condition $\mathcal{M}1$ if both functions are monotone increasing or both are monotone decreasing in D. The pair (f, g) satisfies $\mathcal{M}2$ if one of the functions is monotone increasing and the other is monotone decreasing on D.

Proposition 1. If f, g and $h = f + g \in \mathcal{M}(D)$, then for every $X \subset D$:

$$H(X) = \begin{cases} F(X) + G(X), \text{ if } (f, g) \text{ satisfy } \mathcal{M}1, \\ F(X) \oplus G(X), \text{ if } (f, g) \text{ satisfy } \mathcal{M}2. \end{cases} \tag{3}$$

Proposition 2. If f, g, and $h = f - g$ are monotone on D, then for every $X \subset D$:

$$H(X) = \begin{cases} F(X) - G(X), & \text{if } (f,g) \text{ satisfy } \mathcal{M}1, \\ F(X) \ominus G(X), & \text{if } (f,g) \text{ satisfy } \mathcal{M}2. \end{cases} \tag{4}$$

Proposition 3. Let f and g are such that $|f|$, $|g|$, and $h = f \cdot g$ are monotone. Then for every $X \subset D$,

$$H(X) = \begin{cases} F(X) \cdot G(X), & \text{if } (|f|, |g|) \text{ satisfy } \mathcal{M}1, \\ F(X) \otimes G(X), & \text{if } (|f|, |g|) \text{ satisfy } \mathcal{M}2. \end{cases} \tag{5}$$

Proposition 4. Let f and g are such that $|f|, |g| \in \mathcal{M}(D)$, $g(x) \neq 0$ for $x \in D$, and $h = f/g \in \mathcal{M}(D)$. Then for every $X \subset D$,

$$H(X) = \begin{cases} F(X) : G(X), & \text{if } (|f|, |g|) \text{ satisfy } \mathcal{M}1, \\ F(X) \odot G(X), & \text{if } (|f|, |g|) \text{ satisfy } \mathcal{M}2. \end{cases} \tag{6}$$

The verification of these four propositions is straightforward. As an example we prove the last one.

Verification of Proposition 4. Using (2) for $f/g \in \mathcal{M}(D)$ we can write $H(X) = [(f(x_1)/g(x_1)) \vee (f(x_2)/g(x_2))]$. Now if $|f|$ and $|g|$ satisfy $\mathcal{M}1$, then $(|f(x_1)| - |f(x_2)|)(|g(x_1)| - |g(x_2)|) \geqslant 0$. In this case we see easily that

$$\begin{aligned} H(X) &= [(f(x_1)/g(x_1)) \vee (f(x_2)/g(x_2))] \\ &= [f(x_1) \vee f(x_2)] : [g(x_1) \vee g(x_2)] = F(X) : G(X). \end{aligned}$$

In case $|f|$ and $|g|$ satisfy $\mathcal{M}2$, we have that

$$(|f(x_1)| - |f(x_2)|)(|g(x_1)| - |g(x_2)|) < 0$$

implies

$$\begin{aligned} H(X) &= [(f(x_1)/g(x_1)) \vee (f(x_2)/g(x_2))] \\ &= [f(x_1) \vee f(x_2)] \odot [g(x_1) \vee g(x_2)] = F(X) \odot G(X). \end{aligned}$$

Let us now give some practical applications of Propositions 1 – 4.

Example 1. Find the united extension of the function $h(x) = x^2 + x$, for $x \in \mathcal{R}$.

Solution. The united extension of x is X. By means of Proposition 3 we see that the united extension of $g(x) = x^2 = x \cdot x$ can be calculated for intervals $X \bar{\ni} 0$ (since for such intervals the pair $(|x|, |x|)$ satisfies $\mathcal{M}1$) and is equal to $X \cdot X = X^2$. Applying Proposition 1 we see that the interval extension of $x + x^2$ can be calculated for every interval X that does not contain the points -0.5 and zero in its interior since on such X the functions x, x^2 and $x + x^2$ are monotone. It is given by

$$H(X) = \{x + x^2 : x \in X\} = \begin{cases} X + X^2, & \text{if } X \geqslant 0; \\ X \oplus X^2, & \text{if } X \leqslant -0.5 \text{ or } X \subset [-0.5, 0]. \end{cases}$$

Example 2. For the united extension of the function $h(x) = x - x^2$, we find upon applying Proposition 2, that

$$H(X) = \begin{cases} X - X^2, \text{ if } X \subset [0, +0.5] \text{ or } X \geqslant 0.5; \\ X \ominus X^2, \text{ if } X \leqslant 0. \end{cases}$$

Example 3. Find the united extension of the rational function $h(x) = (2 + x)/(1 + x)$.

Solution. The functions $|2 + x|$ and $|1 + x|$ satisfy $\mathscr{M}1$ in the intervals $(-\infty, -2)$ and $[-1, \infty)$ and satisfy $\mathscr{M}2$ in $[-2, -1]$. We also have $h \in \mathscr{M}((-\infty, -1])$ and $h \in \mathscr{M}([-1, \infty))$. Therefore, in accord with Proposition 4 we may write

$$H(X) = \begin{cases} (2 + X)/(1 + X), & \text{if } \quad X \leqslant -2 \text{ or } X \geqslant -1; \\ (2 + X) \odot (1 + X), & \text{if } -1 \leqslant X \leqslant -2, \end{cases}$$

where we have also used the fact that the united extensions of the functions $2 + x$ and $1 + x$ are $2 + X$ and $1 + X$, respectively. (This example is taken from [2], where it is regretfully noted that the united extension of $(2 + x)/(1 + x)$ cannot be calculated by means of the familiar primitive interval arithmetic operations for $X \subset [1, 2]$).

These examples show that it is possible to outline an algorithm for calculating the united extensions of an arbitrary rational function of one variable in certain intervals.

5. United Extensions of Some Newton-Type Operators

We shall say that f satisfies the N-condition on $D = [d_1, d_2]$ if:

1) there is an x^*, $d_1 < x^* < d_2$, such that $f(x^*) = 0$;
2) f is twice continuously differentiable in D;
3) f' and f'' have constant sign in D;
4) f/f' is monotone increasing in D;
5) f' is not very close to zero in D, i.e. $|f'| \geqslant \lambda > 0$ in D.

Denote by D' the subinterval of D on which $f(x)f''(x) \geqslant 0$; D' is either the interval $[d_1, x^*]$ or the interval $[x^*, d_2]$. Similarly let D'' be the other subinterval of D on which $f(x)f''(x) \geqslant 0$.

Define the Newton operators

$$n(x) = n(f; x) = x - f(x)/f'(x)$$

$$\tilde{n}(x) = \tilde{n}(f; x) = x - f(x)/f'(\tilde{d}),$$

where $\tilde{d}$ is such that $f'(\tilde{d}) = \max_{x \in D}|f'(x)|$. Obviously $\tilde{d} \in \{d_1, d_2\}$.

We shall first calculate the united extension of $\tilde{n}(x)$. To this end note that the functions x, $g(x) = f(x)/f'(\tilde{d})$ and $\tilde{n}(x) = x - g(x)$ are monotone increasing on D (indeed, $\tilde{n}'(x) = 1 - f'(x)/f(\tilde{d}) \geqslant 0$ in D). Then by means of Proposition 2 we obtain for the united extension of $\tilde{n}(x)$:

$$\tilde{N}(X) = X - F(X)/f'(\tilde{d}).$$

We notice that $\tilde{d}$ is an end-point of D (the one that is also end-point of D'). If we recall the definition of the operation A/B for the case $A \ni 0$ (see Table 1) we see that

$\tilde{N}(X)$ can be written as

$$\tilde{N}(X) = X - F(X)/F'(D), \tag{7}$$

where F' is the united extension of f'.

We shall now calculate the united extension of $n(x)$. In order to write down the united extension G of $g(x) = f(x)/f'(x)$ we note that $|f|$ and $|f'|$ satisfy $\mathcal{M}1$ on D' and $\mathcal{M}2$ on D''. Since f/f' is monotone increasing in D we can apply Proposition 4 to obtain

$$G(X) = \begin{cases} F(X)/F'(X) & \text{for } X \subset D', \\ F(X) \odot F'(X) & \text{for } X \subset D'', \end{cases}$$

where F and F' are the united extensions of f and f', respectively.

Now in order to calculate the united extension of $n(x) = x - g(x)$ we first observe that x and $g(x)$ are both monotone increasing in D and that $n(x)$ is monotone in D and D' (since $n'(x) = f(x)f''(x)/f'^2(x)$). Therefore by Proposition 2 we obtain $N(X) = X - G(X)$, for $X \subset D', D''$, that is

$$N(X) = \begin{cases} X - F(X)/F'(X), & \text{for } X \text{ such that } F(X)F''(X) \geqslant 0, \\ X - F(X) \odot F'(X), & \text{for } X \text{ such that } F(X)F''(X) \leqslant 0. \end{cases} \tag{8}$$

where F, F' and F'' are the united extensions of f, f' and f'', respectively.

According to (8) if X is an interval, such that $f(x)f''(x) \geqslant 0$ for all $x \in X$, then

$$N(X) = \{x - f(x)/f'(x) : x \in X\} = X - F(X)/F'(X),$$

where f satisfies the N-conditions on $D \subset X$ (see Fig. 1).

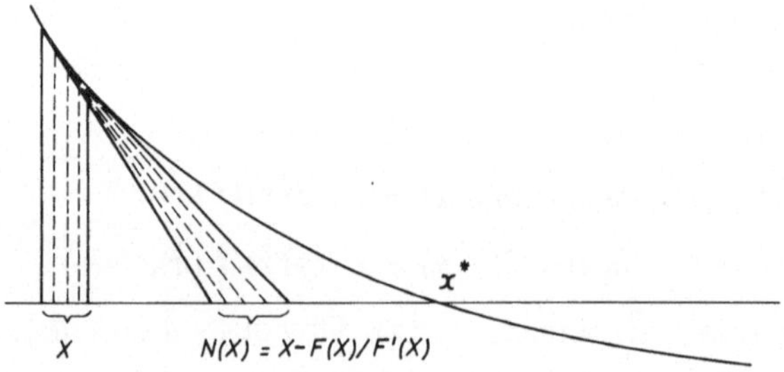

Fig. 1

Assume that the equation $f(x) = 0$ is solved by the iteration procedure $x^{(n+1)} = x^{(n)} - f(x^{(n)})/f'(x^{(n)})$ with $x^{(0)}$ such that $f(x^{(0)})f''(x^{(0)}) > 0$. Suppose we know bounds for $x^{(0)}$, that is, suppose we know (small) interval $X^{(0)}$ containing $x^{(0)}$. Then we can assert that $x^{(1)} \in X^{(1)} = X^{(0)} - F(X^{(0)})/F'(X^{(0)})$ and by induction $x^{(k+1)} \in X^{(k+1)} = X^{(k)} - F^{(k)}/F'(X^{(k)})$. Thus by means of extended interval arithmetic we can calculate an interval for $x^{(k+1)}$ provided we know an interval for $x^{(0)}$.

6. Some Interval Iteration Procedures of Newton Type

As in the previous section we shall assume that f satisfies the N-conditions on D. We shall demonstrate convergency of the interval iteration procedures $X^{(n+1)} = \tilde{N}(X^{(n)})$ and $X^{(n+1)} = N(X^{(n)})$ where $\tilde{N}$ and N are the Newton-type interval operators defined in Sect. 5.

Theorem 3. *Let f satisfy the N-conditions on D and F be the united extension of f. Then the iteration scheme*

$$X^{(0)} = D$$

$$X^{(i+1)} = X^{(i)} - F(X^{(i)}/F'(X^{(0)}), \qquad i = 0, 1, 2, \ldots \tag{9}$$

produces an interval sequence $\{X^{(i)}\}$ converging to x^*, that is $\lim_{i\to\infty}|X^{(i)} - x^*| = 0$.

Proof. As we showed $\tilde{N}(X) = X - F(X)/F'(D)$ is the united extension of $\tilde{n}(x) = x - f(x)/f'(\tilde{d})$. We shall demonstrate that $\tilde{N}(X)$ satisfies $|N(X) - N(Y)| \leqslant q|X - Y|$ with $q < 1$ for every $X, Y \subset D$. Indeed, since $\tilde{N}(X)$ is the united extension of $\tilde{n}(x)$ and $\tilde{n}$ is monotone increasing in D, we can write $\tilde{N}(X) = [\tilde{n}(x_1), \tilde{n}(x_2)]$. Further, using that $|\tilde{n}(u) - \tilde{n}(v)| \leqslant q|u - v|$, $q < 1$ for $u, v \in D$, we have

$$\begin{aligned}|\tilde{N}(X) - \tilde{N}(Y)| &= |[\tilde{n}(x_1), \tilde{n}(x_2)] - [\tilde{n}(y_1), \tilde{n}(y_2)]| \\ &= |[(\tilde{n}(x_1) - \tilde{n}(y_1)) \vee (\tilde{n}(x_2) - \tilde{n}(y_2))]| \\ &= \max\{|\tilde{n}(x_1) - \tilde{n}(y_1)|, |\tilde{n}(x_2) - \tilde{n}(y_2)|\} \\ &\leqslant \max\{q|x_1 - y_1|, q|x_2 - y_2|\} = q|X - Y|.\end{aligned}$$

Thus $\tilde{N}(X)$ satisfies the conditions of Theorem 2 with fix-point $X^* = x^*$. Therefore (9) produces a sequence $X^{(k)} \to x^*$.

Remark 1. Theorem 3 holds true for arbitrary chosen $X^{(0)} \subset D$ (not necessarily $X^{(0)} \ni x^*$).

Remark 2. Theorem 3 follows directly from the theorem in [2].

We shall next consider the interval iteration procedure using the operator $N(X)$.

Theorem 4. *The iteration scheme*

$$\textit{choose } X^{(0)} \subset D$$

$$\textit{let } X^{(i+1)} = X^{(i)} - F(X^{(i)})/F'(X^{(i)}) \tag{10}$$

produces an interval sequence $\{X^{(i)}\}$ that converges to x^* and for the rate of convergence we have

$$|X^{(i)} - x^*| = O(q^{2^i}), \qquad 0 < q < 1.$$

Proof. We shall first consider the case $x^* \in X^{(0)}$. This case is the most interesting for the practice, since then by means of (10) we can obtain a sufficiently narrow interval that contains the zero x^* of $f(x) = 0$.

There are four subcases according to the signs of f' and f'':

a) $f' > 0$ and $f'' < 0$ (Fig. 2a);
b) $f' < 0$ and $f'' > 0$ (Fig. 2b);
c) $f' > 0$ and $f'' > 0$ (Fig. 2c);
d) $f' < 0$ and $f'' < 0$ (Fig. 2d).

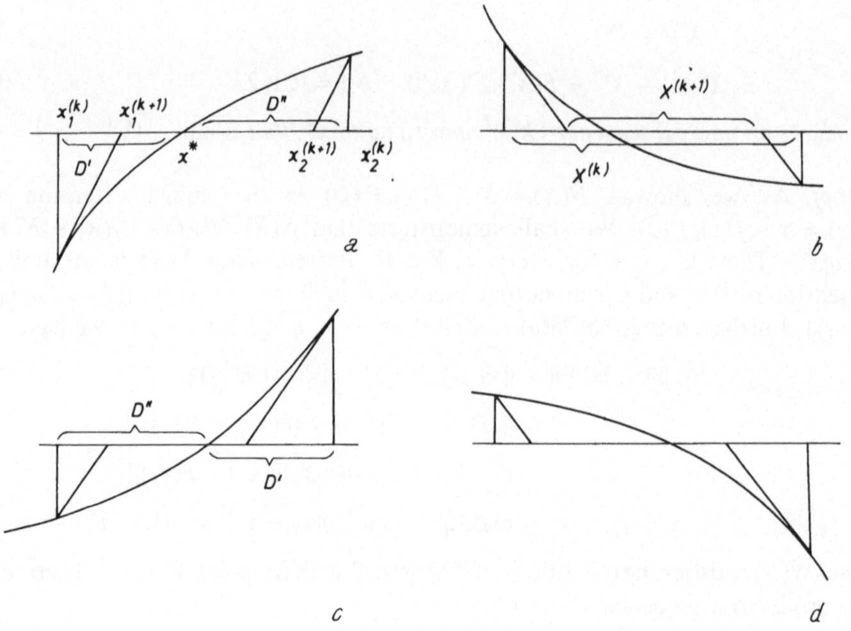

Fig. 2a – d

All subcases are treated similarly; we shall consider in detail the first subcase. $f' > 0$ and $f'' < 0$ imply (see Fig. 2a):

$$F(X^{(k)}) = [f(x_1^{(k)}), f(x_2^{(k)})],$$

$$F'(X^{(k)}) = [f'(x_2^{(k)}), f'(x_1^{(k)})] > 0.$$

Since $F(X^{(k)})$ contains zero and $F'(X^{(k)}) > 0$, we have $F(X^{(k)})/F'(X^{(k)}) = F(X^{(k)})/f'(x_1^{(k)})$. The function $x - f(x)/f'(x_1^{(k)})$ increases in $X^{(k)}$ so that we have

$$x_2^{(k)} - f(x_2^{(k)})/f'(x_1^{(k)}) \geqslant x_1^{(k)} - f(x_1^{(k)})/f'(x_1^{(k)}).$$

Therefore iteration (10) can be written end-point wise:

$$x_1^{(k+1)} = x_1^{(k)} - f(x_1^{(k)})/f'(x_1^{(k)})$$

$$x_2^{(k+1)} = x_2^{(k)} - f(x_2^{(k)})/f'(x_1^{(k)}) \tag{11}$$

(see Fig. 2a).

In subcase b) for the end-point formulation of (10) we obtain again (11), whereas in subcases c) and d) we obtain

$$x_1^{(k+1)} = x_1^{(k)} - f(x_1^{(k)})/f'(x_2^{(k)})$$
$$x_2^{(k+1)} = x_2^{(k)} - f(x_2^{(k)})/f'(x_2^{(k)})$$

Summarizing, for (10) we have

$$X^{(k+1)} = X^{(k)} - F(X^{(k)})/F'(X^{(k)}) = X^{(k)} - F(X^{(k)})/f'(x_j^{(k)}), \tag{10'}$$

where $j = 1$, if $f'f'' < 0$ in $X^{(0)}$ and $j = 2$, if $f'f'' > 0$ in $X^{(0)}$.

This shows that only one computation of f' on each step is required.

We consider now the convergence of the method.

In subcase a) we can write for the width of $F(X^{(k)})$:

$$\begin{aligned} w(F(X^{(k)})) &= f(x_2^k) - f(x_1^k) = (x_2^{(k)} - x_1^{(k)})f'(x_1^{(k)}) + \tfrac{1}{2}(x_2^k - x_1^k)^2 f''(c^k) \\ &= w(X^{(k)})f'(x_1^{(k)}) + \tfrac{1}{2}w^2(X^{(k)})f''(c^{(k)}), \qquad c^{(k)} \in X^{(k)} \end{aligned} \tag{12}$$

Since $f'' < 0$ (12) implies that

$$w(F(X^{(k)})/f'(x_1^{(k)})) < w(X^{(k)}). \tag{13}$$

From (10′) using $w(A - B) = w(A) - w(B)$ and (13) we have

$$\begin{aligned} w(X^{(k+1)}) &= w((X^{(k)}) - F(X^{(k)})/f'(x_1^{(k)})) \\ &= w(X^{(k)}) - w(F(X^{(k)}))/f'(x_1^{(k)}). \end{aligned}$$

Substituting (12) in the above equality we obtain:

$$\begin{aligned} w(X^{(k+1)}) &= w(X^{(k)}) - (w(X^{(k)}f'(x_1^{(k)}) + \tfrac{1}{2}w^2(X^{(k)})f''(c^{(k)}))/f'(x_1^{(k)}) \\ &= -(f''(c^{(k)})/2f'(x_1^{(k)})) \cdot w^2(X^{(k)}), \qquad c^{(k)} \in X^{(k)}. \end{aligned}$$

Subcases b), c), and d) can be treated in a similar way; the results obtained can be summarized as follows:

$$w(X^{(k+1)}) = -\tfrac{1}{2}(f''(c^{(k)})/f'(x_j^{(k)})) \cdot w^2(X^{(k)}),$$

where j is 1 or 2. Choosing $q > 0$ such that $-\frac{1}{2}(f''(c)/f'(d)) \leqslant q$ for all $c, d \in D$ we may write

$$w(X^{(k+1)}) \leqslant q \cdot w^2(X^{(k)}). \tag{14}$$

In view of $x^* \in X^{(k)}$ (14) can be written also as

$$|X^{(k+1)} - x^*| \leqslant 4q|X^{(k)} - x^*|^2 \tag{15}$$

Indeed, using the fact that $x^* \in X^{(k)}$ implies

$$\tfrac{1}{2}w(X^{(k)}) \leqslant |X^{(k)} - x^*| \leqslant w(X^{(k)}),$$

we have

$$|X^{(k+1)} - x^*| \leqslant w(X^{(k+1)}) \leqslant qw^2(X^{(k)}) \leqslant 4q(\tfrac{1}{2}w(X^{(k)}))^2 \leqslant 4q|X^{(k)} - x^*|.$$

Now from (15) and Theorem 2 the proof of Theorem 4 follows for the case $x^* \in X^{(0)}$.

If $x^* \overline{\in} X^{(0)}$, then either $X^{(0)} \subset D'$ or $X^{(0)} \subset D''$. Consider the case $X^{(0)} \subset D'$. In this case (10) can be written in terms of the end-points

$$X^{(k+1)} = \left(x_1^{(k)} - \frac{f(x_1^{(k)})}{f'(x_1^{k})}\right) \vee \left(x_2^{(k)} - \frac{f(x_2^{(k)})}{f'(x_2^{(k)})}\right)$$

which is the usual Newton method starting one time from $x_1^{(k)}$ and another time from $x_2^{(k)}$ (Fig. 3).

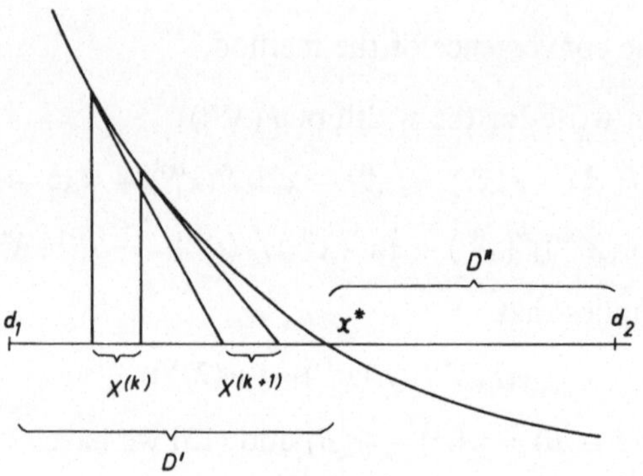

Fig. 3

From the well-known inequalities

$$|x_1^{(k+1)} - x^*| \leqslant q|x_1^{(k)} - x^*|^2, \qquad |x_2^{(k+1)} - x^*| \leqslant q|x_2^{(k)} - x^*|^2$$

it follows

$$\max\{|x_1^{(k+1)} - x^*|, |x_2^{(k+1)} - x^*|\} \leqslant q \max\{|x_1^{(k)} - x^*|^2, |x_2^{(k)} - x^*|^2\}$$
$$= q(\max|x_1^{(k)} - x^*|, |x_2^{(k)} - x^*|)^2,$$

that is

$$|X^{(k+1)} - x^*| \leqslant q|X^{(k)} - x^*|^2,$$

which proves the theorem in this case. Since the case $X^{(0)} \subset D''$ leads after the first iteration to one of the previous cases, the proof of Theorem 4 is complete.

Following Alefeld and Herzberger [1, p. 97] we may modify the iteration scheme (10) as follows:

choose $X^{(0)} \subset D$, $X^{(0)} \ni x^*$, and

let $X^{(k+1)} = Y^{(k)} - F(Y^{(k)})/F'(Y^{(k)})$,

where

$$Y^{(k)} = \begin{cases} [x_1^{(k)}, m(X^{(k)})], & \text{if } f(x_1^{(k)})f(m(X^{(k)}) < 0 \\ m(X^{(k)}), x_2^{(k)})], & \text{if } f(x_1^{(k)})f(m(X^{(k)}) > 0 \\ X^{(k)}, & \text{otherwise} \end{cases} \tag{16}$$

and $m(X)$ denotes the midpoint of X.

In the next section we give a numerical comparison between the interval iteration scheme (10) and Moore's method [5, Sect. 7.2]:

$$\text{choose } X^{(0)} \ni x^*$$

$$\text{let } X^{(k+1)} = (m(X^{(k)}) - F(m(X^{(k)}))/F'(X^{(k)})) \cap X^{(k)} \tag{17}$$

We also give a comparison between the modified interval iteration method (16) and the modified (after Alefeld and Herzberger) Moore's method:

$$\text{choose } X^{(0)} \ni x^*,$$

$$\text{let } X^{(k+1)} = \left(m(X^{(k)}) - \frac{f(m(X^{(k)})}{F'(Y^{(k)})} \right) \cap X^{(k)},$$

where

$$Y^{(k)} = \begin{cases} [x_1^{(k)}, m(X^{(k)})], & \text{if } f(x_1^{(k)}) f(m(X^{(k)})) < 0 \\ [m(X^{(k)}), x_2^{(k)}], & \text{if } f(x_1^{(k)}) f(m(X^{(k)})) > 0 \\ X^{(k)}, & \text{otherwise.} \end{cases} \tag{18}$$

The numerical calculations in the next section are performed by Ivan Nedkov. Thereby a precompiler for extended interval arithmetic created by Nikolai Dushkov has been used.

7. Numerical Results

For the comparison of the methods we use two examples of equations from [1, p. 97–99]. We shall first compare numerically the methods (10) and (17).

Example 1. For the solution of the equation

$$f(x) = x^2(x^2/3 + 2\sin x) - 3/19 = 0$$

with $X^{(0)} = [0.1, 1]$ we obtain the following interval approximations:

a) by means of Moore's method (17):

$$X^{(1)} = [0.000000000000, 0.513607416303]$$

$$X^{(2)} = [0.346173111779, 0.513607416303]$$

$$X^{(3)} = [0.376596765306, 0.405984406594]$$

$$X^{(4)} = [0.392303634719, 0.392471163891]$$

$$X^{(5)} = [0.392379503865, 0.392379510728]$$

$$X^{(6)} = [0.392379507136, 0.392379507136]$$

b) by means of the interval Newton method (10):

$$X^{(1)} = [0.120037044750, 0.680133943621]$$

$$X^{(2)} = [0.161502933620, 0.496973170075]$$

$$X^{(3)} = [0.236010152023, 0.413088339734]$$

$$X^{(4)} = [0.327467643386, 0.393417674107]$$

$$X^{(5)} = [0.381767247666, 0.392382294247]$$

$$X^{(6)} = [0.392089764461, 0.392379507156]$$

$$X^{(7)} = [0.392379289318, 0.392379507136]$$

$$X^{(8)} = [0.392379507136, 0.392379507136]$$

This shows that an accuracy of 10^{-12} is achieved with 6 iterations by means of (17) and with 8 iterations by means of (10).

Example 2. For the solution of the equation

$$p(x) = x(x^9 - 1) - 1$$

with $X^{(0)} = [1, 1.5]$ we obtain the following approximations:

a) by means of (17):

$$X^{(1)} = [0.100000000000, 0.123157901169]$$

$$X^{(2)} = [0.101853906531, 0.110215348995]$$

$$X^{(3)} = [0.107180976833, 0.108476244466]$$

$$X^{(4)} = [0.107564709432, 0.107593118087]$$

$$X^{(5)} = [0.107576603950, 0.107576609732]$$

$$X^{(6)} = [0.107576606608, 0.107576606608]$$

b) by means of (10):

$$X^{(1)} = [0.100260801352, 0.135612883879]$$

$$X^{(2)} = [0.100894156840, 0.123492229604]$$

$$X^{(3)} = [0.102286076683, 0.114352015277]$$

$$X^{(4)} = [0.104657759845, 0.109173023027]$$

$$X^{(5)} = [0.106892552883, 0.107682466778]$$

$$X^{(6)} = [0.107550142704, 0.107577098984]$$

$$X^{(7)} = [0.107576574583, 0.107576606619]$$

$$X^{(8)} = [0.107576606608, 0.107576606608]$$

showing again that Moore's method uses less iterations for a calculation of the zero with an error $\leqslant 10^{-12}$.

We shall next compare the methods (16) and (18) using the same equations as above.

Example 3. For the equation

$$f(x) = x^2(x^2/3 + 2\sin x) - 3/19 = 0$$

with $X^{(0)} = [0.1, 1]$ we obtain:

a) by means of (18):

$$X^{(1)} = [0.100000000000, \ 0.433580540100 \ ]$$

$$X^{(2)} = [0.339533432220, \ 0.433580540100 \ ]$$

$$X^{(3)} = [0.391237711372, \ 0.392469297358 \ ]$$

$$X^{(4)} = [0.392378544687, \ 0.392380226495 \ ]$$

$$X^{(5)} = [0.3923795071359, 0.3923795071364]$$

b) by means of (16):

$$X^{(1)} = [0.164098277476, 0.433580540100]$$

$$X^{(2)} = [0.358156977623, 0.396230123115]$$

$$X^{(3)} = [0.391501480063, 0.392417480786]$$

$$X^{(4)} = [0.392378966757, 0.392379510878]$$

$$X^{(5)} = [0.392379507136, 0.392379507136]$$

Example 4. For the solution of the equation

$$p(x) = x(x^9 - 1) - 1$$

with $X^{(0)} = [1, 1.5]$ we obtain the following approximations:

a) by means of (18):

$$X^{(1)} = [0.100000000000, 0.115390928100]$$

$$X^{(2)} = [0.107452573315, 0.107577227002]$$

$$X^{(3)} = [0.107576435513, 0.107576774994]$$

$$X^{(4)} = [0.107576606608, 0.107576606608]$$

b) by means of (16):

$$X^{(1)} = [0.101360436753, 0.115390928100]$$

$$X^{(2)} = [0.105788147041, 0.107603929621]$$

$$X^{(3)} = [0.107541095006, 0.107576639509]$$

$$X^{(4)} = [0.107576592680, 0.107576606608]$$

$$X^{(5)} = [0.107576606608, 0.107576606608]$$

References

[1] Alefeld, G., Herzberger, J.: Einführung in die Intervallrechnung. Mannheim: Bibliographisches Institut 1974.

[2] Caprani, O., Madsen, K.: Contraction mappings in interval analysis. BIT **15**, 362–366 (1975).

[3] Markov, S.: Extended interval arithmetic. Compt. Rend. Acad. Bulg. Sci. **31**, 163–166 (1978).

[4] Markov, S.: Extended interval arithmetic and some applications. Freiburger Intervall-Berichte 78/4, Institut für Angewandte Mathematik, Univ. Freiburg i. Br. 1978.

[5] Moore, R. E.: Interval Analysis. Englewood Cliffs, N. J.: Prentice-Hall 1966.
[6] Nickel, K.: Intervall-Mathematik. ZAMM **58**, T72 – T85 (1978).

S. M. Markov
Mathematisches Institut
Bulgarische Akademie der Wissenschaften
Postfach 373
Sofia, Bulgaria

Computing, Suppl. 2, 85 – 111 (1980)

Foundations of Finite Precision Rational Arithmetic*

D. W. Matula, Dallas, Texas, and **P. Kornerup,** Aarhus

Abstract

Finite precision fraction number systems are characterized and their number theoretic foundations are developed. Closed approximate rational arithmetic in these systems is obtained by the natural canonical rounding obtained using the continued fraction theory concept of best rational approximation. These systems are shown to be natural finite precision number systems in that they are essentially independent of the apparatus of the representation. The specific fixed-slash and floating-slash fraction number systems are described and their feasibility and convenience for computer implementation are discussed. The foundations of adaptive variable precision are explored. The overall goal is to better understand the inherent mathematical properties of finite precision arithmetic and to provide a most natural and convenient computation system for approximating real arithmetic on a computer.

I. Introduction and Summary

This paper develops the foundations and explores the feasibility of computer number systems composed of limited precision fractions, and coincidently provides insight into a much deeper question: "What are the inherent properties and limitations of finite precision approximate real arithmetic?" Special emphasis is placed on the demonstration that the choices made in specifying fraction number systems are natural and canonical, so that the properties derived help characterize the achievable limits of finite precision computation. Comparison with typical properties of floating-point numeric representation serves to expose accidental and avoidable sources of representation error, such as those attributable to the choice of base.

Approximation by finite precision representation has a variety of implications beyond those tractable by error bounds on the approximations. Thus any practical reliable assessment of the overall accuracy of a scientific computation hosted in a particular finite precision number system requires that the properties and limitations of the number system be rigorously understood. Our development of finite precision fraction number systems explores the following question areas pertinent to the analysis and design of any class of computer number systems:

(1) *Naturalness*: Are these number systems sufficiently natural to allow insight into the inherent properties of finite precision computation? Are they sufficiently natural to suggest canonical implementations on computers made by different manufacturers that would then enhance portability of numeric software?

---

* This research was supported in part by the National Science Foundation under Grant MCS77-21510.

(2) *Complexity*: What is the space complexity of the representation and the time complexity of algorithms for the standard arithmetic operations in these number systems? How cost-effective is arithmetic unit design to realize the arithmetic operator algorithms? How cost-effective is a software multiple precision realization for symbolic and algebraic applications?

(3) *Accuracy*: How numerically accurate is scientific computation hosted in these finite precision number systems?

The fact that approximation of reals by limited precision fractions has many properties superior to those of approximation by limited precision fixed radix representation was clearly and succintly expounded by Khintchin in 1935 in his classic little text *Continued Fractions* (Khintchin [5]). Therein Khintchin noted regarding numeric representation of real numbers that "... continued fractions may, at least in principle, make the same claims for consideration as those which may be made on behalf of decimal fractions, or quite generally for systematic fractions (i.e. those which use any fixed radix system)".[1] Then regarding base dependent representation anomalies he observed "... whilst every systematic fraction is coupled to a definite radix system and therefore unavoidably reflects more the interaction of the radix system and the number, than the absolute properties of the number itself, the continued fraction is completely independent of any radix system and reproduces in a pure form the properties of the number which it represents". Regarding finite precision approximation he first stated that it is essential "... to be able to determine to any prescribed accuracy the approximate value of the number which is being represented, and to be able to do this as simply as possible", and then responded, appealing to the theory of continued fractions as developed in his text, to show "... that the convergents which are yielded by the continued fraction display a property which in a certain extremely simple and important sense, is that of *best approximation*". Khintchine's enthusiasm for continued fractions was dampened only by his unnecessary pessimism regarding the possibilities of computation with continued fractions where he made the unfortunate (and unfounded) claim: "Even the task of finding the continued fraction development of the sum of two continued fractions is extremely complicated and in practice incapable of being carried out". In reality this operation can be performed in time comparable to a floating-point divide operation on arguments of comparable precision [7].

We develop fraction number system foundations from the classical number theoretic literature on continued fractions and Farey fractions, generalizing and evolving additional theory as necessary to interpret and respond to the demands of finite precision numeric computation. Our answers to the question areas of *naturalness*, *complexity*, and *accuracy* are then summarized in ten observations highlighted within Sects. II, III, and IV.

---

[1] The verbatim English quotations are from the translation by Peter Wynn, P. Noordhoff Ltd., Groningen, 1963.

In Sect. II we utilize schematic computer word formats to informally describe two fraction number systems termed fixed-slash: numerator and denominator in separate fixed length fields, and floating-slash: concatenated numerator and denominator packed into a single fixed length field. The monadic operators absolute value, negation, and inversion are shown to be readily and exactly implementable in both systems. Design of the standard dyadic operators add, subtract, multiply, and divide is shown feasible given an appropriate rounding (described in Sect. IV) from fractions with large numerators and denominators to simpler fractions. A verifiable exact rational arithmetic feature for fraction number systems considerably broader in scope than the currently understood concept of exact integer arithmetic on a computer is demonstrated.

Finite precision fraction number systems are formally developed in Sect. III. A theoretically convenient "hyperbolic" fraction number system is shown to closely model the floating-slash system. Then the fixed-slash and hyperbolic systems are shown to be independent of the choice of base and each to possess a single parameter, n, dictating accuracy and range with increasing precision as $n \to \infty$. The concern about loss of representation efficiency due to redundancy (e.g. $\frac{1}{2} = \frac{2}{4} = \frac{3}{6} = \cdots$) is shown to be negligible in that at most one to two bits of an N bit word are lost for any word size N. Material extending the theory of Farey fractions is developed and employed to study the spacing between representable values in fraction number systems. Of primary importance is the result for any two fraction number systems that, locally, one is simply a refinement of the other. Thus any "reasonable" conversion rounding yields a resulting number common to both systems, avoiding the possibility of successive conversion drift and thereby enhancing the study of mathematical software portability. Bounds on the size of the gaps between representable values in fixed-slash and floating-slash number systems are given showing considerable gap size variability with relatively simple fractions always farther removed from their closest neighbors. In spite of this "microscopic" variability, desirable "macroscopic" spacing is noted in that the fixed-slash numbers become uniformly dense on the unit interval as $n \to \infty$, and the hyperbolic systems (closely resembling floating-slash number systems) approach log uniform density on the positive real line as $n \to \infty$.

Closed arithmetic in fraction number systems is realized by the natural canonical rounding obtained using the continued fraction theory concept of best rational approximation. Foundations for the rounding and the distribution of rounding error are explored in Sect. IV. The rounding is shown to provide a natural adaptive variable single-to-double precision environment regarding absolute rounding error over the unit interval in fixed-slash and similarly regarding relative rounding error over the reals in floating-slash. The adaptive precision favors roundings to simple fractions, so that the simple fractions effectively form a single precision subsystem imbedded within an overall double precision environment.

From the theorems and observations developed we conclude that while fraction number systems possess the inherent limitations of finite precision arithmetic, they also possess several natural and intriguing deeply rooted features that could serve both to characterize and realize the achievable goals of approximate real arithmetic.

II. Fixed-Slash and Floating-Slash Number Systems

Fixed-slash and floating-slash approximate rational arithmetic may be informally characterized by specifying the format for the representable fractions along with a rounding procedure for approximating the results of arithmetic operations that are not exactly representable within the precision constraints of the system. The fixed-slash fraction representation of a rational number contains: (i) a sign field, (ii) two identical integer fields for representing the numerator and denominator magnitudes, and (iii) an optional status flag field. Each of the integer fields will be assumed to allow integer values up to a maximum of n. The flag field may be used to indicate the status of the fraction as (*a*) an exact number, or (*b*) an approximation (resulting either from approximate input data or the result of a computation in which rounding approximation was employed). A diagram of the representation of a fixed-slash number is then given by

```
                                                 optional
[sign] [numerator field] [denominator field] [status flag]
       |← fixed length →| |← fixed length →|
```

The full set of representable fractions in the *fixed-slash*(n) *number system* includes reducible and irreducible finite and infinite fractions and is given by

$$\mathrm{FXS}(n) = \left\{ \pm \frac{p}{q} \,\middle|\, 0 \leqslant p, q \leqslant n \right\}.$$

Note that the monotonically increasing sequence of fixed-slash(n) representable irreducible fractions between zero and unity is well known in classical number theory (Hardy and Wright [3]) as the Farey series $\mathscr{F}_n$. This and other important number theoretic aspects of fixed-slash numbers will be developed in the formal treatment of Sects. III and IV.

A hardware oriented binary fixed-slash(n) implementation would likely have the sign, numerator, and denominator fields all packed into one (or two) machine words with typical orders such as $n = 2^{15} - 1$ for a 32 bit word or $n = 2^{31} - 1$ for a 64 bit word. An additional bit is then available for the exact/rounded flag if desired. The feasibility and hardware efficiencies possible for such systems were studied in Matula and Kornerup [11] and Kornerup and Matula [7]. Alternatively, a software oriented fixed-slash(n) implementation would likely allow the numerator and denominator fields to each be multiple precision integers, where then the unused sign bit for the denominator could be used as the flag bit location for exact/rounded status of the result. In this case much higher values of n should be considered feasible with implementation costs comparable to the level of multiple precision integers utilized. It is important to note that the single parameter n of a fixed-slash(n) system determines both the range of allowed values and the spacing between successive representable values and does so in a canonical manner *independent* of the numeric base of the number representation.

The floating-slash representation (Matula [10]) of a rational number contains the following three fields: a sign field, a slash position indicator field, and a fraction field. The fraction field is assumed to be composed of k subfields each capable of representing an integer in the range $[0, \beta - 1]$. The numerator and denominator are each assumed to be in base β representation and are placed in concatenated form in the fraction field. The break point or "slash position" is indicated by an integer in the range $[1, k - 1]$ stored in the slash position field. A diagram of the representation of a floating-slash number is then given by

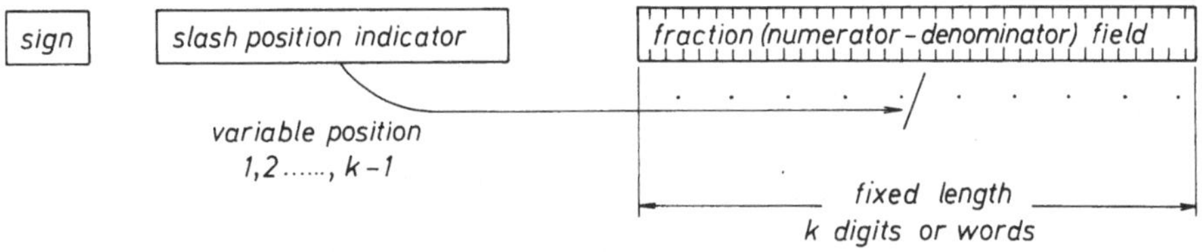

Fig. 1. Format of floating-slash representation

The set of representable fractions in the *floating-slash*(β, k) *number system* with $\beta, k \geqslant 2$ is then given by

$$\mathrm{FLS}(\beta, k) = \left\{ \pm \frac{p}{q} \,\middle|\, p, q \geqslant 1, \left\lfloor \frac{\log p}{\log \beta} \right\rfloor + \left\lfloor \frac{\log q}{\log \beta} \right\rfloor \leqslant k - 2 \right\}$$
$$\cup \left\{ \pm \frac{0}{q}, \frac{p}{0} \,\middle|\, 1 \leqslant p, q \leqslant \beta^{k-1} - 1 \right\}.$$

A hardware oriented floating-slash(β, k) implementation would likely have the three fields packed in a binary word ($\beta = 2$) with values of k typically between 26 and 120 (Matula and Kornerup [11]). At least $\lceil \log_2 k \rceil$ bits would be needed for the slash position indicator field, and another bit may be explicitly dedicated for the exact/rounded status flag if desired. A software implementation would typically have multiple precision integers for numerators and denominators utilizing at most k words/bytes in total. Thus parameter values such as $\beta = 2^8, 2^{16}$ or 2^{32} and small values for k, e.g. $2 - 12$, should be feasible at implementation costs comparable to k-level multiple precision integer arithmetic. An additional word could then be used for the sign, slash position indicator, and optional exact/rounded status flag.

In both fixed-slash(n) and floating-slash(β, k) number systems the monadic arithmetic operators absolute value, negation, and inverse are readily and exactly implemented without overflow or underflow as follows:

Monadic Operator	*Implementation*
Absolute Value	Sign $\leftarrow +$
Negation	Change Sign
Inverse	Swap Numerator and Denominator Fields, Adjust Slash Position Indicator

Each of the dyadic algebraic operators $+, -, \times, \div$ applied to a pair of fixed-slash(n) numbers yields an exact result representable as a fixed-slash$(2n^2)$ number, and similar dyadic operator application to a pair of floating-slash(β, k) numbers yields an exact result representable as a floating-slash$(\beta, 3k)$ number (Matula and Kornerup [11]). Implementation of closed approximate rational arithmetic for fixed- and floating-slash number systems thus requires mappings Φ_n: $\mathrm{FXS}(2n^2) \rightarrow$

FXS(n) and $\Phi_{(\beta,k)}$: FLS(β, $3k$) $\rightarrow$ FLS(β, k) to effect roundings of the intermediate results to representable fractions. Denoting such roundings by Φ we implement the dyadic operators as follows:

Dyadic Operator Implementation (*Rounded Arithmetic*)

$$\frac{p}{q} \otimes \frac{r}{s} = \Phi\left(\frac{pr}{qs}\right)$$

$$\frac{p}{q} \oslash \frac{r}{s} = \Phi\left(\frac{ps}{qr}\right)$$

$$\frac{p}{q} \oplus \frac{r}{s} = \Phi\left(\frac{ps + qr}{qs}\right)$$

$$\frac{p}{q} \ominus \frac{r}{s} = \Phi\left(\frac{ps - qr}{qs}\right)$$

A canonical specification of the mappings Φ_n and $\Phi_{(\beta,k)}$ is critical to achieving unique approximate arithmetic in these systems. Fortunately, there is a clear natural choice provided by the theory of convergents and "best rational approximation". The considerable theory behind the rounding based on best rational approximation is developed in Sect. IV. It is sufficient at this point to note that if the computed result can be reduced by eliminating the greatest common divisor to a fraction exactly representable in the system, then the rounding produces this exact irreducible fraction as the rounded result.

Observation 1 (Exact Rational Computation). Both fixed-slash and floating-slash number systems allow extensive exact rational expression evaluation when the arguments are relatively simple fractions. This feature includes and is much more comprehensive than the exact integer arithmetic feature implicit in fixed-point and floating-point number systems. Furthermore, exact/rounded status flags may be conveniently maintained for each fraction variable by the hardware to allow verifiable assurance that any particular computational result was derived exactly. ∎

Both fixed-slash and floating-slash representation have ranges for representable finite values that are considerably limited in comparison to ranges employed in typical hardware floating-point systems of comparable word size. An extended range feature can be employed in floating-slash representation by allowing the slash position field to assume larger values implicitly scaling up the numerator (implicit denominator unity), or scaling up the denominator (implicit numerator unity) as discussed in Matula and Kornerup [11]. Since such extended floating-slash number systems only add representable values with magnitudes between the otherwise largest finite fraction and $\frac{1}{0}$, and between the otherwise smallest non-zero fraction and $\frac{0}{1}$, the following theory will still be appropriate to the range of extended floating-slash numbers representable with slash positions internal to the fraction field.

III. Finite Precision Fraction Number Systems

From the point of view of computer architecture, a natural finite precision number system must have a convenient computer word format for the representation of its members. This property was amply demonstrated for fixed-slash and floating-slash representation in the previous section. From the complementary point of view of approximate real arithmetic, a natural finite precision number system must provide in some sense a best approximation system which is not compromised by the artifacts of the representation. In other words we want to avoid situations that occur, for example, in floating-point representation, where the artifact of the choice of base interacts with the inherent finite precision limitation and implants base dependent anomalies on top of the natural limitations of finite precision.

Our purpose in this and the next section is to show that the sets of representable real values characterized by the fixed-slash and floating-slash number systems have a natural structure derived from foundations deep in number theory, in particular the theories of Farey fractions, continued fractions, and best rational approximation. In view of the sign magnitude feature of the fixed-slash and floating-slash number systems it is sufficient to limit our formal treatment to the region of nonnegative reals.

Formally, a *fraction*, denoted p/q or $\frac{p}{q}$, is an ordered pair composed of a nonnegative integer *numerator* p, and a nonnegative integer *denominator* q, which are not both zero. The quotient of p/q is the rational number determined by the ratio of p to q for $q \neq 0$, and is taken to be positive infinity when $q = 0$. The numerator and denominator of an *irreducible* fraction must have a greatest common divisor (gcd) of unity, other fractions being termed *reducible*. Two fractions are equal, denoted $p/q = r/s$, if $qr = ps$ ($p/q = r/s$ does not necessarily imply identical numerators and denominators).

In order to formulate a notion of precision for fraction number systems, we say that the fraction p/q is *simpler than* the fraction r/s if and only if $p \leqslant r$, $q \leqslant s$, and at least one inequality is strict. To investigate the range and spacing of numeric values of the representable fractions of a fraction number system we then define the *simple chain* F to be a finite ordered set

$$F = \left\{\frac{0}{1} = \frac{p^{(1)}}{q^{(1)}}, \frac{p^{(2)}}{q^{(2)}}, \ldots, \frac{p^{(k-1)}}{q^{(k-1)}}, \frac{p^{(k)}}{q^{(k)}} = \frac{1}{0}\right\}$$

of irreducible fractions ordered by monotone increasing numeric value of their quotients where all irreducible fractions simpler than any member of the chain are also in the chain, i.e. F is closed under the simpler than relation over irreducible fractions. Thus $F = \{\frac{0}{1}, \frac{1}{3}, \frac{1}{2}, \frac{2}{3}, \frac{1}{1}, \frac{3}{2}, \frac{2}{1}, \frac{3}{1}, \frac{1}{0}\}$ is a simple chain, however $F = \{\frac{0}{1}, \frac{1}{3}, \frac{1}{2}, \frac{1}{1}, \frac{3}{2}, \frac{2}{1}, \frac{1}{0}\}$ is not a simple chain, since $\frac{3}{1} \notin F$ is simpler than $\frac{3}{2} \in F$.

Any reasonable computer word format characterizing a fraction number system that provides for a separate representation of the numerator and denominator would almost certainly allow any fraction simpler than some representable fraction to also be representable. Thus analysis of the properties of simple chains should allow us to determine the properties attainable via any such characterization of a finite precision fraction number system.

The fixed-slash and floating-slash number systems are thus specific examples determining simple chains given formally by

Fixed-slash. $n \geqslant 1$, chain ordered by increasing value,

$$\mathrm{FXS}^*(n) = \left\{ \frac{p}{q} \,\middle|\, p, q \leqslant n, \gcd(p,q) = 1 \right\}. \tag{1}$$

Floating-slash. $\beta \geqslant 2$, $k \geqslant 2$, chain ordered by increasing value,

$$\mathrm{FLS}^*(\beta,k) = \left\{ \frac{0}{1} \right\} \cup \left\{ \frac{p}{q} \,\middle|\, p, q \geqslant 1, \gcd(p,q) = 1, \right.$$

$$\left. \lfloor \log_\beta p \rfloor + \lfloor \log_\beta q \rfloor \leqslant k - 2 \right\} \cup \left\{ \frac{1}{0} \right\}. \tag{2}$$

$\lfloor \log_\beta p \rfloor$ and $\lfloor \log_\beta q \rfloor$ denote integer portions of the base β logarithms and render floating-slash representation base dependent. To study the base independent features of floating-slash representation it is desirable to introduce the

Hyperbolic Chain. $n \geqslant 1$, chain ordered by increasing value,

$$H^*(n) = \left\{ \frac{p}{q} \,\middle|\, pq \leqslant n, \gcd(p,q) = 1 \right\} \tag{3}$$

from which we then note comparability of floating-slash chains to hyperbolic chains "within one digit", since

$$H^*(\beta^{k-1} - 1) \subset \mathrm{FLS}^*(\beta,k) \subset H^*(\beta^k - 1). \tag{4}$$

The hyperbolic chain $H^*(n)$ determined by (3) may be considered to represent a "pure floating-slash" system analogous to the inherently pure fixed-slash system as specified by equation (1). From the definitions of $\mathrm{FXS}^*(n)$ and $H^*(n)$ two immediate observations should be noted.

Observation 2 (Independence of Representation). The representable values of $\mathrm{FXS}^*(n)$ and $H^*(n)$ are determined solely by integral valued bounds on the numerator and denominator and are independent of how the numerator and denominator values are represented, be they radix representation of arbitrary base or other representation methods. ❙

Observation 3 (Single Parameter Specification of Accuracy and Range). Note that any real $\alpha \geqslant 0$ will fall in some gap between consecutive members of $\mathrm{FXS}^*(n)$ (or of $H^*(n)$), so $p/q \leqslant \alpha < p'/q'$, and will, with increasing n, eventually have both gap bounds finite and then have their difference, $(p'q - q'p)/qq'$, monotonically approach zero as $n \to \infty$. Thus the parameter n acts in a natural manner both to extend the range and increase the available accuracy of the representation of the reals by finite precision fractions. ❙

These two observations indicate desirable properties of the representable values of our fixed-slash and floating-slash number systems. One cost of the representation is in the redundancy. The relatively simple fractions have many equivalent reducible representations in fixed- and floating-slash word formats that degrades the efficiency of the representation. Fortunately, the following classical number theoretic results on the proportion of irreducible fractions (see Knuth [6, p. 314], for a proof of (5) and Dickson [2, p. 283] for a discussion of (6)) shows that the net cost of this redundancy is negligible.

Theorem 1. *For fixed-slash*:

$$\lim_{n\to\infty} \frac{\left|\left\{\frac{p}{q}\,\middle|\, p, q \leqslant n, \gcd(p,q) = 1\right\}\right|}{(n+1)^2 - 1} = \frac{6}{\pi^2} = .6079\ldots. \tag{5}$$

For hyperbolic chains (Dirichlet, see [2, p. 283]):

$$\lim_{n\to\infty} \frac{\left|\left\{\frac{p}{q}\,\middle|\, pq \leqslant n, \gcd(p,q) = 1\right\}\right|}{\left|\left\{\frac{p}{q}\,\middle|\, pq \leqslant n\right\}\right|} = \frac{6}{\pi^2} = .6079\ldots. \tag{6}$$

Thus approximately 60% of the representable fractions are irreducible. To appreciate that this representation efficiency is acceptable consider that the storage capacity of a computer is effectively measured in words (or bits). For a numeric word format to attain 60% of the possible bit patterns as distinct values is to lose less than one bit (actually $\log_2(\pi^2/6) = 0.718$ bits) per word independent of word length.

The representation efficiency of floating-slash is also affected by any unutilized portion of the range in the slash position field. In Matula and Kornerup [11] this feature was investigated in detail for binary floating-slash number systems. With suitable architectural encodings it was demonstrated that the total representation efficiency loss due to both reducible fractions and slash position encoding can be held under two bits per word. In fact for popular word sizes such as 32 or 64 bits the loss is essentially only one bit per word. Thus we state:

Observation 4 (Representation Efficiency). The formats of fixed-slash and floating-slash representation are efficient in storage space utilization. With suitable binary encodings the representation loss is never more than one or two bits for any length word or multi-word encoding (e.g. a k-bit representation yields approximately at least 2^{k-2} distinct representable real values). ∎

The members of the simple chains $\mathrm{FXS}^*(5) = \{\frac{0}{1}, \frac{1}{5}, \frac{1}{4}, \frac{1}{3}, \frac{2}{5}, \frac{1}{2}, \frac{3}{5}, \frac{2}{3}, \frac{3}{4}, \frac{4}{5}, \frac{1}{1}, \frac{5}{4}, \frac{4}{3}, \frac{3}{2}, \frac{5}{3}, \frac{2}{1}, \frac{5}{2}, \frac{3}{1}, \frac{4}{1}, \frac{5}{1}, \frac{1}{0}\}$ and $H^*(18) = \{\frac{0}{1}, \frac{1}{18}, \frac{1}{17}, \ldots, \frac{1}{5}, \frac{2}{9}, \frac{1}{4}, \frac{2}{7}, \frac{1}{3}, \frac{2}{5}, \frac{1}{2}, \frac{3}{5}, \frac{2}{3}, \frac{3}{4}, \frac{1}{1}, \frac{4}{3}, \frac{3}{2}, \frac{5}{3}, \frac{2}{1}, \frac{5}{2}, \frac{3}{1}, \frac{7}{2}, \frac{4}{1}, \frac{9}{2}, \frac{5}{1}, \frac{6}{1}, \ldots, \frac{18}{1}, \frac{1}{0}\}$ from $\frac{0}{1}$ to $\frac{1}{1}$ are shown to scale in Fig. 2.

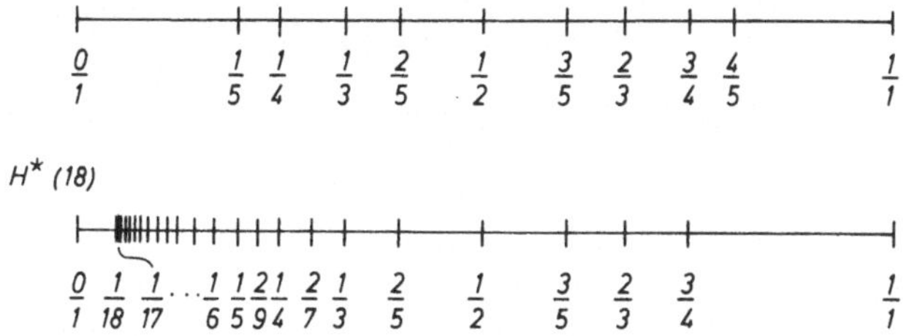

Fig. 2. Spacing of values of FXS*(5) and $H^*(18)$ over the unit interval

The spacing between representable values is clearly quite irregular in both cases. For the relationship between a choice of precision level and the accuracy guarantee of an approximation of that precision to be meaningful, it is necessary to investigate the nature and size of the gaps between representable values in finite precision fraction number systems.

For the finite distinct fractions $p/q < r/s$, the size of the interval $[p/q, r/s]$ is given by the expression $(rq - ps)/qs$ and will be a minimum relative to the size of the denominator when $rq - ps = 1$. We say the fractions p/q and r/s are *adjacent* whenever $|rq - ps| = 1$. Note for p/q adjacent to r/s it follows immediately that both p/q and r/s are irreducible and, excluding the pair $\frac{0}{1} < \frac{1}{0}$, that one of them is simpler than the other.

By direct computation the chains FXS*(5) and $H^*(18)$ of Fig. 2 are both seen to have the property that consecutive fractions are adjacent. We shall show in fact that consecutive fractions of *any* simple chain are adjacent.

The content of the following Theorems 2 through 5 is at least implicit in the classical 19th century literature on Farey series [see 2, p. 156–158]. We provide our own concise self-contained development to establish a rigorous foundation for fraction number systems. We extract as the essential element of Farey theory the concept we term "adjacency" of fractions, and guide our development accordingly, starting with three equivalent definitions of adjacency.

Theorem 2. *For the fractions $p/q, r/s$, each of the following three properties implies the other two and serves as an equivalent definition of adjacency.*

(i) *Determinant form*: $|rq - ps| = 1$,

(ii) *Difference form*: $|r/s - p/q| = 1/qs$ *for* $qs \neq 0$, *and* $|rq - ps| = 1$ *for* $qs = 0$,

(iii) *Interval form*: p/q, r/s *are both irreducible and both simpler than any other fraction within the interval bounded by* p/q *and* r/s.

Proof. Clearly (i) and (ii) are equivalent. To prove the equivalence of (iii) and (i) we shall first derive some properties about the mediant of the adjacent fractions p/q, r/s, where, for arbitrary fractions $p/q, r/s$, the *mediant* is defined as the fraction $(p + r)/(q + s)$.

Theorem 3. *The mediant* $(p + r)/(q + s)$ *of the adjacent fractions* $p/q < r/s$ *satisfies*:

(i) *Irreducibility*: $(p + r)/(q + s)$ *is irreducible,*

(ii) *Quotient ordering*: $p/q < (p + r)/(q + s) < r/s$,

(iii) *Precision ordering*: p/q *and* r/s *are both simpler than* $(p + r)/(q + s)$,

(iv) *Adjacency*: $(p + r)/(q + s)$ *is adjacent to* p/q *and* r/s,

(v) *Simplicity*: $p/q < t/u < r/s$ *implies* $(p + r)/(q + s)$ *is simpler than or identical to* t/u.

Proof. Property (iii) is immediate. Note then that

$$\frac{p+r}{q+s} - \frac{p}{q} = \frac{pq + rq - pq - ps}{(q+s)q} = \frac{1}{(q+s)q} > 0,$$

establishing both that $p/q < (p + r)/(q + s)$ and that p/q and $(p + r)/(q + s)$ are adjacent. Also for finite r/s

$$\frac{r}{s} - \frac{p + r}{q + s} = \frac{rq + rs - ps - rs}{s(q + s)} = \frac{1}{s(q + s)} > 0,$$

so that in all cases

$$\frac{p + r}{q + s} < \frac{r}{s}$$

and they are likewise adjacent fractions. This proves (iv) and (ii) and (i) then follows from (iv), since adjacent fractions must be irreducible. To obtain (v) let $p/q < t/u < r/s$. Then for finite r/s

$$\frac{qr - ps}{qs} = \frac{ur - ts}{us} + \frac{qt - up}{qu},$$

hence

$$\frac{1}{qs} \geqslant \frac{1}{us} + \frac{1}{qu},$$

so $u \geqslant q + s$. If $r/s = 1/0$, then $q = 1$, so also $u \geqslant q + s = 1$. Noting $s/r < u/t < q/p$, we similarly obtain $t \geqslant p + r$, proving (v). □

Pairs of adjacent fractions are nested as shown in the following.

Corollary 3.1. *Let $p/q < t/u < r/s$ where p/q is adjacent to r/s. Then t/u cannot be adjacent to any fraction outside the interval $[p/q, r/s]$.*

Proof. If t/u is adjacent to v/w where either $v/w < p/q$ or $r/s < v/w$, then by Theorem 3 (v) and (i), $t \geqslant p + r \geqslant p + 1$, and similarly either $p \geqslant v + t \geqslant t$, a contradiction, or $r \geqslant t + v \geqslant t + 1$, a contradiction. □

The relation between the pair of adjacent fractions $p/q, r/s$ and their mediant $(p + r)/(q + s)$ is shown to be of even greater fundamental importance by the following converse relation.

Theorem 4. *Every finite non-zero irreducible fraction is the mediant of precisely one pair of adjacent fractions.*

Proof. It is sufficient to consider $0/1 < p/q < 1/0$, where p/q is irreducible. Form the sequence $F_i(p/q)$ as follows for $i = 0, 1, 2, \ldots, N(p, q)$.

$$F_0\left(\frac{p}{q}\right) = \left[\tfrac{0}{1}, \tfrac{1}{0}\right]$$

$$F_1\left(\frac{p}{q}\right) = \left[\tfrac{0}{1}, \tfrac{1}{1}, \tfrac{1}{0}\right]$$

and $F_i(p/q)$, $i \geqslant 2$, is formed recursively from $F_{i-1}(p/q)$ by inserting the mediant in the interval between the two consecutive fractions of $F_{i-1}(p/q)$ that contains the quotient of p/q. If the mediant inserted to form $F_i(p/q)$ is p/q, the process terminates with $N(p,q) = i$, otherwise it continues. Note from Theorem 3 that each F_i is a monotonically increasing sequence of consecutive adjacent fractions, and before termination p/q always falls in a unique interval between two consecutive adjacent fractions. By Theorem 3 (iii), (v) this procedure must terminate with $N(p,q) \leqslant \max\{p,q\}$, determining an adjacent pair r/s, t/u for which p/q is the mediant. It then follows from Corollary 3.1 and Theorem 3 that this pair of adjacent fractions, r/s, t/u, is unique. □

For the finite non-zero irreducible fraction t/u, the unique pair of adjacent fractions $p/q < r/s$, of which t/u is the mediant by Theorem 4, are termed the *parents* of t/u. Note that one parent is less than t/u, the other parent greater than t/u, and both parents are simpler than t/u. We now complete the proof of Theorem 2.

Proof (Equivalence of (i) and (iii) in Theorem 2). From Theorem 3 (v) the mediant $(p+r)/(q+s)$ is the simplest fraction falling between the adjacent fractions $p/q, r/s$, so property (iii) of the statement of Theorem 2 follows from (i). To show that property (iii) implies (i), assume $p/q < r/s$ irreducible with no fraction simpler than either of $p/q, r/s$ in the open interval $(p/q, r/s)$. Let p^*/q^* be the larger parent of p/q, so $p/q < p^*/q^*$, and r^*/s^* the smaller parent of r/s, so $r^*/s^* < r/s$. The parents are simpler fractions and thus cannot fall in the open interval $(p/q, r/s)$ by assumption, so $r^*/s^* \leqslant p/q < r/s \leqslant p^*/q^*$. Now $r^*/s^*, r/s$ are adjacent and $p/q, p^*/q^*$ are adjacent by Theorem 3 (iv), and both inequalities cannot be strict by Corollary 3.1. So one of the fractions $p/q, r/s$ is a parent of the other, hence they are adjacent. Thus property (iii) implies (i) and the proof of Theorem 2 is complete. □

For any $n \geqslant 1$, the fractions of the simple chain FXS*(n) between $\frac{0}{1}$ and $\frac{1}{1}$ are well known in number theory as the *Farey series* $\mathscr{F}_n$. It is a classic theorem of number theory that consecutive fractions of a Farey series are adjacent (Hardy and Wright [3, Ch. III]). From the equivalent definitions of adjacency given in Theorem 2, an immediate generalization of this fundamental property of Farey series is obtained.

Theorem 5. *Consecutive fractions of any simple chain are adjacent.*

Proof. Let $p/q, p'/q'$ be consecutive fractions of any simple chain F. Now F contains all irreducible fractions simpler than either p/q or p'/q', and none of these fall between p/q and p'/q' since F is ordered by increasing quotient value. By Theorem 2 it follows that p/q and p'/q' are adjacent. □

Theorems 2 through 5 provide a foundation for the investigation of the nature of the gaps and the spacing between representable values of a fraction number system.

Observation 5 (Canonical Representation Gaps). A gap between two consecutive representable values $p/q, p'/q'$ of a simple chain has important properties derived from the bounds $p/q, p'/q'$ of the gap which may be interpreted independently of the overall charàcterization of the fraction number system. Specifically, the gap size is $1/qq'$, which is as small as possible relative to the size of the product qq'. Furthermore, no fraction simpler than $(p+p')/(q+q')$ falls in the gap. ▮

The canonical nature of the representation gaps is significant relevant to considerations of (i) input-output conversion and (ii) portability of computations in alternative fraction number systems. Note first from the definition of simple chains that the following is immediate.

Lemma 6. *The union $F_1 \cup F_2$ and the intersection $F_1 \cap F_2$ of the simple chains F_1 and F_2 are also simple chains.*

Our next theorem is best appreciated by contrast with a problem area in base conversion for fixed-point and/or floating-point radix representation. In finite precision radix representation it is possible for two differently based systems to yield approximately the same accuracy although the representable values are "out-of-phase". For example, there are 501 values between one-half and unity representable by three decimal digit numbers at increments of 1/1000, and 513 values between one-half and unity representable by ten bit binary numbers at increments of 1/1024, where only 5 values between one-half and unity are exactly representable in both systems. Although such pairs of systems are of roughly the "same precision", iterated conversion between such out-of-phase systems can cause undesirable accumulated error (Matula [9]). The next theorem shows that two fraction number systems cannot have any such phase distortion anomalies, since locally one system is always a refinement of the other.

Theorem 7. *For any two simple chains F_1 and F_2, any two consecutive members of the intersection $F_1 \cap F_2$ are consecutive members of either F_1 or F_2 (or both).*

Proof. $F_1 \cap F_2$ is a simple chain, so the consecutive members $p/q, p'/q'$ of $F_1 \cap F_2$ are adjacent. If either F_1 or F_2 contained any fraction of the open interval $(p/q, p'/q')$, they would then by Theorem 3 (v) also have to contain $(p+p')/(q+q')$. But not both F_1 and F_2 can contain $(p+p')/(q+q')$, so p/q and p'/q' must be consecutive in at least one of F_1, F_2. □

Corollary 7.1. *Let F_1 and F_2 be simple chains and $\Phi: F_1 \to F_2$ any mapping satisfying the following:*

$$\Phi(p/q) = p/q \qquad \text{for } p/q \in F_1 \cap F_2,$$

$$\left.\begin{array}{l}\Phi(p/q) = r/s \\ \text{or} \\ \Phi(p/q) = r'/s'\end{array}\right\} \qquad \text{if } r/s < p/q < r'/s' \text{ for consecutive } r/s, r'/s' \in F_2.$$

Then

$$\Phi(p/q) \in F_1 \cap F_2 \text{ for all } p/q \in F_1.$$

Proof. Suppose $p/q \in F_1$, $p/q \notin F_2$, and $r/s < p/q < r'/s'$ for consecutive $r/s, r'/s'$ of the intersection $F_1 \cap F_2$. By Theorem 7, $r/s, r'/s'$ are then consecutive members of F_2, hence the corollary. □

Observation 6 (Rounding Stability and Portability). For any "reasonable" rounding procedure, the rounding of a value in one fraction number system to a value in another fraction number system will, by Corollary 7.1, always yield a value common to both systems. Thus iterated conversion between two fraction number systems is stable after the first conversion; (i.e. there is no successive conversion drift as may occur in floating-point base conversion (Matula [9]). Since the spacing in $F_1 \cap F_2$ is locally no worse than the worst spacing in either of the simple chains F_1 or F_2, comparison of a computation hosted in F_1 to a computation hosted in F_2 is much easier to model. Hence portability studies are more tractable. ∎

The results of Theorems 2 through 5 also allow for an analysis of the variation in the size of the gaps between representable values of our fraction number systems. For the fixed-slash simple chains we are primarily interested in the absolute (rather than relative) size of gaps over the interval [0, 1].

Lemma 8. *For the interval* [0, 1] *and any* $n \geqslant 2$, *the size of the gaps between consecutive representable values of* FXS*(n) *varies from a minimum of* $1/n(n-1)$ *to a maximum of* $1/n$.

Proof. For $n \geqslant 2$,

$$\mathrm{FXS}^*(n) = \left\{\frac{0}{1}, \frac{1}{n}, \frac{1}{n-1}, \ldots, \frac{n-2}{n-1}, \frac{n-1}{n}, \frac{1}{1}, \ldots\right\},$$

and the gap size $1/n(n-1)$ is obtained between the consecutive pairs of fractions $1/n, 1/(n-1)$ and $(n-2)/(n-1), (n-1)/n$. The gap size $1/n$ is obtained between the consecutive pairs $0/1, 1/n$ and $(n-1)/n, 1/1$. By Theorems 2 and 5 we have for the maximum gap size the upper bound

$$\max\left\{\frac{1}{qs}\,\middle|\, 1 \leqslant q, s \leqslant n; q + s \geqslant n + 1\right\} = \frac{1}{n} \qquad \text{for } n \geqslant 2. \tag{7}$$

For the minimum gap size we note in addition to Theorems 2 and 5 that $0/1 \leqslant p/q < r/s \leqslant 1/1$ and $qr - ps = 1$ allows $q = s$ only if $q = s = 1$. Thus for $n \geqslant 2$ we obtain the following lower bound on the minimum gap size,

$$\min\left\{\frac{1}{qs}\,\middle|\, 1 \leqslant q, s \leqslant n; q \neq s; q + s \geqslant n + 1\right\} = \frac{1}{n(n-1)} \qquad \text{for } n \geqslant 2. \ \square$$

Thus the spacing between representable values of FXS*(n) varies between $O(1/n^2)$ and $O(1/n)$, which we might loosely interpret as a variation between single and double precision. The average gap size is readily available in the limit for large n from Theorem 1 as $(\pi^2/6)(1/n^2)$, so "most gaps" are of the smaller double precision size. It should be noted that a gap is of large size only when one of the bounding fractions is relatively simple.

For floating-slash chains and hyperbolic chains we shall now show that the relative spacing between consecutive fractions exhibits this same single-to-double precision variation. The relative size of the gap between the finite adjacent fractions $p/q < r/s$ is defined by

$$\mathrm{relgap}\left(\frac{p}{q}, \frac{r}{s}\right) = \int_{p/q}^{r/s} \frac{1}{x}\,dx, \tag{8}$$

from which we obtain

$$\mathrm{relgap}\left(\frac{p}{q}, \frac{r}{s}\right) = \mathrm{relgap}\left(\frac{s}{r}, \frac{q}{p}\right) = \log\left(1 + \frac{1}{ps}\right). \tag{9}$$

We also note the strict bounds

$$\operatorname{relgap}\left(\frac{p}{q},\frac{r}{s}\right) > \left(\frac{r}{s}-\frac{p}{q}\right)\Big/\frac{r}{s} = \frac{1}{rq} = \frac{1}{ps+1},$$

$$\operatorname{relgap}\left(\frac{p}{q},\frac{r}{s}\right) < \left(\frac{r}{s}-\frac{p}{q}\right)\Big/\frac{p}{q} = \frac{1}{ps}. \tag{10}$$

For $n \geqslant 9$, note that $\lfloor\sqrt{n}\rfloor/(\lfloor\sqrt{n}\rfloor - 1), (\lfloor\sqrt{n}\rfloor - 1)/(\lfloor\sqrt{n}\rfloor - 2)$ are consecutive adjacent fractions of $H^*(n)$, and from the preceeding bounds (10)

$$\min_{H^*(n)}\left\{\operatorname{relgap}\left(\frac{p}{q},\frac{r}{s}\right)\right\} \leqslant \operatorname{relgap}\left(\frac{\lfloor\sqrt{n}\rfloor}{\lfloor\sqrt{n}\rfloor-1},\frac{\lfloor\sqrt{n}\rfloor-1}{\lfloor\sqrt{n}\rfloor-2}\right)$$

$$< \frac{1}{\lfloor\sqrt{n}\rfloor(\lfloor\sqrt{n}\rfloor-2)}. \tag{11}$$

Then since $\lceil\sqrt{n}+1\rceil/\lceil\sqrt{n}\rceil \notin H^*(n)$,

$$\max_{H^*(n)}\left\{\operatorname{relgap}\left(\frac{p}{q},\frac{r}{s}\right)\right\} > \frac{1}{\lceil\sqrt{n}\rceil+1}.$$

Thus a single-to-double precision variation from $O(1/n^{1/2})$ to $O(1/n)$ is observed in the relative spacing for the hyperbolic chains, which clearly also pertain to floating-slash chains by formula (4). The following theorem shows that the variation exhibited above is essentially the extreme of the variation of the relative gap size.

Theorem 9. *For any $n \geqslant 2$ the relative gap size between the consecutive pair $p/q < r/s$ of finite non-zero fractions of $H^*(n)$ satisfies*

$$\frac{1}{n} < \operatorname{relgap}\left(\frac{p}{q},\frac{r}{s}\right) < \frac{2}{\sqrt{n}}. \tag{12}$$

Proof. Now $pq \leqslant n$ and $rs \leqslant n$ so $pqrs \leqslant n^2$. Hence $ps(ps+1) \leqslant n^2$, so $ps \leqslant n-1$ since ps is integral. Then the left-hand inequality of (12) follows from the lower bound of (10).

For the right-hand inequality we may assume $1/1 \leqslant p/q < r/s$ since

$$\operatorname{relgap}\left(\frac{s}{r},\frac{q}{p}\right) = \operatorname{relgap}\left(\frac{p}{q},\frac{r}{s}\right).$$

If $p/q = 1/1$, the inequality is immediate, so we further assume $1/1 < p/q$, hence it follows that $p + r - 2 \geqslant q + s$. Then

$$(p+r-1)^2 > (p+r)(p+r-2) \geqslant (p+r)(q+s) > n,$$

so $p + r - 1 > \sqrt{n}$.

Utilizing the upper bound of (10),

$$\operatorname{relgap}\left(\frac{p}{q},\frac{r}{s}\right) = \operatorname{relgap}\left(\frac{p}{q},\frac{p+r}{q+s}\right) + \operatorname{relgap}\left(\frac{p+r}{q+s},\frac{r}{s}\right)$$

$$< \frac{1}{p(q+s)} + \frac{1}{(p+r)s}$$

$$= \frac{1}{(p+r)q-1} + \frac{1}{(p+r)s} < \frac{2}{\sqrt{n}}. \square$$

Since about 60% of the representable fractions in both fixed-slash and floating-slash are irreducible, it would be possible to obtain uniform gap sizes of order $1/n^2$ in fixed-slash and uniform relative gap sizes of order $1/n$ in floating-slash if the representable values were appropriately spaced. Note that the role played by n^2 in fixed-slash representation should be similar to the role played by n in floating-slash representation since the fixed-slash bound of n pertains separately to both the numerator and denominator whereas the floating-slash bound relates to the whole fraction. Theorem 2(ii) and the bounds of formula (10) show that the resulting gap sizes can be considerably larger when one of the bounding consecutive fractions is relatively simple. The following two theorems (Matula and Kornerup [13]) are stated here (without proof) to indicate that this extreme variation in spacing is limited to the neighborhoods of particular relatively simple fractions and does not alter the asymptotic density of representable real values over any interval as $n \to \infty$.

Theorem 10. *The representable values of* FXS*(n) *become uniformly dense on the unit interval as* $n \to \infty$.

Theorem 11. *The representable values of* $H^*(n)$ *become log uniformly dense on the positive real line as* $n \to \infty$.

Observation 7 (Gap Size Variability and Representable Value Uniformity over Intervals). The absolute gap size in fixed-slash number systems over the unit interval and the relative gap size in floating-slash number systems over the non-negative reals can vary in size from single-to-double precision level, with the lower precision extremes limited to the immediate neighborhood of particular simple fractions. Over the interval [0, 1] fixed-slash number systems approach uniform density of double precision order as $n \to \infty$, which is desirable for problems where absolute error must be kept small. Over the non-negative reals the hyperbolic chains (closely associated with floating-slash number systems) approach log uniform density of double precision order as $n \to \infty$, which is desirable for the control of relative error growth. ∎

The presence of a large variation in gap size and its effect on a computation is one of the most important issues in finite precision rational arithmetic. The obvious disadvantages of this variable precision feature and the not-so-obvious advantages are intimately connected with the choice of rounding procedure employed, and a detailed treatment of this issue is the main content of the next section.

IV. Approximate Rational Arithmetic

From the point of view of computer architecture, the unary operators absolute value, negation, inverse and the dyadic algebraic operators add, subtract, multiply, divide, and any necessary rounding must be convenient to implement in an efficient manner on any operands chosen from the finite precision number system. From the complementary point of view of approximate real arithmetic, the results of the arithmetic operations must (when approximation is necessary) yield in some sense a best approximation to the true result on the same operands.

It was noted in Sect. II that the unary operators are always exact for fixed-slash and floating-slash arithmetic systems. Furthermore all dyadic operators, through elementary integer arithmetic on numerators and denominators, yield exact results representable in higher precision fixed-slash and floating-slash number systems. Thus the fundamental issue underlying the structure of approximate rational arithmetic is the mathematical nature and implementation feasibility of a rounding algorithm which rounds fractions with relatively large numerators and denominators to simpler fractions that are good approximations. We shall now show that a canonical choice for the rounding is available with foundations derived from the theory of continued fractions and best rational approximation.

Utilizing the notation $[a_0, a_1, \ldots]$ for the *continued fraction*

$$a_0 + \cfrac{1}{a_1 + \cfrac{1}{a_2 + \ddots}}, \qquad a_i \geqslant 0$$

where the *partial quotients* a_i are assumed to be integral, any non-negative rational number p/q has a finite expansion

$$\frac{p}{q} = [a_0, a_1, \ldots, a_m]$$

which is unique (*canonical*) with the added requirements $a_0 \geqslant 0$; $a_i \geqslant 1$, $i = 1, \ldots, m-1$; and $a_m \geqslant 2$ where $m \geqslant 1$. The truncated continued fractions

$$\frac{p_i}{q_i} = [a_0, a_1, \ldots, a_i], \qquad i = 0, 1, \ldots, m$$

yield rational numbers which form a sequence of continued fraction approximations of p/q (called the *convergents*) whose properties can be summarized from classical material on continued fractions (Hardy and Wright [3, Ch. X]; Khintchine [5]) as follows.

Theorem 12. *The convergents* $p_i/q_i = [a_0, a_1, \ldots, a_i]$ *of* $p/q = [a_0, a_1, \ldots, a_m]$ *for* $i = 0, 1, \ldots, m$ *satisfy the following properties*:

(i) *Recursive ancestry*: *With* $p_{-2} = 0$, $p_{-1} = 1$, $q_{-2} = 1$ *and* $q_{-1} = 0$,

$$p_i = a_i p_{i-1} + p_{i-2},$$

$$q_i = a_i q_{i-1} + q_{i-2},$$

(ii) *Irreducibility*:

$$\gcd(p_i, q_i) = 1,$$

(iii) *Adjacency*:

$$q_i p_{i-1} - p_i q_{i-1} = (-1)^i,$$

(iv) *Alternating convergence*:

$$\frac{p_0}{q_0} < \frac{p_2}{q_2} < \cdots < \frac{p_{2j}}{q_{2j}} < \cdots \leqslant \frac{p}{q} \leqslant \cdots < \frac{p_{2j-1}}{q_{2j-1}} < \cdots < \frac{p_1}{q_1},$$

(v) *Best rational approximation*:

$$\frac{r}{s} \neq \frac{p_i}{q_i}, \qquad s \leqslant q_i \Rightarrow \left|\frac{r}{s} - \frac{p}{q}\right| > \left|\frac{p_i}{q_i} - \frac{p}{q}\right|,$$

(vi) *Quadratic convergence*:

$$\frac{1}{q_i(q_{i+1} + q_i)} < \left|\frac{p_i}{q_i} - \frac{p}{q}\right| \leqslant \frac{1}{q_i q_{i+1}} \qquad \textit{for } i \leqslant m - 1.$$

Since the partial quotients a_i of the expression $p/q = [a_0, a_1, \ldots, a_m]$ are the quotients obtained from the standard Euclidean GCD-Algorithm applied to p, q, we may simply extend the Euclidean Algorithm by incorporating computation of p_i and q_i in parallel with computation of the a_i to achieve an algorithm for determination of all convergents of p/q.

Algorithm EC (Euclidean-Convergent Algorithm). For any $p \geqslant 0$, $q \geqslant 1$, let

$$b_{-2} = p; \quad p_{-2} = 0; \quad q_{-2} = 1;$$

$$b_{-1} = q; \quad p_{-1} = 1; \quad q_{-1} = 0.$$

Determine a_i as the quotient, and b_i as the non-negative remainder of the division of b_{i-2} by b_{i-1}, so

$$b_{i-2} = b_{i-1} \cdot a_i + b_i,$$

and compute

$$p_i = p_{i-1} \cdot a_i + p_{i-2},$$

$$q_i = q_{i-1} \cdot a_i + q_{i-2},$$

for $i = 0, 1, \ldots$ until $i = m$, where $b_{m-1} \neq 0$, $b_m = 0$. ∎

For any rational $x = p/q = [a_0, a_1, \ldots, a_m]$ or irrational $x = [a_0, a_1, \ldots]$, the convergents $p_0/q_0, p_1/q_1, p_2/q_2, \ldots$ represent successively more accurate approximations to x, where every convergent is simpler than any subsequent convergent. Thus for any simple chain F there is a largest indexed convergent $p_i/q_i \in F$ such that no subsequent convergent is in F, and this fact provides the basis for a natural rounding procedure.

Let $\pm F$ be the set of all signed fractions corresponding to fractions of a simple chain F. For any simple chain F, the mapping Φ_F: Reals $\rightarrow \pm F$ is defined for every real number x, where $p_0/q_0, p_1/q_1, p_2/q_2, \ldots$ are the convergents to $|x|$, by

$$\Phi_F(x) = \begin{cases} p_m/q_m & \text{if } x = p_m/q_m \in F, \\ p_i/q_i & \text{if } x > 0, \quad p_i/q_i \in F, \quad p_{i+1}/q_{i+1} \notin F, \\ -\Phi_F(-x) & \text{if } x < 0. \end{cases} \tag{13}$$

Suppose, for example, we wish to round $\frac{277}{642}$ to a fraction limited to two decimal digits in numerator and denominator. From Algorithm EC we determine $\frac{277}{642} = [0, 2, 3, 6, 1, 3, 3]$, and that the convergents are $\frac{0}{1}, \frac{1}{2}, \frac{3}{7}, \frac{19}{44}, \frac{22}{51}, \frac{85}{197}, \frac{277}{642}$. Hence we obtain the "rounded value"

$$\Phi_{\mathrm{FXS}^*(99)}\left(\frac{277}{642}\right) = \frac{22}{51}.$$

Theorem 13. *For any simple chain F the mapping Φ_F:* Reals $\to \pm F$ *satisfies the following three properties for all real x, y:*

(i) *Monotonic*: $x < y \Rightarrow \Phi_F(x) \leqslant \Phi_F(y)$,

(ii) *Antisymmetric*: $\Phi_F(-x) = -\Phi_F(x)$,

(iii) *Fixed points*: $|x| = p/q \in F \Rightarrow \Phi_F(x) = x$.

The preceeding theorem is stated without proof, noting that the proof may be obtained from standard results in the theory of continued fractions. Note also that the properties stated in Theorem 13 are those characterizing an "optimal rounding" in the sense of Kulisch [8].

If $p/q, p'/q'$ are consecutive fractions of the simple chain F, it follows from Theorem 13 that there is a real number $p/q \leqslant y \leqslant p'/q'$ such that $\Phi_F(x) = p/q$ for $p/q \leqslant x < y$ and $\Phi_F(x) = p'/q'$ for $y < x \leqslant p'/q'$, where $\Phi_F(y)$ must be one of $p/q, p'/q'$. The following theorem states that this "splitting point" y is always the mediant, and exhibits the interval mapped into $p/q \in F$ under Φ_F.

Theorem 14. *Let $p/q, p'/q', p''/q''$ be consecutive fractions of a simple chain F. Then*

$$\Phi_F(x) = \frac{p'}{q'} \quad \textit{for} \quad \frac{p + p'}{q + q'} < x < \frac{p' + p''}{q' + q''},$$

and the only other values mapped to p'/q' are determined by:

$$\Phi_F\left(\frac{p + p'}{q + q'}\right) = \frac{p'}{q'} \quad \textit{if } \frac{p'}{q'} \textit{ is simpler than } \frac{p}{q},$$

$$\Phi_F\left(\frac{p' + p''}{q' + q''}\right) = \frac{p'}{q'} \quad \textit{if } \frac{p'}{q'} \textit{ is simpler than } \frac{p''}{q''}.$$

Again we simply note without proof that Theorem 14 follows in a straightforward manner from the theory of continued fractions. In view of the prominence of the mediant as exhibited in Theorem 14, we term the mapping Φ_F: Reals $\to \pm F$ *mediant rounding* to $\pm F$.

A detailed study of a binary hardware implementation of the mediant rounding algorithm was pursued in Kornerup and Matula [7]. In summary, it was shown that with suitable parallelism the rounding execution time would be of the order of one or two floating-point divide instructions on comparably sized arguments. A multiple precision software implementation of fixed-slash and/or floating-slash was discussed in Matula and Kornerup [11] as being of value for symbolic computation studies. In this case the software implementation of the rounding algorithm is clearly comparable in execution time requirements to that of the GCD-Algorithm as customarily utilized for the elimination of common factors in numerator and denominator in implementations for applications in symbolic and algebraic computation.

Observation 8 (Canonical Rounding). For any number system based on the representable fractions of a simple chain, there is a natural and canonical rounding procedure that unambiguously determines the appropriate rounded value. The mediant rounding procedure determines as the unique rounded value the fraction which is a best rational approximation in the number theoretic sense. The approximate arithmetic utilizing this rounding is thus canonically specified with a foundation deep in number theory. Hardware implementation of the rounding is competitive with a floating-point divide instruction in time and logic requirements. ∎

In Lemma 8 we showed that the size of the gaps between consecutive representable fractions of the fixed-slash chain FXS*(n) over the interval [0, 1] varied between $1/n(n-1)$ and $1/n$. Utilizing the definition of mediant rounding we are now able to assess the approximation errors introduced by computations in such a variable precision environment. Let Φ_n denote mediant rounding, Φ_F, for the case where F is the fixed-slash chain FXS*(n), hence with $p_0/q_0, p_1/q_1, \ldots, p_m/q_m$ the convergents to $|x|$:

For Fixed-Slash

$$\Phi_n(x) = \begin{cases} p_m/q_m & \text{if } x \geqslant 0, \quad x = p_m/q_m \text{ with } p_m, q_m \leqslant n, \\ p_i/q_i & \text{if } x > 0, \quad p_i, q_i \leqslant n \text{ and } \max\{p_{i+1}, q_{i+1}\} > n, \\ -\Phi_n(-x) & \text{if } x < 0. \end{cases} \tag{14}$$

From Theorem 14 we are then able to derive the following bounds on the portion of the unit interval rounding to p/q.

Corollary 14.1. *For* $p/q \in \mathrm{FXS}^*(n)$, $0 < p/q < 1$,

$$\frac{1}{(n+q)q} \leqslant \sup\left\{x \,\middle|\, \Phi_n(x) = \frac{p}{q}\right\} - \frac{p}{q} \leqslant \frac{1}{(n+1)q},$$

$$\frac{1}{(n+q)q} \leqslant \frac{p}{q} - \inf\left\{x \,\middle|\, \Phi_n(x) = \frac{p}{q}\right\} \leqslant \frac{1}{(n+1)q}. \tag{15}$$

Proof. For p/q, p'/q' consecutive fractions of FXS*(n), with $0/1 < p/q$ and $p'/q' \leqslant 1/1$, we obtain from Theorems 14, 3 and 2

$$\sup\left\{x \,\middle|\, \Phi_n(x) = \frac{p}{q}\right\} - \frac{p}{q} = \frac{p+p'}{q+q'} - \frac{p}{q} = \frac{1}{q(q+q')}.$$

Now $q \leqslant n$, $q' \leqslant n$, and $q + q' \geqslant n + 1$, so

$$\frac{1}{q(n+q)} \leqslant \frac{1}{q(q+q')} \leqslant \frac{1}{q(n+1)},$$

and the other inequality is obtained similarly. □

Utilizing Corollary 14.1 we observe that the approximation error in rounding to $p/q \in \mathrm{FXS}^*(n)$ for $0/1 \leqslant p/q \leqslant 1/1$ is at most $1/n$, but is much more likely to be of the order $1/n^2$, since q is usually of order n. It is important to note that it is the simpler fractions ($q \ll n$) that have the wider rounding intervals, and the extent of the variation is very pronounced, from single, $O(n^{-1})$, to double, $O(n^{-2})$, precision. We may therefore visualize the fixed-slash system as *an adaptive variable single-to-double precision arithmetic system*. The relatively simple fractions exist as a natural imbedded single-precision subsystem. Intermediate computations that overflow and require rounding are effectively computed in a near double-precision background in fixed-slash. When a true simple fractional result would be encountered in the actual exact computation, the single-precision rounding interval implicitly associated with the simpler fraction will almost certainly capture the approximate result whose accumulated error has been effectively growing in a double-precision environment, thus providing the possibility of recovering the exact value.

To measure the extent to which recovery of exactness is likely, it is necessary to investigate the possibility that an arbitrary rounded computation could incur a much larger than double-precision error due to the variable precision environment. An important measure of this likelihood is provided by the distribution of rounding error $|X - \Phi_n(X)|$, under uniform distribution of numbers to be rounded. This distribution was initially investigated in Matula and Kornerup [12], and the following result is repeated here for completeness.

Theorem 15. *For the random variable X chosen uniformly on* $[0, 1]$ *the expected value and the variance of the rounding error $|X - \Phi_n(X)|$ are respectively*:

$$\mathrm{Exp}(|X - \Phi_n(X)|) = \frac{6}{\pi^2}\frac{\log n}{n^2} + O\left(\frac{1}{n^2}\right),$$

$$\mathrm{Var}(|X - \Phi_n(X)|) = \frac{\zeta(2)}{\zeta(3)} \cdot \frac{2}{3n^3} + O\left(\frac{\log^2 n}{n^4}\right), \tag{16}$$

where $\zeta(X)$ is the Riemann Zeta-function ($\zeta(2) = \pi^2/6$).

Proof. For $0 < X < 1$ and $1 \leqslant p < q \leqslant n$ we obtain from Corollary 14.1

$$\frac{2}{q(n+q)} \leqslant \mathrm{Prob}\left\{\Phi_n(X) = \frac{p}{q}\right\} < \frac{2}{nq}.$$

The average error rounding to p/q, denoted $\varepsilon(p/q)$ is also obtained from Corollary 14.1 as

$$\frac{1}{2q(n+q)} \leqslant \varepsilon\left(\frac{p}{q}\right) < \frac{1}{2nq},$$

and since

$$\frac{1}{q^2n^2} - \frac{1}{q^2(n+q)^2} = \frac{1}{q^2n^2} - \frac{(n-q)^2}{q^2(n^2-q^2)^2} \leqslant \frac{2}{qn^3}$$

we obtain

$$\operatorname{Exp}(|X - \Phi_n(X)|) = \frac{1}{n^2} \sum_{\substack{1 \leqslant p \leqslant q \leqslant n \\ \gcd(p,q)=1}} \frac{1}{q^2} + O\left(\frac{1}{n^2}\right)$$

Utilizing the Euler φ function and following the proof procedure employed in Hardy and Wright [3, p. 268] one obtains

$$\sum_{\substack{1 \leqslant p \leqslant q \leqslant n \\ \gcd(p,q)=1}} \frac{1}{q^2} = \sum_{q=1}^{n} \frac{\varphi(q)}{q^2} = \frac{\log n}{\zeta(2)} + O(1)$$

from which the first part of the theorem is proved.

The variance can be found from the second moment which may be similarly expressed as

$$\operatorname{Exp}(|X - \Phi_n(X)|^2) = \frac{2}{3n^3} \sum_{\substack{1 \leqslant p \leqslant q \leqslant n \\ \gcd(p,q)=1}} \frac{1}{q^3} + O\left(\frac{\log n}{n^4}\right)$$

and utilizing that

$$\sum_{q=1}^{n} \frac{\varphi(q)}{q^3} = \sum_{q=1}^{\infty} \frac{\varphi(q)}{q^3} + O\left(\sum_{q=n+1}^{\infty} \frac{\varphi(q)}{q^3}\right) = \frac{\zeta(2)}{\zeta(3)} + O\left(\frac{1}{n^2}\right). \square$$

The standard deviation of $|X - \Phi_n(X)|$ over $[0, 1]$ is then asymptotically proportional to $\sqrt{n}/n^2$, hence it dominates the mean. The Chebychev inequality would then give a very pessimistic estimate regarding the distribution about the mean. The following theorem provides a direct bound on the distribution of error for X uniform on $[0, 1]$. The bound is stated for the probability that the error is greater than one part in n^α for $1 \leqslant \alpha \leqslant 2$, so α provides a measure from single ($\alpha = 1$) to double ($\alpha = 2$) precision.

Theorem 16. *If X is chosen uniformly on* $[0, 1]$, *then for any* α, $1 \leqslant \alpha \leqslant 2$, *and the fixed-slash rounding* Φ_n:

$$\operatorname{Prob}\left\{|X - \Phi_n(X)| > \frac{1}{n^\alpha}\right\} \leqslant 2n^{\alpha-2}. \tag{17}$$

Proof. For $0 \leqslant X \leqslant 1$ with $\Phi_n(X) = p/q$, from Corollary 14.1

$$\operatorname{Prob}\{|X - \Phi_n(X)| \leqslant n^{-\alpha}\} \geqslant \operatorname{Prob}\left\{|X - \Phi_n(X)| \leqslant \frac{1}{n\lceil n^{\alpha-1}\rceil}\right\} \geqslant \operatorname{Prob}\{q \geqslant n^{\alpha-1}\}$$

hence

$$\operatorname{Prob}\{|X - \Phi_n(X)| > n^{-\alpha}\} \leqslant \operatorname{Prob}\{q \leqslant \lfloor n^{\alpha-1} \rfloor\}.$$

But

$$\begin{aligned}\operatorname{Prob}\{q \leqslant k\} &= \operatorname{Prob}\{p = 0\} + \operatorname{Prob}\{p = q = 1\} + \operatorname{Prob}\{1 \leqslant p < q \leqslant k\} \\ &\leqslant \frac{2}{n} + \sum_{\substack{1 \leqslant p < q \leqslant k \\ \gcd(p,q)=1}} \operatorname{Prob}\left\{\Phi_n(X) = \frac{p}{q}\right\}\end{aligned}$$

$$< \frac{2}{n} + \sum_{q=2}^{k} \sum_{p=1}^{q-1} \frac{2}{nq}$$

$$< \frac{2}{n} \cdot \sum_{q=1}^{k} \sum_{p=1}^{q} \frac{1}{q} = \frac{2k}{n};$$

which with $k = \lfloor n^{\alpha-1} \rfloor$ proves the theorem. □

From (15) and (17) note that the distribution function of $|X - \Phi_n(X)|$ is always greater than are equal to the distribution function of the random variable Y specified by

$$\text{Prob}\{Y > x\} = \begin{cases} 1 & \text{for } 0 \leqslant x \leqslant 2/n^2, \\ 2/xn^2 & \text{for } 2/n^2 \leqslant x < 1/n, \\ 0 & \text{for } 1/n \leqslant x. \end{cases}$$

Then for any fixed n

$$\text{Exp}(Y) = \frac{2}{n^2} + \int_0^{1/n} \text{Prob}\{Y > x\}\,dx = \frac{2}{n^2} + \int_{2/n^2}^{1/n} \frac{2}{xn^2}\,dx$$

$$= 2\frac{\log n}{n^2} + O\left(\frac{1}{n^2}\right). \tag{18}$$

From (16) and (18) we conclude that the bound of Theorem 16 is of the right order. We further conjecture that $\text{Prob}\{|X - \Phi_n(X)| > 1/n^\alpha\}$ has an order with leading term $(6/\pi^2)n^{\alpha-2}$.

Some numeric examples using Theorem 16 illustrate the extent to which near double precision accuracy (i.e. α near 2) is likely. Consider first a system with $n = 2^{31} - 1$ (e.g. a fixed-slash representation using a 32-bit word for numerator and sign and a 32-bit word for the denominator ignoring the sign). We consider the probability that the absolute error is worse than one part in 10^t (indicating about t digits of accuracy) and note the corresponding value of α.

Prob{rounding error > 10^{-18}}	< 0.43	$\alpha = 1.93$
Prob{rounding error > 10^{-15}}	$< 4.3 \times 10^{-4}$	$\alpha = 1.61$
Prob{rounding error > 10^{-11}}	$< 4.3 \times 10^{-8}$	$\alpha = 1.18$
Prob{rounding error > $10^{-9.3}$}	$= 0$	$\alpha = 1$

Now consider the case $n = 2^{127} - 1$ corresponding to multiple precision range for numerator and denominator. Note in this case that the likelihood of operating close to double-precision, e.g. α near 2, increases dramatically.

Prob{rounding error > 10^{-75}}	< 0.069	$\alpha = 1.96$
Prob{rounding error > 10^{-70}}	$< 6.9 \times 10^{-7}$	$\alpha = 1.83$
Prob{rounding error > 10^{-60}}	$< 6.9 \times 10^{-17}$	$\alpha = 1.57$
Prob{rounding error > 10^{-45}}	$< 6.9 \times 10^{-32}$	$\alpha = 1.18$
Prob{rounding error > $10^{-38.5}$}	$= 0$	$\alpha = 1$

For scientific computation we are generally more concerned with the relative rounding error where X ranges over all real values. For this case first note the following result for the interval $[0, 1]$ proved in Matula and Kornerup [12].

Theorem 17. *If X is chosen uniformly on $[0, 1]$, then for any α, $1 \leqslant \alpha \leqslant 2$,*

$$\text{Prob}\left\{\frac{|X - \Phi_n(X)|}{X} > \frac{1}{n^\alpha}\right\} \leqslant 2n^{\alpha-2}(\log\lceil n^{2-\alpha}\rceil + \tfrac{1}{2}). \tag{19}$$

Observe then that

$$\left|\frac{1}{X} - \Phi_n\left(\frac{1}{X}\right)\right| \Big/ \left(\frac{1}{X}\right) = \frac{|X - \Phi_n(X)|}{\Phi_n(X)},$$

so the bound of Theorem 17 is also of the right order for X in the interval $[1, \infty)$ chosen so that $1/X$ is uniform on $[0, 1]$.

Numeric computations using inequality (19), as we utilized (17) in the preceding examples, gives nearly as high an order for the probability that the relative rounding error in fixed-slash is close to the double precision level. Despite this good "global" behavior, it should be noted that the fixed-slash relative errors for roundings to fractions in the neighborhood of zero are systematically much larger than for roundings to fractions in the neighborhood of unity. Thus scaling could unduly influence the relative rounding error behavior in fixed-slash, introducing a feature that is undesirable for many scientific computations.

It was noted in the statement of Theorem 11 that the representable values of $H^*(n)$ approach log uniform density on the positive real line as $n \to \infty$. Thus the relative rounding error in floating-slash computation should be essentially free of any systematic bias that would generally alter the results of a computation under scaling. The behavior of the relative rounding error in floating-slash computation has several natural and desirable features for hosting scientific computation as indicated in the following results for roundings to $H^*(n)$.

Theorem 18. *For any finite non-zero $p/q \in H^*(n)$ with $n \geqslant 2$, the relative size of the interval rounding to p/q is bounded as follows:*

$$\frac{1}{\sqrt{npq} + pq} \leqslant \frac{\sup\{x|\Phi_{H^*(n)}(x) = p/q\} - p/q}{p/q} \leqslant \frac{1}{\sqrt{npq} - 1}, \tag{20}$$

$$\frac{1}{\sqrt{npq} + pq + 1} \leqslant \frac{p/q - \inf\{x|\Phi_{H^*(n)}(x) = p/q\}}{p/q} \leqslant \frac{1}{\sqrt{npq}}, \tag{21}$$

and the order is given by

$$\frac{\sup\{x|\Phi_{H^*(n)}(x) = p/q\} - \inf\{x|\Phi_{H^*(n)}(x) = p/q\}}{p/q} = \frac{2}{\sqrt{npq}} + \frac{c}{n} \tag{22}$$

where $|c| \leqslant 4$.

Proof. Let $p/q < p'/q'$ be consecutive finite non-zero fractions of $H^*(n)$, so by Theorem 14,

$$\frac{\sup\{x|\Phi_{H^*(n)}(x) = p/q\} - p/q}{p/q} = \frac{1}{p(q+q')}. \tag{23}$$

Now $pq' = qp' - 1$ and $p'q' \leqslant n$, so

$$pq' < \sqrt{pq'qp'} \leqslant \sqrt{npq},$$

which with (23) yields the left-hand side of (20). Also $p(q+q') + 1 = q(p+p')$ and $(p+p')(q+q') > n$, so

$$p(q+q') + 1 > \sqrt{p(q+q')q(p+p')} > \sqrt{npq},$$

which with (23) yields the right-hand side of (20). Inequality (21) follows similarly, and (22) then follows directly from (20) and (21). □

From Theorem 18 note that the relative size of the interval rounding to $p/q \in H^*(n)$ will be (i) of the smaller order $1/n$ when the product pq is close to n, and (ii) of the larger order $1/n^{1/2}$ when p/q is a relatively simple fraction. Specifically, for any fixed k as $n \to \infty$, all fractions $p/q \in H^*(k)$ have rounding intervals whose relative size is of order $1/n^{1/2}$ in $H^*(n)$.

To determine the expected size of the relative error for rounding to $H^*(n)$ first consider that both empirical evidence and some theoretical arguments (Benford [1], Pinkham [14], Hamming [4]) suggest we investigate the case where X is distributed in a log uniform manner over the range $[1/n, n]$, i.e. X is distributed over $[1/n, n]$ so that

$$\text{Prob}\{X \leqslant y\} = \frac{1}{2\log n}\int_{1/n}^{y} \frac{dx}{x}.$$

From Theorem 18 the probability that X rounds to the irreducible finite non-zero fraction p/q is then of order

$$\frac{1}{2\log n}\,\frac{2}{\sqrt{npq}},$$

and the relative rounding error in this case has average order $1/(2\sqrt{npq})$. Hence the expected relative error introduced by the rounding $\Phi_{H^*(n)}$ with X chosen log uniformly over the interval $[1/n, n]$ is of order

$$\frac{1}{2\log n} \sum_{\substack{1 \leqslant pq \leqslant n \\ \gcd(p,q)=1}} \frac{1}{npq} \leqslant \frac{1}{2\log n} \sum_{q=1}^{n} \frac{1}{nq} \sum_{p=1}^{\lfloor n/q \rfloor} \frac{1}{p} < \frac{\log n}{2n}, \tag{24}$$

so that the smaller end of the range $[1/n, 1/n^{1/2}]$ of relative rounding errors is obtained as the expected behavior. Furthermore, for X chosen in the same log uniform manner, arguments similar to the proof of Theorem 16 yield the following bound on the relative error distribution for the rounding $\Phi_{H^*(n)}$ for suitably large n with $\frac{1}{2} \leqslant \alpha \leqslant 1$,

$$\text{Prob}\left\{\frac{|X - \Phi_{H^*(n)}(X)|}{X} > n^{-\alpha}\right\} < 4\alpha n^{\alpha - 1}. \tag{25}$$

It is instructive to summarize and compare the properties of the absolute rounding error in fixed-slash over $[0, 1]$ with the relative rounding error in floating-slash over the positive reals.

Observation 9 (Adaptive Variable Precision). In the fixed-slash system FXS*(n) the size of the interval rounding to p/q for $0 < p/q < 1$ is of order $1/qn$ and varies from single precision, $O(1/n)$, to double precision, $O(1/n^2)$. The expected absolute rounding error for X chosen uniformly on the unit interval is $O(\log n/n^2)$. Thus the simple fractions ($q \ll n$) essentially constitute an embedded single precision subsystem within a near double precision environment in FXS*(n). The bound n applies separately to both numerator and denominator in FXS*(n) and hence should be compared with the role of $n^{1/2}$ in $H^*(n)$. For a floating-slash system as interpreted through the ideal system $H^*(n)$, the relative size of the interval rounding to p/q for $0 < p/q < \infty$ is of order $1/(npq)^{1/2}$ in $H^*(n)$, and thus varies from single precision, $O(1/n^{1/2})$, to double precision, $O(1/n)$. The expected relative rounding error for X chosen log uniformly over $[1/n, n]$, is $O(\log n^{1/2}/n)$. Thus the simple fractions ($pq \ll n$) essentially constitute an embedded single precision relative error subsystem within a near double precision relative error environment in $H^*(n)$. ∎

The adaptive variable precision feature summarized in Observation 9 suggests that fixed-slash and floating-slash systems have the intriguing potential to recover exact results in certain computations. Our final observation treats this issue where we caution that the breadth and significance for applications of this potential to recover exact results requires both further analysis and empirical testing on typical real world problems.

Observation 10 (Recovery of Exactness). A moderate length approximate rational computation in either fixed-slash or floating-slash arithmetic will have an accumulated error governed with high probability by a near double precision error bound. If the exact result of the same rational computation is a rather simple fraction, the implicit single precision rounding interval associated with this simple fraction is very likely to contain the approximate near double precision computed result prior to the final rounding, so that the final rounding then recovers the exact simple fractional result. ∎

In the definitions and theorems of this paper we have attempted to provide a firm foundation for approximate rational arithmetic. In the observations we have sought to interpret from the foundation the natural principles governing the application of fixed-slash and floating-slash computation as an apparatus for hosting approximate real arithmetic on a computer. We believe the depth of number theoretic support for these systems has indeed provided insight into the inherent properties of finite precision computation.

References

[1] Benford, F.: The law of anomalous numbers. Proc. Am. Phil. Soc. **78**, 551–572 (1938).
[2] Dickson, L. E.: History of the theory of numbers, Vol. 1 (Reprint). Chelsea Publ. Co. 1971.
[3] Hardy, G. H., Wright, E. M.: An introduction to the theory of numbers, 4th ed. Oxford: Clarendon Press 1960.
[4] Hamming, R. W.: On the distribution of numbers. Bell Sys. Tech. Jour. **49**, 1609–1625 (1970).

[5] Khintchin, A. Ya: Continued fractions. (Translated from Russian by P. Wynn.) Groningen: P. Noordhoff Ltd. 1963.
[6] Knuth, D. E.: The art of computer programming, Vol. 2: Seminumerical algorithms. Reading: Addison-Wesley 1969.
[7] Kornerup, P., Matula, D. W.: A feasibility analysis of fixed-slash rational arithmetic, in: Proceedings of the 4th IEEE Symposium on Computer Arithmetic, IEEE Catalog No. 78CH1412-6C, 1978, pp. 39 – 47.
[8] Kulish, U.: An axiomatic approach to rounded computations. Numerische Mathematik **18**, 1 – 17 (1971).
[9] Matula, D. W.: A formalization of floating-point numeric base conversion. IEEE Trans. on Comp. **C-19**, 681 – 692 (1970).
[10] Matula, D. W.: Fixed-slash and floating-slash rational arithmetic, in: Proceedings of the 3rd IEEE Symposium on Computer Arithmetic, IEEE Catalog No. 75CH1017-3C, 1975, pp. 90 – 91.
[11] Matula, D. W., Kornerup, P.: A feasibility analysis of binary fixed-slash and floating-slash number systems, in: Proceedings of the 4th IEEE Symposium on Computer Arithmetic, IEEE Catalog No. 78CH1412-6C, 1978, pp. 29 – 38.
[12] Matula, D. W., Kornerup, P.: Approximate rational arithmetic systems: analysis of recovery of simple fractions during expression evaluation, in: Proceedings of EUROSAM 79 (Lecture Notes in Computer Science, Vol. 72), pp. 383 – 397. Berlin-Heidelberg-New York: Springer 1979.
[13] Matula, D. W., Kornerup, P.: In preparation.
[14] Pinkham, R. S.: On the distribution of first significant digits. Ann. Math. Stat. **32**, 1223 – 1230 (1961).

Dr. D. W. Matula
Department of Computer Science
and Engineering
Southern Methodist University
Dallas, Texas, U.S.A.

Dr. P. Kornerup
Computer Science Department
Aarhus University
Aarhus, Denmark

Computing, Suppl. 2, 113–120 (1980)

Interval Methods for Nonlinear Systems

R. E. Moore*, Madison, Wisconsin

Abstract

Interval methods provide computational tests for the existence or non-existence of a solution to a given system of nonlinear equations in an n-dimensional rectangle. Tests are also provided for the convergence of certain iterative methods within suitable regions (safe starting regions). Using bisection procedures, we can search an arbitrary n-dimensional rectangle for a safe starting region. Various bisection rules are discussed.

1. Computational Tests for Existence, Non-Existence, and Convergence

We will discuss methods for finding sequences of vectors which converge to a solution of a system of equations

$$f_1(x_1, x_2, \ldots, x_n) = 0$$
$$f_2(x_1, x_2, \ldots, x_n) = 0$$
$$\ldots$$
$$f_n(x_1, x_2, \ldots, x_n) = 0. \tag{1.1}$$

We can write (1.1) in vector form

$$f(x) = 0 \tag{1.2}$$

Recently (Hansen [3, p. 23], Nickel [15], Moore [10, 11], Alefeld [1]), computational tests for the existence of a solution in an n-dimensional rectangle have been found which are simpler than the well-known Kantorovich conditions (Kantorovich [6], Rall [16]). In addition, tests for convergence (of Newton-type iterative methods) which are also simpler than the Kantorovich conditions have been found (Moore [11, 13]).

In this section, we will discuss a number of such tests and also some simple tests for non-existence of a solution in an n-dimensional rectangle (Moore [8, 9], Moore and Jones [14]). We begin with the tests for non-existence.

We assume that f and f' are continuous on some open set of R^n and have inclusion monotonic interval extensions F and F'.

We denote an n-dimensional rectangle by $X = (X_1, X_2, \ldots, X_n)$ or $B = (B_1, B_2, \ldots, B_n)$, etc., where $X_i = [\underline{X}_i, \bar{X}_i]$ and $B_i = [\underline{B}_i, \bar{B}_i]$ are closed bounded intervals of real numbers. Thus X and B are interval vectors.

* NSF Grant MCS78–03824.

(1.3) *First Non-Existence Test*: If $0 \notin F_i(X)$ for some i, then (1.2) has no solution in X.

Comments: If we can evaluate $F_i(X)$, then $F_i(X) = [a, b]$ for some real numbers (machine numbers) $a \leqslant b$. If $a > 0$ or $b < 0$, then $0 \notin F_i(X)$. Since $f(x) \in F(X)$ for all $x \in X$, the condition $0 \notin F_i(X)$ implies that $f_i(x)$ cannot vanish for any x in X.

(1.4) *Second Non-Existence Test*: Suppose that $y \in X$ and that Z is an interval matrix containing the inverses of all real matrices A such that $A_{ij} \in F'_{ij}(X)$. If $\{y - Zf(y)\} \cap X$ is empty, then there is no solution (to (1.2)) in X.

Comments: To find such a Z, we "invert" the interval matrix $F'(X)$; see, e.g., Alefeld and Herzberger [2]. If $\{y - Zf(y)\}_i \cap X_i$ is empty for any i then there is no solution in X. Note that $[a, b] \cap [c, d]$ is empty if $b < c$ or $a > d$.

(1.5) *Third Non-Existence Test*: Suppose that $y \in X$ (for instance y might be chosen as the midpoint of X) and that Y is any non-singular real matrix. Let $K(X) = y - Yf(y) + \{I - YF'(X)\}(X - y)$, where I is the $n \times n$ identity matrix. If $K(X) \cap X$ is empty, then there is no solution in X (to (1.2)).

Comments: This test follows from the fact that if X contains a solution, then so does $K(X)$. (For the previous test, if X contains a solution, then so does $y - Zf(y)$).

Several tests for existence have been derived using the Schauder fixed point theorem and the fact that an interval vector is a convex, compact set in R^n: see Hansen [3], Nickel [15], Alefeld [1], Moore [10, 12, 13]. We will be content with two of these here.

(1.6) *First Test for Existence*: If $K(X) \subseteq X$, then X contains a solution of (1.2), with $K(X)$ as defined in (1.5).

Comments: For Y, we may choose an approximate inverse of the midpoint matrix of $F'(X)$ – that is, of the real matrix with elements $m(F'(X)_{ij})$ – or, alternatively, of the real matrix $F'(m(X))$ – or *any* non-singular matrix. If y is the midpoint of X, and if X is not too wide and if $f(y)$ is small, and if $I - YF'(X)$ has a small norm, then $K(X)$ will be contained in X. The norm of $I - YF'(X)$ will be small if X is not too wide and if Y is chosen as indicated. A comparison of the test (1.6) with the Kantorovich conditions for existence of a solution has been made by L. B. Rall [16]. Note that (1.6) does not require an explicit bound on the norm of the second Fréchet derivative nor an explicit Lipschitz constant for the first Fréchet derivative, nor an explicit bound on the norm of the inverse of a Fréchet derivative.

Another test for existence is based on first re-writing the system (1.1) or (1.2) in "fixed point form"

$$x = g(x). \tag{1.7}$$

This can always be done in a large number of ways. If the resulting interval function g has an inclusion monotonic interval extension G, then we have the following.

(1.8) *Second Test for Existence*: If $G(X) \subseteq X$ and if $w(G(V)) \leqslant cw(V)$ for some $c < 1$ and for all $V \subseteq X$, then X contains a solution of (1.2).

Comments: Here, $w(V)$ denotes the *width* of V; thus, $w(V) = \max w(V_i)$, where $V = (V_i, V_2, \ldots, V_n)$ and $w(V_i) = \bar{V}_i - \underline{V}_i$. Similarly, $w(G(V))$ denotes the width of $G(V)$. In the rest of this paper we will only be concerned with the first test for existence. The second test suffers the weakness of not providing a means for determining whether such $c < 1$ exists. In a particular application it may be possible to do so "analytically".

For convenience in our discussion of convergence tests, we repeat here the interval operator used in (1.5) – which was introduced by Krawczyk [7] – namely,

$$K(X) = y - Yf(y) + \{I - YF'(X)\}(X - y) \tag{1.9}$$

where y is any real vector in X (for example $m(X)$, the midpoint of X) and where Y is any non-singular real matrix.

Consider the iterative algorithm:

$$x^{(k+1)} = x^{(k)} - Yf(x^{(k)}),\ Y \text{ non-singular, (if } Y = f'(x^{(0)})^{-1}, \text{ this is called "the simplified Newton method").} \tag{1.10}$$

(1.11) *First Test for Convergence* (Moore [11]): Suppose that X is an n-cube, $X = y + E$ with $E_i \equiv [-1, 1]r, r > 0$, and that

$$\|K(X) - y\| = \| - Yf(y) + \{I - YF'(X)\}E\| < r \quad \text{(max norm)}, \tag{1.12}$$

then (1.10) converges to the unique solution x^* in X from any $x^{(0)}$ in X. In particular, (1.10) converges from $x^{(0)} = y = m(X)$.

Comments: This test can be used for any algorithm of the form (1.10) with Y fixed throughout the iterations, as long as Y is non-singular. In particular, Y could be chosen as a non-zero multiple of the identity matrix, or as the diagonal matrix with elements $\{\partial f_i/\partial x_i\}^{-1}$ evaluated at $x^{(0)}$ (if possible). The norm in (1.11) can be evaluated computationally using interval arithmetic, thus the test can be carried out on a computer for a given n-cube, X.

Consider next the class of iterative algorithms of the form:

$$x^{(k+1)} = x^{(k)} - Y^{(k)}f(x^{(k)}) \tag{1.13}$$

where $Y^{(0)} = Y$ is non-singular and where $\|I - Y^{(k)}F'(X)\| \leqslant \|I - Y^{(k-1)}F'(X)\|$.

The ordinary Newton method for n-dimensional systems is included if we "update" $Y^{(k)}$ only when $\|I - Y^{(k)}F'(X)\|$ does not increase. Note that we can compute this norm as

$$\|I - Y^{(k)}F'(X)\| = \max_i \sum_p |\delta_{ip} - \sum_j Y_{ij}^{(n)}F'(X)_{jp}|.$$

Recall that $|[a, b]| = \max(|a|, |b|)$ for an interval $[a, b]$ and δ_{ip} is one, for $i = p$, and zero, for $i \neq p$.

(1.14) *Second Test for Convergence* (Moore [11]): The first test for convergence (1.11) applies to the algorithm (1.13). If (1.12) holds, then (1.13) converges to the unique solution of (1.2) in X from any $x^{(0)}$ in X, in particular from $x^{(0)} = y = m(X)$.

Comments: Note that we need only make the test (1.12) once, at the beginning for $y = x^{(0)}$. In the ordinary Newton method, we would take $Y^{(k)} = f'(x^{(k)})^{-1}$. In (1.13) we are requiring that we leave $Y^{(k)} = Y^{(k-1)}$ if it should turn out that

$$\|I - Y^{(k)}F'(X)\| > \|I - Y^{(k-1)}F'(X)\|. \tag{1.15}$$

Note also that the test for convergence (1.12) also implies existence and local uniqueness in X, if it is satisfied.

Consider, finally, the interval algorithm (Krawczyk [7], Moore [10], Hansen [4]):

$$X^{(k+1)} = X^{(k)} \cap K(X^{(k)}) \tag{1.16}$$

where $K(X)$ is given by (1.9) and where we put $y = m(X^{(k)})$ and we take, for $Y^{(k)}$, an approximation to $m(F'(X^{(k)}))^{-1}$, unless (1.15) holds, in which case we leave $Y^{(k)} = Y^{(k-1)}$.

(1.17) *Third Test for Convergence* (Moore [11]): The first test for convergence (1.11) applies to the algorithm (1.16). If (1.12) holds, then (1.16) converges to the unique solution x^* of (1.2) in X from $X^{(0)} = X$. Furthermore, $x^* \in X^{(k+1)} \subseteq X^{(k)}$ for every k.

Comments: For the interval algorithm (1.16), we obtain a nested sequence of n-dimensional rectangles containing the exact solution. Thus, if we stop the iteration at any particular k, we have bounds on the solution components. In rounded interval arithmetic, we can iterate until we obtain *numerical convergence*: $X^{(k+1)} = X^{(k)}$. This will always happen in a finite number of iterations.

(1.18) *Fourth Test for Convergence* (Moore [10, 11]): We have existence of a solution in X if $K(X) \subseteq X$, by (1.6). Whether or not X is an n-cube, we will have convergence of (1.16) from $X^{(0)} = X$, and (1.10), (1.13) from any $x^{(0)}$ in X if, in addition to (1.6), we have $\|I - Y^{(0)}F'(X)\| < 1$.

2. Bisection in n Dimensions

Given a nonlinear system of equations (1.1) and an iterative algorithm, the problem remains to *find a starting vector* from which the algorithm converges to a solution. Often it is necessary to start very close to a solution in order to obtain convergence.

In Sect. 1, we have given tests which may be applied to a trial region X in order to determine, computationally, whether there is: (1) no solution in X (non-existence test), (2) a solution in X (existence test), (3) convergence to a solution from $x^{(0)}$ in X (convergence test), using certain algorithms: (1.10), (1.13), (1.16).

In this section we discuss a procedure for searching an *arbitrary* n-dimensional rectangle B in order to determine one of two things: (1) there is no solution anywhere in B; *or* (2) there is a solution in a certain subrectangle X in B and a chosen one of the algorithms converges to that solution from X. Such a region X will be called a *safe starting region*.

The basic idea is to carry out a "depth first" search using bisections, while "stacking" the halves not chosen first in the bisection process. In this way, we reach a small region which is a safe starting region; or we reach a small region which we can exclude as not containing a solution. In the latter case, we next examine the last region which was put on the stack. If the stack becomes empty without our having found a safe starting region, then there is no solution in the initial region B. The region B should be chosen *large* enough so that it has some chance of containing a solution. We might even apply the search procedure to each of a sequence of larger and larger regions.

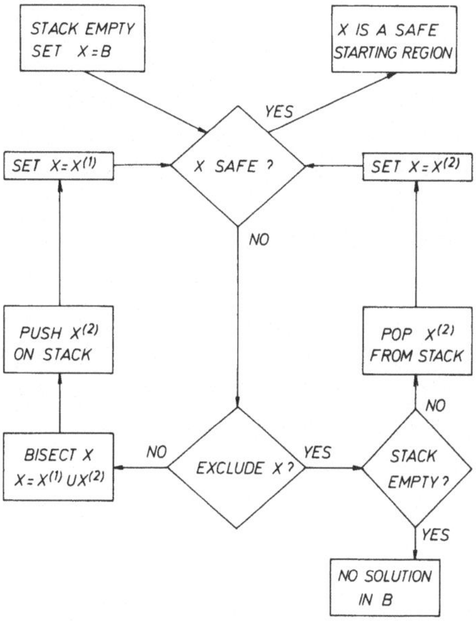

Fig. 1. The search procedure

Fig. 1 shows a logical flow chart for the search procedure. The search procedure can be used with *any* iterative algorithm for which there are tests for existence and convergence for a given trial region X. For example, we can use the search procedure with the algorithm (1.10) with the tests (1.6) and (1.11); or the algorithm (1.13) with the tests (1.6) and (1.11); or the algorithm (1.16) with the test (1.11) or with the tests (1.6) and (1.18). We could also use the search procedure with the algorithm

$$x^{(k+1)} = g(x^{(k)}) \tag{2.1}$$

and the test (1.8), if we have some means for determining that such a $c < 1$ exists.

We can exclude X for *any* iterative algorithm used with the search procedure if the non-existence test (1.3) is satisfied. If we are using the tests (1.6) or (1.11) then we will be computing $K(X)$ and we can use the non-existence test (1.5) in addition to (1.3) without any additional computational effort.

"Push" and "pop" refer to the operations of adding a region to the stack or removing a region from the stack (respectively), with a "last-in, first-out" ordering.

For the *bisection* of a region X, we must have rules for choosing: (1) a coordinate direction in which to bisect the region X, (2) which half to search next and which half to stack for possible search later. We will discuss a number of such *bisection rules.*

By the *bisection of X in direction i,* we mean that we cut X in half in coordinate direction x_i and represent X as the union of two half regions $X^{(1)}$ and $X^{(2)}$. We have $X^{(1)} = (X_1^{(1)}, \ldots, X_n^{(1)})$ and $X^{(2)} = (X_1^{(2)}, \ldots, X_n^{(2)})$ with $X_j^{(1)} = X_j^{(2)} = X_j$ for all j except $j = i$. The ith component interval of X is cut in half as $X_i = [\underline{X}_i, m(X_i)] \cup [m(X_i), \bar{X}_i]$ with $X_i^{(1)}$ chosen as one of these halves and $X_i^{(2)}$ the other.

Note that if we cannot even carry out one of the required tests (perhaps because we cannot evaluate $F(X)$ – machine overflow, X not in the domain of F, etc. – or perhaps we cannot find a suitable Y for use in $K(X)$, for whatever reason) then we have not satisfied the test and we take the "no" exit from the test box for "X SAFE?" or "EXCLUDE X?" in Fig. 1. Thus, there is no restriction against the singularity of the Jacobian of the system nor is it even necessary that the functions f_i are bounded or even defined in all of B.

Thus the search procedure is not terminated by an attempt to compute an undefined quantity. It could happen, because of finite precision machine arithmetic, that we cannot bisect further some small region X. In this case we can print such a region along with a message that it may still contain a solution and then exclude it and proceed with the search. We could later continue the search on any such small regions remaining, using higher precision arithmetic.

Note that we only bisect a region if it might still contain a solution but has not been determined to be a safe starting region.

Bisection Rules I (Moore [13]): Make a cyclic choice of coordinate direction in successive bisections; select the left half of the bisected interval for $X^{(1)}$.

Comments: By the "left half", we mean put $X_i^{(1)} = [\underline{X}_i, m(X_i)]$. Coordinate directions are chosen as $1, 2, \ldots, n, 1, 2, \ldots$. After pn bisections, we have $w(X) = 2^{-p} w(B)$.

Bisection Rules II (Moore [13]): Bisect in a coordinate direction x_i in which X_i is of maximum width; select a half toward which the Newton method points from $m(X)$.

Comments: Again we will have $w(X) = 2^{-p} w(B)$ after np bisections. The choice of a half "toward which the Newton method points" means we will take $[\underline{X}_i, m(X_i)]$ for $X_i^{(1)}$ if $-[Yf(m(X))]_i < 0$, or $[m(X_i), \bar{X}_i]$ if $-[Yf(m(X))]_i > 0$, or either if $-[Yf(m(X))]_i = 0$. If we cannot compute $Yf(m(X))$, we select the left half of X as in Bisection Rules I. If the Newton steps from the midpoints always point to a half containing a solution, then the sequence of midpoints will converge to a solution and we will likely find a safe starting region in a relatively small number of bisections.

Bisection Rules III (Moore [13]): When using (1.5) along with (1.3) for non-existence tests to decide whether to exclude a particular region X (and using (1.6) as an existence test), we will have computed $K(X)$. If X is not safe and if we cannot exclude X, then we will arrive at the box requiring a bisection of X in the logical flow chart of Fig. 1. However, since, if X contains a solution, so does $K(X)$, it follows that we can replace X by $K(X) \cap X$ (which is non-empty – otherwise X would have been excluded). That is, we may replace $X^{(1)}$, chosen by Bisection Rules II, by $X^{(1)} \cap K(X)$ and $X^{(2)}$ by $X^{(2)} \cap K(X)$. It could happen that $X^{(2)} \cap K(X)$ is empty in which case we do not push anything onto the stack.

Comments: After *np* bisections, we will have a region X of width $w(X) \leqslant 2^{-p}w(B)$. We can use the fourth test for convergence, namely: X is a safe starting region for the algorithms (1.10), (1.13), and (1.16) if $K(X) \subseteq X$ and $\|I - Y^{(0)}F'(X)\| < 1$ are both satisfied. We illustrate the use of Bisection Rules III with an example. Consider the system

$$\begin{aligned} f_1(x_1, x_2) &= x_1^2 + x_2^2 - 1 = 0 \\ f_2(x_1, x_2) &= x_1 - x_2^2 \qquad = 0. \end{aligned} \tag{2.2}$$

We search the initial region $B = ([0,1],[0,1])$ for a safe starting region for the system (2.2). After *two* bisections, we obtain the safe starting region $X = ([0.5, 0.98125], [0.5625, 1])$. Starting at the midpoint of $K(X)$ in this region, the ordinary Newton method produced, in two iterations, an approximate solution accurate to at least eight decimal places in x_1 and x_2. The Kantorovich conditions *cannot* be satisfied at $x^{(0)} = (0.5, 1.0)$ even though the Newton method converges from there.

In fact, for the relatively simple system (2.2) we can find the exact set of $x^{(0)}$ at which the Kantorovich conditions can be satisfied. This is shown in Fig. 2, along with some safe starting regions found by interval methods, for comparison.

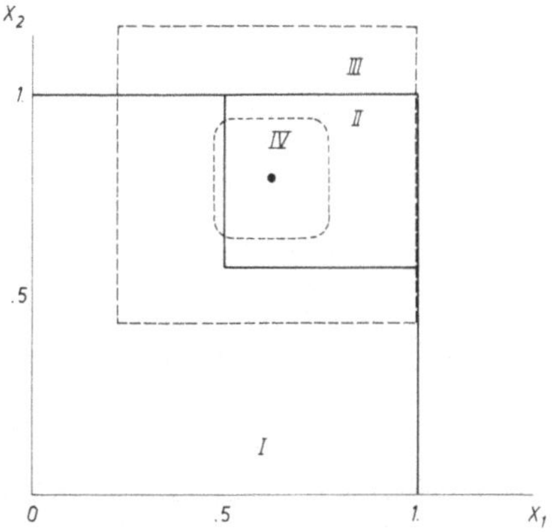

Fig. 2. Safe starting regions

The following regions are indicated in Fig. 2:

I: An arbitrary choice for B.
II: A safe starting region found in B.
III: The largest X such that $K(X) \subseteq X$ and $r_0 < 1$.
IV: $\{x^{(0)}\}$ at which the Kantorovich conditions can be satisfied.

Bisection Rules IV (Moore [13]): Bisect in a coordinate direction x_i in which $|[Yf(m(X))]_i|$ is maximum and choose a half toward which the Newton method points from $m(X)$. (See comments on Bisection Rules II).

Comments: It can be proved (Moore [13]) that if X contains a solution and if Y is non-singular, then Bisection Rules IV will select a half of X which still contains a solution providing that $\|I - YF'(X)\|$ is sufficiently small.

A number of other bisection rules have been studied (Moore and Jones [14], Jones [5]). Further research and computational experience is needed to determine which are the most efficient bisection rules. For a certain system of dimension 9, a solution was sought in the unit 9-cube. Jones [5] found a safe starting region in 168 bisections (in direction i with $w(X_i) = w(X)$ and *random* choice of half) with 73 "backtracks" (popping 73 regions from the stack). Note that 63 cyclic bisections of the unit 9-cube will produce a smaller 9-cube of width less than 0.01. The safe starting region found by Jones (starting with a slightly larger B than the unit 9-cube) had width about 0.11. The systematic testing of a partition of the unit 9-cube into cubes of width 0.01 would require testing 10^{18} such smaller cubes! The bisection search procedure – if it can keep going in more or less the right direction – can be vastly more efficient.

References

[1] Alefeld, G.: Intervallanalytische Methoden bei nichtlinearen Gleichungssystemen. Technische Universität Berlin, Fachbereich Mathematik (3), Nr. 43. 1978.

[2] Alefeld, G., Herzberger, J.: Einführung in die Intervallrechnung. Mannheim: Bibliographisches Institut 1974.

[3] Hansen, E. (ed.): Topics in Interval Analysis, p. 23. Oxford University Press 1969.

[4] Hansen, E. (ed.): Interval Forms of Newton's Method. Computing **20**, 153–163 (1978).

[5] Jones, S. T.: Searching for Solutions of Finite Nonlinear Systems – An Interval Approach. Ph. D. Thesis, University of Wisconsin-Madison, 1978.

[6] Kantorovich, L. V.: Functional analysis and applied mathematics. Uspehi Mat. Nauk **3**, 89–185 (1948). (In Russian.)

[7] Krawczyk, R.: Newton-Algorithmen zur Bestimmung von Nullstellen mit Fehlerschranken. Computing **4**, 187–201 (1969).

[8] Moore, R. E.: Interval Arithmetic and Automatic Error Analysis in Digital Computing. Ph. D. Dissertation, Stanford University, 1962.

[9] Moore, R. E.: Interval Analysis. Englewood Cliffs, N.J.: Prentice-Hall 1966.

[10] Moore, R. E.: A test for existence of solutions to nonlinear systems. SIAM J. Numer. Anal. **14**, 611–615 (1977).

[11] Moore, R. E.: A computational test for convergence of iterative methods for nonlinear systems. SIAM J. Numer. Anal. **15**, 1194–1196 (1978).

[12] Moore, R. E.: Bounding sets in function spaces with applications to nonlinear operator equations. SIAM Rev. **20**, 492–512 (1978).

[13] Moore, R. E.: Methods and Applications of Interval Analysis. Series on Studies in Applied Mathematics (Ames, W. F., ed.). Philadelphia: SIAM Publications 1979.

[14] Moore, R. E., Jones, S. T.: Safe starting regions for iterative methods. SIAM J. Numer. Anal. **14**, 1051–1065 (1977).

[15] Nickel, K.: On the Newton method in interval analysis. MRC Report #1136, Mathematics Research Center, University of Wisconsin-Madison, 1971.

[16] Rall, L. B.: A comparison of the existence theorems of Kantorovich and Moore. MRC Report #1944, Mathematics Research Center, University of Wisconsin-Madison, 1979. [SIAM J. Numer. Anal. **17** (1980).]

Prof. Dr. R. E. Moore
Computer Sciences Department
University of Wisconsin-Madison
1210 West Dayton Street
Madison, WI 53706, U.S.A.

Computing, Suppl. 2, 121 – 129 (1980)

Algorithms for Multiplication with Given Precision

W. Oberaigner, Innsbruck

Abstract

The general problem is to find for given lattices X, Y a function $f\colon X \to Y$ and a rounding T_Y in Y, a rounding T_X in X, or a family of roundings in X with the property $T_Y f(x) = T_Y f(T_X x)$ for any $x \in X$. We present a general scheme of an algorithm solving this problem and then concretize the algorithm-scheme for the multiplication of fixed-point and floating-point numbers. The resulting algorithm is based on a formula for the multiplication of polynomials and we discuss the advantages of the algorithm as well as the results of a simulation of the algorithm.

1. Introduction

We first present the problem in a general situation: Let $X, Y\ (\neq \emptyset)$ be lattices, $f\colon X \to Y$ a function and $T_Y\colon\ Y \to Y$ a rounding in Y (see [2]); we are not interested in $f(x)$ for any argument $x \in X$ but in the rounded picture $T_Y f(x)$. It is then natural to ask whether it is necessary to have an exact argument if we are rounding the picture; or in other words: does there exist a rounding T_X in X so that

$$\bigwedge_{x \in X} T_Y f(x) = T_Y f(T_X x) \tag{1}$$

If we take as example $Y = \mathbb{R}$, T_Y the classical rounding of the real to the entire numbers, then we can construct a set X, arguments $x \in X$ and a rounding T_X in X with $f(x) = z + 0.5$, $z \in Z$, $f(T_X x) < z + 0.5$ and $f(x) - f(T_X x)$ very small if f is "bad" enough. So we have $T_Y f(x) \neq T_Y f(T_X x)$ in spite of a very small difference $f(x) - f(T_X x)$. For this reason it is in general not sufficient to work only with one rounding and we look for a family of roundings in X $(T_X^i)_{i \in I}$ ($I \neq \emptyset$ index-set) with the property:

$$\bigwedge_{x \in X} \bigvee_{i_x \in I} T_Y f(x) = T_Y f(T_X^{i_x} x) \tag{2}$$

We can write Eq. (2) in a more general way when taking into consideration that $f(T_X^i x)$ is nothing else than an approximation of $f(x)$, therefore we can replace $f(T_X^i x)$ by $\mathrm{appr}_f^i(x)$ in Eq. (2).

We present now a very general scheme for an algorithm for the computation of $T_Y f(x)$ without having in general $f(x)$. The language is an ALGOL-like one we are sure that everybody can easily interpret.

Algorithm-Scheme. (Let $x \in X$ be given)

1. $p^0 := 0$; $i := 1$;

2. $p^i(x) := \text{appr}_f^i(x, p^{i-1}(x))$;
3. if $T_Y f(x) \neq T_Y p^i(x)$ then begin $i := i + 1$; goto 2.; end
4. STOP

Remark. When concretizing this algorithm-scheme we must solve two problems: namely

(P1) give a detailed description of the approximation function appr_f^i

(P2) say in which way one can verify the equation

$$T_Y f(x) = T_Y p^i(x)$$

to hold without knowing $f(x)$.

2. Definitions, Assumptions

We restrict ourselves for the rest of the paper to the multiplication of fixed-point or floating-point numbers. Let the basis B ($B \in N$, $B > 1$) and the number of digits of the fixed-point numbers resp. of the mantissa of the floating-point numbers n ($n \in N$) be arbitrary but fixed in the following. Then we define as usual the fixed-point numbers by

$$F_n(B) := F_n := \left\{ \pm 0.x_1 \ldots x_n \,\middle|\, \bigwedge_{j=1(1)n} x_j \in \{0, \ldots, B-1\} \right\} \tag{3}$$

(the notation $\bigwedge_{j=1(1)n}$ is an abbreviation for $\bigwedge_{j \in \{1, \ldots, n\}}$) and the floating-point numbers with n-digit mantissa and exponents between eu and eo ($eu, eo \in Z$, $eu \leqslant eo$) by

$$\text{Gl}(n, eu, eo; B) := \left\{ x \,\middle|\, \bigvee_{m \in F_n(B)} \bigvee_{e \in \{eu, \ldots, eo\}} x = m * B^e \wedge m_1 \neq 0 \right\} \tag{4}$$

where for any $x \in F_n(B)$ and any $j \in \{1, \ldots, n\}$ x_j denotes the jth digit of x (see for example [4] and N resp. Z denotes the set of natural resp. entire numbers).

Now we denote for arbitrary but in the following fixed r ($r \in N$, $r \geqslant 2$ will denote the number of factors)

$$X_F := \mathop{\times}_{j=1}^{r} F_n(B), \quad X_G := \mathop{\times}_{j=1}^{r} \text{Gl}(n, eu, eo; B), \quad X := X_F \quad \text{or} \quad X := X_G$$

(if the distinction is not relevant)

$$Y := F_{r*n}(B) \quad \text{or} \quad Y := \text{Gl}(r*n, r*eu, r*eo; B)$$

$$f\colon X \to Y\colon x \mapsto f(x) := f(x^1, \ldots, x^r) := \prod_{j=1}^{r} x^j \tag{5}$$

Furthermore we denote by TK_l the classical rounding to l digits which is defined by

$$TK_l\colon F_m(B) \to F_l(B)$$

$$x \mapsto TK_l(x) := (\text{entier}(x * B^l + 0.B_2)) * B^{-l}$$

$$B_2 := \text{entier}((B-1)/2)+1$$

$$x_{m+1} := 0 \tag{6}$$

for $m, l \in N$, $m \geqslant l$.

We will do the further computation only for the classical rounding but we want to emphasize the fact that the restriction to this special kind of rounding has no substantial reasons; all of the following results can by easy computation be transferred to e.g. directed monotonic roundings (see [4]) or other types of roundings.

Regarding problem (P2) we can give an obvious necessary condition for $TK_l f(x) = TK_l p^i(x)$ to hold (let $(p^i(x))_{i \in I}$ be generated by any approximation)

$$\bigwedge_{x \in X} \bigwedge_{i \in I} TK_l f(x) = TK_l p^i(x) \Rightarrow |f(x) - p^i(x)| < B^{-l} \tag{7}$$

The sufficient condition for this equation will be based on the maximal error of the approximation; we will use the following abbreviations

$$\bigwedge_{x \in X} \bigwedge_{i \in I} \text{err}^i(x) := |f(x) - p^i(x)| \tag{8}$$

$$\bigwedge_{i \in I} \text{errmax}^i := \max\{\text{err}^i(x) | x \in X\} \tag{9}$$

$$\bigwedge_{u,v,w \in N} \text{card}(u,v,w) := \left|\left\{(z^1,\ldots,z^v) \,\middle|\, \bigwedge_{t=1(1)v} z^t \in N \wedge 1 \leqslant z^t \leqslant w \wedge \sum_{t=1}^{v} z^t = u\right\}\right|$$

$$= \sum_{t=0}^{v} (-1)^t * \binom{v}{t} * \binom{u - t*w - 1}{v-1} \tag{10}$$

3. Concretizing the Algorithm-Scheme

We assume without loss of generality that all factors be nonnegative, so we use the abbreviations

$$X_F^+ := \{x | x \in X_F \wedge x \geqslant 0\},\ X_G^+ := \{x | x \in X_G \wedge x \geqslant 0\},\ X^+ := \{x | x \in X \wedge x \geqslant 0\} \tag{11}$$

3.1. Pre-Rounding

According to Eq. (2) we set for a family $(l^i)_{i \in N}$ with the property

$$\bigwedge_{i \in N} 1 \leqslant l^i \leqslant l^{i+1} \leqslant n,\ l^i \in N \tag{12}$$

$$\bigwedge_{x \in X_F^+} \bigwedge_{i \in N} p^i(x) := \text{appr}^i(x) := \text{appr}^i(x^1,\ldots,x^r) := \prod_{j=1}^{r} TK_{l^i}(x^j) \tag{13}$$

For the case $l = n$ which will be the most important one in applications we get the implication (7) $\Rightarrow \bigwedge_{i \in N} l^i \geqslant n$ which holds also if l^i is dependent of j; but this means exact multiplication. For this reason we will not go into further detail for this kind of approximation.

3.2. Polynomial-Formula

The following approximation is based on a well-known formula for the product of polynomials. We first present the idea for fixed-point numbers and then generalize the method for floating-point numbers.

3.2.1. Fixed-Point Factors

We can write the factors $x^j \in F_n(B)$, $x^j \geqslant 0$ in the following way

$$\bigwedge_{j=1(1)r} x^j := 0.x_1^j \ldots x_n^j = \sum_{t=1}^{n} x_t^j * B^{-t} \tag{14}$$

and get therefore

$$\bigwedge_{x \in X_F^+} f(x) := \prod_{j=1}^{r} x^j = \sum_{t=r}^{r*n} \left(\sum_{s^1 + \cdots + s^r = t} x_{s^1}^1 * \cdots * x_{s^r}^r \right) * B^{-t} \tag{15}$$

We can now define another kind of approximation for a family of natural numbers $(l^i)_{i \in N_0}$ with the property

$$l^0 := 0, \bigwedge_{i \in N_0} l^i \leqslant l^{i+1} \leqslant r*n \tag{16}$$

by recursion:

$$\bigwedge_{x \in X_F^+} p^0(x) := p^0 := 0$$

$$\bigwedge_{i \in N} \bigwedge_{x \in X_F^+} p^i(x) := \mathrm{appr}^i(x, p^{i-1}(x))$$

$$:= p^{i-1}(x) + \sum_{t=l^{i-1}+1}^{l^i} \left(\sum_{s^1 + \cdots + s^r = t} x_{s^1}^1 * \cdots * x_{s^r}^r \right) * B^{-t} \tag{17}$$

For any $x \in X_F^+$ $(p^i(x))_{i \in N_0}$ converges to $TK_l f(x)$ in the sense that there exists an index $i_x \in N$ with the property $TK_l f(x) = TK_l p^{i_x}(x)$ as one sees immediately by putting $l^1 := r*n$.

We obtain by easy computation

$$\bigwedge_{i \in N_0} \bigwedge_{x \in X_F^+} \mathrm{err}^i(x) = \sum_{t=l^i+1}^{r*n} \left(\sum_{s^1 + \cdots + s^r = t} x_{s^1}^1 * \cdots * x_{s^r}^r \right) * B^{-t} \tag{18}$$

and

$$\bigwedge_{i \in N_0} \mathrm{errmax}^i = \sum_{t=l^i+1}^{r*n} \mathrm{card}(t, r, n) * (B-1)^r * B^{-t} \tag{19}$$

and we must choose l^1 great enough to have $\mathrm{errmax}^1 < B^{-l}$. We can now formulate a sufficient condition for $TK_l f(x) = TK_l p^i(x)$ to hold with the abbreviations

$$\bar{B} := \mathrm{entier}(B/2), \qquad \bigwedge_{i \in N_0} em^i := (\mathrm{errmax}^i)_{l+1} + 1 \tag{20}$$

$$\bigwedge_{i \in N_0} \bigwedge_{x \in X_F^+} TK_l f(x) \neq TK_l p^i(x) \Rightarrow (em^i \leqslant \bar{B} \wedge p^i(x)_{l+1} < B_2 \wedge p^i(x)_{l+1} + em^i \geqslant B_2)$$
$$\vee\ (em^i > \bar{B} \wedge (p^i(x)_{l+1} < B_2 \vee p^i(x)_{l+1} \geqslant B_2 \wedge p^i(x)_{l+1} + em^i - B \geqslant B_2)) \tag{21}$$

The proof is done by solving some simple inequalities.

We now describe an algorithm which gives as result $TK_l(x^1 * \cdots * x^r)$ without calculating (if it is not necessary) the exact product $x^1 * \cdots * x^r$.

Algorithm 1.

1. read r, n, l;
   compute errmax^i, $1 \leqslant i \leqslant r * n$ according Eq. (19) for $(l^i)_{i \in N_0}$ with
   $\bigwedge_{i \in N_0} l^i := i;$
   find $a \in N$ minimal with the property errmax$^a < B^{-l}$;

2. read $\tilde{x}^1, \ldots, \tilde{x}^r \in F_n(B)$, $\tilde{x}^j = \text{sign}^j * x^j$, $\text{sign}^j \in \{1, -1\}$, $x^j \geqslant 0$ for $1 \leqslant j \leqslant r$;
   if there exists $j \in \{1, \ldots, r\}$ with $x^j = 0$ then
     begin sign := 1;
     prod := 0;
     goto 5.; end
   $\text{sign} := \prod_{j=1}^{r} \text{sign}^j;$
   $ue := b := 0;$

3. for $i := a(-1)1$ do
     begin
     $\tilde{p}^i := \sum_{s^1 + \cdots + s^r = i} x^1_{s^1} * \cdots * x^r_{s^r} + ue;$
     $p^i := (\tilde{p}^i)_0;$
     $ue := (\tilde{p}^i - p^i)/B;$ end

4. $em := (\text{errmax}^{a+b})_{l+1} + 1;$
   if $(em \leqslant \bar{B} \wedge p^{l+1} < B_2 \wedge p^{l+1} + em \geqslant B_2) \vee$
     $\vee\ (em > \bar{B} \wedge (p^{l+1} < B_2 \vee p^{l+1} \geqslant B_2 \wedge p^{l+1} + em - B \geqslant B_2))$ then
     begin $b := b + 1;$
     $\tilde{p}^{a+b} := \sum_{s^1 + \cdots + s^r = a+b} x^1_{s^1} * \cdots * x^r_{s^r};$
     $p^{a+b} := (\tilde{p}^{a+b})_0;$
     $ue := (\tilde{p}^{a+b} - p^{a+b})/B;$
     for $i := a + b - 1(-1)1$ do while $ue \neq 0$
       begin $\tilde{p}^i := p^i + ue;$
       $p^i := (\tilde{p}^i)_0;$
       $ue := (\tilde{p}^i - p^i)/B;$ end
     goto 4.; end
   $\text{prod} := TK_l(0.p^1, \ldots, p^l p^{l+1});$

5. print sign * prod;
   if there are factors for multiplication then goto 2.;
   STOP

Remarks. (1) Algorithm 1 uses the family $(l^i)_{i \in N_0}$ with $l^0 := 0, l^1$ minimal with the property $\text{errmax}^1 < B^{-l}, \bigwedge_{i \geq 2} l^i := l^1 + i - 1$; the approximation stops as soon as the sufficient condition (21) holds.

(2) Algorithm 1 was simulated on a CDC 3300, the results are shown in the following table:

Table 1

n	r	l	number of examples	S	em^a
5	5	5	80	57%	3
6	4	6	100	52%	10
8	4	8	50	63%	3
8	4	12	20	36%	3
12	4	8	30	87%	3

S is defined by

$$S := \sum_{t = i_x + 1}^{r*n} \text{card}(t, r, n) \Big/ \sum_{t=1}^{r*n} \text{card}(t, r, n) * 100$$

and is a measure for the saving of multiplications of the kind $x^1 * \cdots * x^r$ compared with calculating the exact product. Another advantage of Algorithm 1 is in our opinion the fact that one can choose r, n, l arbitrarily.

(3) In the case $p^i(x)_{l+1} = B_2 - 1 \wedge em^i = 1$ it is eventually better to do a more detailed computation to avoid the computation of the exact product.

(4) Algorithm 1 is independent of a fixed word-length; one must be able to realize digit-multiplication and addition, the maximal value of the sum is given by

$$\text{card}(\text{entier}((n - 1)/2) + 1, r, n) * (B - 1)^r.$$

Example. For the case $r = 2$ we obtain the simple formulas

$$x^1 * x^2 = \sum_{t=2}^{2*n} \left(\sum_{s=1}^{t-1} x_s^1 * x_{t-s}^2 \right) * B^{-t} \tag{15'}$$

$$\text{card}(u, 2, n) = \binom{u-1}{1} - 2 * \binom{u-n-1}{1} \quad \text{for } u \leq 2*n \tag{10'}$$

Let us put

$$B = 10,\ n = 6,\ l = 6;$$

therefore we compute $a = 9$ and we put

$$x^1 = 0.563407, \qquad x^2 = 0.730595.$$

Then we compute with $l^1 = 9$

$$\text{prod}^1 = 0.411622336, \qquad em^1 = 4, \qquad p_7^1 = 3$$

and as $(4 \leqslant 5 \wedge 3 < 5 \wedge 3 + 4 \geqslant 5)$ is true we must compute a second step with $l^2 = 10$:

$$\text{prod}^2 = 0.4116223371, \qquad em^2 = 1, \qquad p_7^1 = 3$$

and as $((1 \leqslant 5 \wedge 3 < 5 \wedge 3 + 1 \geqslant 5) \vee (1 > 5 \wedge (3 < 5 \vee 3 \geqslant 5 \wedge 3 + 1 - 10 \geqslant 5)))$ is false we have

$$TK_6(0.563407 * 0.730595) = 0.411622.$$

Remark. In the above example we have computed 2 steps although the comparation with the exact product shows that already the first step would have given the correct result. The reason for this is on the one hand the factor-independent condition (21) and on the other hand the "bad" values of em^1 and p_{l+1}^1 in the above example.

3.2.2. Floating-Point Numbers

The generalization of the above method to floating-point numbers is obvious:

$$\text{for} \quad x = (x^1, \ldots, x^r) = (m^1 * B^{e^1}, \ldots, m^r * B^{e^r}) \in X_G^+, \quad \bigwedge_{j=1(1)r} m^j = 0.m_1^j \ldots m_n^j$$

we derive from (15)

$$\prod_{j=1}^{r} x^j = \sum_{t=r}^{r*n} \left(\sum_{s^1 + \cdots + s^r = t} m_{s^1}^1 * \cdots * m_{s^r}^r \right) * B^{e^1 + \cdots + e^r - t} \tag{22}$$

and therefore we define for a family of natural numbers $(l^i)_{i \in N_0}$ with the property (16) the approximation by

$$\bigwedge_{x \in X_G^+} p^0(x) : p^0 := 0$$

$$\bigwedge_{i \in N} \bigwedge_{\substack{x \in X_G^+, \\ x = (m^1 * B^{e^1}, \ldots, m^r * B^{e^r})}} p^i(x) := \text{appr}^i(x, p^{i-1}(x)) := p^{i-1}(x)$$

$$+ \sum_{t = l^{i-1}+1}^{l^i} \left(\sum_{s^1 + \cdots + s^r = t} m_{s^1}^1 * \cdots * m_{s^r}^r \right) * B^{e^1 + \cdots + e^r - t} \tag{23}$$

As the approximation error of (23) is based only on the multiplication of the mantissas we define the maximal error by

$$\bigwedge_{i \in N_0} \bigwedge_{e^1, \ldots, e^r \in \{eu, \ldots, eo\}} \text{flerrmax}^i(e^1, \ldots, e^r)$$

$$:= \max \left\{ \text{err}^i(x^1, \ldots, x^r) \,\middle|\, \bigwedge_{j=1(1)r} x^j = m^j * B^{e^j}, \right.$$

$$\left. m^j \in F_n(B), m^j \geqslant 0 \right\} \tag{24}$$

where $\text{err}^i(x)$ is defined by (8) and we obtain with errmax^i of Eq. (19)

$$\bigwedge_{i \in N_0} \bigwedge_{e^1,\ldots,e^r \in \{eu,\ldots,eo\}} \text{flerrmax}^i(e^1,\ldots,e^r) = B^{e^1+\cdots+e^r} * \text{errmax}^i \tag{25}$$

We can formulate the sufficient condition for $TK_l f(x) = TK_l p^i(x)$ to hold in exactly the same way as in (21) when replacing the definition of em^i in (20) by

$$\bigwedge_{i \in N_0} em^i(e^1,\ldots,e^r) := em^i$$

$$:= (\text{errmax}^i)_{e^1+\cdots+e^r+l+1} + 1, \qquad e^1,\ldots,e^r \in \{eu,\ldots,eo\} \tag{26}$$

(we drop the arguments $e^1,\ldots,e^r$ if there is no doubt about the meaning).

Remarks. (1) We did not take into consideration the problem of over- or underflow of exponents when assuming that the addition of exponents is done without error.

(2) We can generalize Algorithm 1 by replacing (17) by (23), and errmax^i resp. (20) by flerrmax^i resp. (26) to floating-point numbers. As this is only a simple transferring of statements we will not formulate the algorithm for floating-point numbers.

4. Conclusion

We developed an algorithm for the "rounding-exact" (see Eqs. (1), (2)) product of fixed-point and floating-point numbers which is based on a well-known formula for the product of polynomials. The main advantages of this algorithm are in our opinion first the fact that

- the basis of the fixed-point resp. floating-point numbers
- the number of digits of the fixed-point numbers resp. of the mantissa of the floating-point numbers
- the number of the factors
- the precision-parameter l

are all free parameters of our algorithm and second the saving of digit-multiplications which is for some examples summarized in Table 1. Although we restricted ourselves to the classical rounding the transferring of the results to other types of roundings can be done by easy computation. The algorithm seems therefore to be well-suited for implementation provided that one wishes to have the rounding-exact product for all factors.

References

[1] Albrecht, R.: Grundlagen einer Theorie gerundeter algebraischer Verknüpfungen in topologischen Vereinen. Computing, Suppl. 1, pp. 1–14. Wien-New York: Springer 1977.

[2] Albrecht, R.: Rundungen und Approximationen in geordneten Mengen. (This volume.)

[3] Bohlender, G.: Produkte und Wurzeln von Gleitkommazahlen. Computing, Suppl. 1, pp. 33–46. Wien-New York: Springer 1977.

[4] Kulisch, U.: Grundlagen des Numerischen Rechnens. Mannheim-Wien-Zürich: Bibliographisches Institut 1976.

[5] Kulisch, U.: Über die beim numerischen Rechnen mit Rechenanlagen auftretenden Räume. Computing, Suppl. 1, pp. 107–120. Wien-New York: Springer 1977.

[6] Lang, S.: Algebra. London: Addison-Wesley 1971.
[7] Pichat, M.: Correction d'une somme en arithmétique à virgule flottante. Numer. Math. **19**, 400 – 406 (1972).
[8] Schmid, H.: Decimal Computation. New York: J. Wiley 1974.
[9] Spaniol, O.: Arithmetik in Rechenanlagen: Logik und Entwurf. Stuttgart: Teubner 1976.
[10] Stummel, F.: Rounding error analysis of numerical algorithms. (This volume.)
[11] Wilkinson, J. H.: Rundungsfehler. Berlin-Heidelberg-New York: Springer 1969.

Dr. W. Oberaigner
Institut für Informatik und
Numerische Mathematik
Universität Innsbruck
Innrain 52
A-6020 Innsbruck
Austria

Computing, Suppl. 2, 131 – 140 (1980)

Unrestricted Algorithms for Generating Elementary Functions*

F. W. J. Olver, College Park, Maryland

Abstract

An "unrestricted" algorithm for generating a mathematical function is a computational algorithm in which the user may demand any accuracy for arguments of any magnitude. Two interesting mathematical problems arise in the construction of such algorithms for the elementary functions. First, an efficient method is needed to determine realistic a priori error bounds that is applicable when the number of arithmetical operations is unbounded. Secondly, the free parameters associated with the algorithm need to be optimized in order to minimize the total computing time. The first problem is solved by application of a recently-developed logarithmic form of interval analysis. The second is solved by asymptotic methods.

1. Introduction and Summary

An unrestricted algorithm for generating a mathematical function is one in which the user may demand any accuracy in the computed value of the function for any value(s) of the argument(s). Reliability is therefore of overriding importance. However, certain free parameters present themselves in the construction of the algorithms as a rule, and an associated problem is to reduce the computing time by optimum choice of these parameters.

We assume at the outset that the basic arithmetic operations of addition, subtraction, multiplication, and division may be carried out to arbitrary precision. The efficient implementation of these procedures depends on the availability of comprehensive software packages, such as those described in [3], [4], [9][1]. The algorithm, however, is independent of such considerations, as well as the choice of programming language.

Several multiprecision algorithms for generating the elementary mathematical functions are already available; see, for example, [1], [2], [8]. However, none are unrestricted algorithms in the sense just described, since they lack strict error analyses. Furthermore, the free parameters that occur are optimized uniformly only with respect to restricted ranges of the argument.

---

* This research was supported by the U.S. Army Research Office, Durham, under Grant DAAG 29-77-G-0003, and the National Science Foundation under Grant MCS 78-02111.

[1] From the present standpoint there is a defect common to these multiprecision packages: only one word length is allocated to the exponent of a floating-point number. To implement an unrestricted algorithm it is just as important to be able to store an arbitrary number of digits in front of the decimal (or binary) point as it is to store an arbitrary number of digits after this point.

There are two reasons for seeking unrestricted algorithms. First, such an algorithm furnishes a constructive numerical definition of the corresponding function. Hence in the construction of restricted algorithms for everyday use, an unrestricted algorithm may be used to generate any of the more familiar forms of approximation, such as polynomials, rational functions, spline functions, or Chebyshev-series expansions. It may also be used in the systematic testing of such algorithms. Secondly, it is believed that in the course of solving analytical and numerical problems that arise in the construction of unrestricted algorithms, new light may be shed upon the procedures of numerical approximation. This belief has already been realized in one respect: a new form of error analysis has been found that is well suited to the determination of *a priori* error bounds with floating-point arithmetic. Existing procedures, such as interval analysis, proved to be too cumbersome for the present purpose.

This paper describes joint work of C. W. Clenshaw of the University of Lancaster and the author. Functions that have been treated successfully so far include the exponential, sine, and cosine functions. For the most part, the paper is confined to the exponential function. In Sect. 2 the new methods of error analysis for floating-point arithmetic operations are described briefly. The method to be used for computing the exponential function is explained in Sect. 3, and the associated errors are analyzed in Sect. 4. Optimization of the free parameters by asymptotic methods is discussed in Sect. 5, and the final form of the algorithm is described in the concluding section, Sect. 6. Further details will be found in [5].

2. Relative Precision

Let a and $\bar{a}$ be any nonzero real or complex numbers. Then $\bar{a}$ is said to be an approximation to *a of relative precision* α if

$$a = \bar{a}e^u, \qquad \text{where} \qquad |u| \leqslant \alpha. \tag{2.01}$$

In symbols, we write

$$a \simeq \bar{a}; \quad \mathrm{rp}(\alpha). \tag{2.02}$$

If the variables are real, then (2.01) implies that $\ln \bar{a}$ approximates $\ln a$ within an interval of length 2α. Similarly, if the variables are complex, then with appropriate choice of branches $\ln \bar{a}$ approximates $\ln a$ within a disk of diameter 2α.

For small values of α (which is the usual situation in practice) and real variables, the definition of relative precision agrees with the conventional definition of relative error α to within $O(\alpha^2)$. However, the new definition enjoys many useful properties that do not apply with the conventional definition. Let us assume that (2.02) holds. Then we have:

Symmetry $$\bar{a} \simeq a; \quad \mathrm{rp}(\alpha). \tag{2.03}$$

Inclusion $$a \simeq \bar{a}; \quad \mathrm{rp}(\delta), \quad \forall \delta > \alpha. \tag{2.04}$$

Powering $$a^p \simeq \bar{a}^p; \quad \mathrm{rp}(|p|\alpha). \tag{2.05}$$

Next, assume also that

$$b \simeq \bar{b}; \quad \mathrm{rp}(\beta).$$

Then we have:

Multiplication $$ab \simeq \bar{a}\bar{b}; \quad \mathrm{rp}(\alpha + \beta). \tag{2.06}$$

Division $$a/b \simeq \bar{a}/\bar{b}; \quad \mathrm{rp}(\alpha + \beta). \tag{2.07}$$

Lastly, assume also that

$$\bar{a} \simeq \bar{\bar{a}}; \quad \mathrm{rp}(\delta).$$

Then we have:

Accumulation $$a \simeq \bar{\bar{a}}; \quad \mathrm{rp}(\alpha + \delta). \tag{2.08}$$

Of the six properties (2.03) to (2.08), only (2.04) is shared with the conventional definition.

Rules for addition and subtraction naturally are more complicated. Suppose first that a and b are real and of the same sign. Then by elementary analysis it follows that

$$a + b \simeq \bar{a} + \bar{b}; \quad \mathrm{rp}\left\{\ln\left(\frac{\bar{a}e^{\alpha} + \bar{b}e^{\beta}}{\bar{a} + \bar{b}}\right)\right\}. \tag{2.09}$$

By use of the well-known inequalities

$$e^{\eta} - 1 \leqslant \eta/(1 - \eta), \quad \forall \eta \in (-\infty, 1); \qquad \ln(1 + \eta) \leqslant \eta, \quad \forall \eta \in (-1, \infty),$$

and the inclusion property (2.04), we may simplify (2.09) to

$$a + b \simeq \bar{a} + \bar{b}; \quad \mathrm{rp}\left\{\frac{\bar{a}\alpha + \bar{b}\beta}{(\bar{a} + \bar{b})(1 - \alpha - \beta)}\right\}, \tag{2.10}$$

provided that $\alpha + \beta < 1$. However, this simplification is not always advantageous. Next, in consequence of the symmetry property (2.03) we have a *dual* form of (2.09), given by

$$a + b \simeq \bar{a} + \bar{b}; \quad \mathrm{rp}\left\{\ln\left(\frac{ae^{\alpha} + be^{\beta}}{a + b}\right)\right\}. \tag{2.11}$$

This dual form is especially useful for calculating *a priori* bounds, because exact analytic or algebraic bounds may be available for the true values a and b, as opposed to purely numerical values for the approximations $\bar{a}$ and $\bar{b}$.

In the general case in which a, $\bar{a}$, b, and $\bar{b}$ are real or complex, we have

$$a + b \simeq \bar{a} + \bar{b}; \quad \mathrm{rp}\{-\ln(1 - \kappa)\}, \tag{2.12}$$

where

$$\kappa = \frac{|\bar{a}|(e^{\alpha} - 1) + |\bar{b}|(e^{\beta} - 1)}{|\bar{a} + \bar{b}|}, \tag{2.13}$$

provided that $\bar{a} + \bar{b} \neq 0$ and $0 \leqslant \kappa < 1$. A special case of this result is the subtraction of positive real numbers. Again, (2.12) and (2.13) can be simplified to get rid of the logarithm and exponentials with a resulting loss of precision that is of the second order of small quantities. There is also a dual form of result obtained by replacing $\bar{a}$ and $\bar{b}$ in (2.13) by a and b, respectively.

Further results pertaining to relative precision will be found in [7].

3. Method for Computation of e^x

The problem that we set ourselves is the computation of e^x to rp(α), where x and α are any given real numbers such that $-\infty < x < \infty$ and $\alpha > 0$. The method that we employ has been used before and comprises three stages. First, the absolute value of x is halved repeatedly in order to produce a reduced argument t; thus

$$|x| = 2^m t = Mt, \tag{3.01}$$

where m is a nonnegative integer and $M = 2^m$. Secondly, an approximation f_0 for e^t is found from the $(n + 1)$st partial sum of its Maclaurin expansion, given by

$$f_0 = 1 + \frac{t}{1!} + \frac{t^2}{2!} + \cdots + \frac{t^n}{n!}. \tag{3.02}$$

Thirdly, the required approximation F to e^x is obtained by successive squaring; thus

$$f_{j+1} = f_j^2, \qquad j = 0, 1, \ldots, m - 1, \tag{3.03}$$

and

$$F = f_m^{\pm 1}, \tag{3.04}$$

the upper or lower sign being taken according as $x \geqslant 0$ or $x < 0$.

Clearly m and n are parameters that are at our disposal in implementing the algorithm. A third parameter governs the number of guarding figures to be retained in the computations. Two conditions govern the choice of the three parameters. First, by hypothesis the stored value $\bar{F}$, say, of F must satisfy

$$e^x \simeq \bar{F}; \quad \text{rp}(\alpha). \tag{3.05}$$

Fulfillment of this condition forms the subject of Sect. 4. Secondly, the total computing time is to be as short as possible. This aspect is discussed in Sect. 5.

4. Error Analysis

Two unavoidable types of error are incurred in the computations described in the previous section. First, there is the *analytic* or *truncation* error caused by taking only a finite value for n in (3.02). We denote this error in rp form by β, so that

$$e^t \simeq f_0; \quad \text{rp}(\beta). \tag{4.01}$$

By using the elementary inequality

$$e^t \leqslant \left(1 + \frac{t}{1!} + \frac{t^2}{2!} + \cdots + \frac{t^n}{n!}\right)\exp\left\{\frac{t^{n+1}}{(n+1)!}\right\}, \qquad t \geqslant 0,$$

we see that we may set

$$\beta = t^{n+1}/(n + 1)!. \tag{4.02}$$

Secondly, there are the *abbreviation* errors introduced by rounding or chopping. For simplicity, we assume that for each prescribed number of word lengths of the multiprecision software all operations (other than input and output) that are accompanied by an abbreviation introduce errors bounded by the same rp. For a given computer the value of this common rp can be assessed by methods similar to those described in [7, §4].

The first task is to assess how much precision is lost during the computation of the truncated Maclaurin expansion (3.02). We denote by γ the rp bound for all abbreviation errors incurred in this process. Since the number of word lengths is at our disposal, γ may be regarded as the explicit form of the third parameter associated with the algorithm (the other two being m and n). We shall assume that (3.02) is evaluated by nested multiplication, that is, we construct the sequence $\{a_j\}$, $j = n, n-1, \ldots, 0$, defined by $a_n = 1$ and

$$a_{j-1} = 1 + (ta_j/j). \tag{4.03}$$

Then

$$f_0 = a_0. \tag{4.04}$$

We denote the stored value of a_j by $\bar{a}_j$ and suppose that

$$a_j \simeq \bar{a}_j; \quad \text{rp}(\delta_j), \tag{4.05}$$

where $\delta_n = 0$ and the remaining δ_j are to be bounded.

The value of t is found from $|x|$ by division by the exact number 2^m. Continuing to add bars to symbols in order to distinguish stored values, we have

$$t \simeq \bar{t}; \quad \text{rp}(\gamma).$$

In forming $\overline{ta_j/j}$ two further abbreviations take place, one with multiplication and one with division; hence

$$ta_j/j \simeq \overline{ta_j/j}; \quad \text{rp}(\delta_j + 3\gamma).$$

The next step is to add unity to $\overline{ta_j/j}$ and abbreviate the result to form $\bar{a}_{j-1}$. Using the dual form (2.11) of the addition rule and comparing the result with (4.05), we perceive that

$$\delta_{j-1} = \gamma + \ln\left\{\frac{1 + (ta_j/j)e^{\delta_j + 3\gamma}}{1 + (ta_j/j)}\right\}.$$

Applying (4.03), we obtain

$$\delta_{j-1} = \gamma + \ln\left\{1 + \frac{ta_j}{ja_{j-1}}(e^{\delta_j + 3\gamma} - 1)\right\} \leqslant \gamma + \ln\left\{1 + \frac{t}{j}(e^{\delta_j + 3\gamma} - 1)\right\}, \tag{4.06}$$

the last step following from the easily-verified inequality $a_{j-1} \geqslant a_j$. From this result and the starting value $\delta_n = 0$ we are able to deduce a realistic bound for δ_0. The first step is to replace (4.06) by a linear inequality. Since

$$\ln\left\{1 + \frac{t}{j}(e^v - 1)\right\} = \frac{t}{j}v + O(v^2), \qquad v \to 0,$$

it is natural to linearize by trying to choose a constant k such that

$$\ln\left\{1 + \frac{t}{j}(e^{\delta_j + 3\gamma} - 1)\right\} \leqslant k\frac{t}{j}(\delta_j + 3\gamma), \quad \forall j,$$

in the expectation that a value of k can be found that is only slightly greater than unity. This approach leads to the following rigorous result, the proof of which will be found in [5, §3]:

Lemma. *Let k be an arbitrary number in the open interval* $(1, \infty)$ *such that*

$$\gamma \leqslant (k-1)/(4ke^{kt}). \tag{4.07}$$

Then

$$\delta_0 \leqslant (4e^{kt} - 3)\gamma. \tag{4.08}$$

In order to apply this result we could specify k in an arbitrary manner, e.g. $k = 1.01$. This leads to bounds that are reasonably satisfactory in practice. However, in the process of optimizing the parameters described in the next section, we select automatically the best possible value of k in all circumstances. It should be noted, in passing, that the right-hand side of (4.08) exceeds γ only slightly when kt is small (which is the practical situation); this confirms the satisfactory nature of the error analysis.

After summation of the truncated Maclaurin expansion, the remaining stage of the algorithm is to square the sum m times. Assuming that the conditions of the lemma are satisfied, we have from (4.01), (4.04), (4.05), and (4.08)

$$e^t \simeq \bar{f}_0; \quad \mathrm{rp}\{\beta + (4e^{kt} - 3)\gamma\}.$$

If the working rp were maintained at γ throughout the squarings, then by using the power rule (2.05) and allowing for the introduction of an abbreviation error of rp(γ) at each step we would obtain

$$e^{|x|} \equiv e^{Mt} \simeq \bar{f}_m; \quad \mathrm{rp}\{M\beta + M(4e^{kt} - 3)\gamma + (M-1)\gamma\}.$$

However, owing to the systematic loss of relative accuracy caused by the squaring process, the overall computing time can be reduced by supposing that one binary figure is discarded after each squaring. In practice this is implemented by discarding one word length every B steps, where B is the number of binary digits in each word. The corresponding result (which is verifiable by induction) is found to be

$$e^{|x|} \simeq \bar{f}_m; \quad \mathrm{rp}\{M\beta + M(4e^{kt} - 3)\gamma + mM\gamma\}.$$

There is one further possible abbreviation error in the calculation; this is incurred in the formation of the reciprocal of $\bar{f}_m$ when x is negative. Assuming this division is carried out to the precision used in forming $\bar{f}_m$, that is, rp($M\gamma$), we may assert that in all cases

$$e^x \simeq \bar{F}; \quad \mathrm{rp}\{M\beta + (4e^{kt} + m - 2)M\gamma\}. \tag{4.09}$$

Hence the stored value $\bar{F}$ of F represents e^x to within rp(α), provided that

$$M\{\beta + (4e^{kt} + m - 2)\gamma\} \leqslant \alpha. \tag{4.10}$$

This is the desired condition. It is valid whenever the conditions of the lemma are satisfied.

A similar algorithm and error analysis can be constructed for the cosine and sine functions. In effect, we replace x by ix and use complex rp analysis. We find that the stored value of e^{ix} represents the true value to within rp(α), provided that

$$M\left[\beta + \left\{7\cosh\left(\frac{kt}{\sqrt{1-\frac{1}{2}t^2}}\right) + 2m - 4\right\}\gamma\right] \leqslant \alpha, \tag{4.11}$$

where the quanties m, M, t, k, β, and γ essentially have the same meaning as before. In this case we also have to restrict $t < \sqrt{2}$.

5. Optimization

In order to implement the algorithm discussed in Sects. 3 and 4 in an efficient manner, the parameters m, n, and γ should be chosen in such a way that the total computing time T is as short as possible. First, we need a measure of T.

A convenient basic unit of time is that required to multiply two multiprecision numbers. If the number of word lengths is W, then this unit is proportional to W^{ω}, where ω is a constant whose value depends on the software being used. In any case, however, we may assume that $1 < \omega \leqslant 2$; compare [6, pp. 258 et seq.].

In the evaluation of (3.02) all calculations are carried out to rp(γ). Hence W is proportional to c, where

$$c = \log_2(1/\gamma). \tag{5.01}$$

For each of the n steps of the recurrence process (4.03) there is one full multiplication, one division by an integer, and one addition of an integer. From the standpoint of timing we shall suppose that this is equivalent to λ full multiplications, where $\lambda \equiv \lambda(c)$ is a function of c that depends on the software, and will need to be specified by the user. Clearly $\lambda(c) > 1$, and in practice $\lambda(c)$ is likely to be close to unity when c is large. For our purposes, however, we need only the following assumptions:

$$\lambda(c) = O(1), \quad \lambda'(c) = o(1/c), \quad c \to \infty.$$

These conditions are satisfied by all known multiprecision packages.

Whatever the actual form of $\lambda(c)$, the computing time of the whole recurrence process is proportional to $nc^{\omega}\lambda(c)$. Next, as noted at the end of Sect. 4, the $(j+1)$st step of the squaring process (3.03) is carried out to rp($2^j\gamma$), and there are m such steps altogether. Hence our measure of the total computing time is defined to be

$$T \equiv nc^{\omega}\lambda(c) + \sum_{j=0}^{m-1} (c-j)^{\omega}. \tag{5.02}$$

The relevant relations that connect the variables are given by

$$M = 2^m, \quad |x| = Mt, \quad t^{n+1}/(n+1)! = \beta, \tag{5.03}$$

the inequalities (4.07) and (4.10), and Eq. (5.01). For any chosen pair of values of m

and n, the largest permissible value of γ (and hence the smallest possible value of T) will be found by solving the set of six equations given by (5.01), (5.03), and the extreme forms of (4.07) and (4.10):

$$\gamma = (k-1)/(4ke^{kt}), \qquad \alpha = M\{\beta + (4e^{kt} + m - 2)\gamma\}. \tag{5.04}$$

These six equations contain eight unknowns, namely m, M, t, n, β, γ, c, and k. In order to minimize the expression (5.02) in a reasonably simple manner, we treat all unknowns, including m, M, and n, as continuous variables. If we regard two of the variables, for example, n and γ, as independent and the rest as depending on them, then T is minimized when

$$\partial T/\partial n = \partial T/\partial \gamma = 0, \tag{5.05}$$

these two equations completing the set of eight required to determine the eight unknowns.

The exact solution of these equations is impossible, and numerical methods would not help owing to the lack of an adequate first approximation. Instead, we solve them asymptotically. We introduce the variable

$$\xi \equiv \log_2|x| + \log_2(1/\alpha), \tag{5.06}$$

and seek asymptotic approximations to m, n, and γ for large ξ, since it is only when $|x|$ is large, or α is small, or both events occur, that it becomes important to minimize T.

The required asymptotic approximations for the parameters may be verified to be

$$m = \log_2|x| + \{\xi\lambda(\xi)\}^{1/2} - \tfrac{1}{2}\log_2\xi + O(1), \tag{5.07}$$

$$n = \{\xi/\lambda(\xi)\}^{1/2}\{1 + O(\xi^{-1/2})\}, \tag{5.08}$$

$$\log_2(1/\gamma) = \xi + \{\xi\lambda(\xi)\}^{1/2} + O(\log_2\xi), \tag{5.09}$$

as $\xi \to \infty$, uniformly with respect to unbounded values of both $|x|$ and $1/\alpha$. These results are proved in [5, §4 and Appendix A]. Actually only one of the estimates (5.07), (5.08), and (5.09) is needed, because in order to guarantee final accuracy the other two quantities have to be chosen to satisfy strictly the inequalities (4.07) and (4.10). The estimate that we actually use is (5.07).

6. Implementation

In order to apply the estimate (5.07) we need to be able to calculate logarithms to base 2 for arguments of any size. Fortunately, this particular problem is solvable in an easy manner, because we need calculate the logarithms in (5.07) only to within the tolerance of the uniform error term. Let a be any nonzero real number, and represent it in normalized form

$$a = 2^{d(a)} \times \mu(a),$$

where $d(a)$ is the exponent of a in base 2, and $\frac{1}{2} \leqslant |\mu(a)| < 1$. Then

$$\log_2|a| = d(a) - \vartheta(a),$$

where $0 < \vartheta(a) \leqslant 1$. In consequence $d(a)$, which is easily available, is an adequate approximation to $\log_2|a|$. Therefore, for example, we replace the definition (5.06) of ξ by

$$\xi = d(x) + d(1/\alpha) - 1. \tag{6.01}$$

We conclude by describing briefly the six steps that comprise the final algorithm. Further details will be found in [5, §5 and Appendix B].

Evaluation of Parameters

(1) Calculate ξ from (6.01). If $\xi \leqslant 0$, take $m = 0$. If $\xi > 0$, compute the largest integer, l, say, such that $l^2 \leqslant 4\xi\lambda(\xi)$, and then choose m to be the greater of zero and

$$d(x) + \operatorname{int}[\tfrac{1}{2}\{l - d(\xi) + 3\}], \tag{6.02}$$

where int denotes, as usual, the integer part of the succeeding quantity in square brackets. [The expression (6.02) is obtained from the asymptotic estimate (5.07).]

(2) Take n to be the smallest nonnegative integer that satisfies

$$n\{2m - 5 - 2d(x)\} + (2n + 3)d(n + 1) \geqslant 2\{d(x) + d(1/\alpha) + d(m + 4)\} + 4. \tag{6.03}$$

[This guarantees that sufficient terms are taken in the Maclaurin expansion of e^t.] The value of n is found by trial and error, an initial guess being obtainable from (5.08).

(3) Set

$$\log_2(1/\gamma) = m + 1 + d(1/\alpha) + d(m + 4). \tag{6.04}$$

This determines the initial working precision. Suppose that C denotes the total of number of bits that corresponds to $\mathrm{rp}(\gamma)$ for the particular computer being used. Then the initial number of word lengths in the multiprecision software is taken to be W, where W is the integer that satisfies

$$C \leqslant WB < B + C,$$

and B again denotes the number of bits in each word.

[Formula (6.04) ensures that the condition (4.10) is satisfied. It should be noted that the terms $m + d(1/\alpha)$ represent the "obvious" minimum initial accuracy that must be used in order to realize a final rp of α after m squarings. The remaining terms $1 + d(m + 4)$ represent guarding bits, the number of which grows only logarithmically with m.]

Computation of e^x

(4) Compute $t = |x|/2^m$ and evaluate the sum $\bar{f}_0$ of the first $n + 1$ terms of the Maclaurin expansion of e^t by nested multiplication.

(5) Compute $\bar{f}_m$ from $\bar{f}_0$ by m successive squarings. After $B + C - WB$ squarings have been completed, the number of word lengths in the software is reduced to $W - 1$; thereafter one word length is discarded after every B squarings.

(6) Compute $\bar{F}$ from (3.04) without any further change of wordlength. This is the required value of e^x, correct to rp(α).

References

[1] Brent, R. P.: Fast multiple-precision evaluation of elementary functions. J. Assoc. Computing Machinery **23**, 242–251 (1976).

[2] Brent, R. P.: Multiple-precision zero-finding methods and the complexity of elementary function evaluation. In: Analytic Computational Complexity (Traub, J. F., ed.), pp. 151–176. New York: Academic Press 1976.

[3] Brent, R. P.: A Fortran multiple-precision arithmetic package. ACM Trans. Math. Software **4**, 57–70 (1978).

[4] Brent, R. P.: Algorithm 524. MP, A Fortran multiple-precision arithmetic package. ACM Trans. Math. Software **4**, 71–81 (1978).

[5] Clenshaw, C. W., Olver, F. W. J.: An unrestricted algorithm for the exponential function. SIAM J. Numer. Anal. (In press.)

[6] Knuth, D. E.: The Art of Computer Programming, 2: Seminumerical Algorithms. Reading, Mass.: Addison-Wesley 1969.

[7] Olver, F. W. J.: An new approach to error arithmetic. SIAM J. Numer. Anal. **15**, 368–393 (1978).

[8] Schonfelder, J. L., Keech, M. S.: Arbitrary precision elementary functions in ALGOL 68. (To be published.)

[9] Wyatt, W. T., jr., Lozier, D. W., Orser, D. J.: A portable extended precision arithmetic package and library with Fortran precompiler. ACM Trans. Math. Software **2**, 209–231 (1976).

Dr. F. W. J. Olver
Institute for Physical Science and Technology
University of Maryland
College Park, MD 20742, U.S.A.

Computing, Suppl. 2, 141 – 156 (1980)

Applications of Software for Automatic Differentiation in Numerical Computation*

L. B. Rall, Madison, Wisconsin

Abstract

This paper describes construction and applications of software for automatic differentiation. Examples from actual experience are cited to illustrate the theory.

1. Introduction

At the foundations of numerical computation lie the means by which calculations are actually done. Today, this means not only the computing machines (hardware), but also the programs (software) available to the user. This paper describes software for differentiation of functions and some of its applications. Differentiation is somewhat similar to washing dishes, a dull, uninteresting task best left to a machine. Two methods are given for automatic differentiation, and their use in various programs for applications is described. As the discussion is limited to actual experience at the Mathematics Research Center over the past 15 years, no mention is made of much fine software which has doubtless been developed elsewhere.

At the end of the paper, some conclusions concerning the past usefulness of automatic differentiation will be given, together with some predictions of possible future trends.

2. Differentiation by List Processing

An approach to formula translation and differentiation which is conceptually simple is provided by a list processing technique. Rather than use a general purpose list processing language, however, it was considered more appropriate to develop software for the specific purpose of evaluating and differentiating formulas. This technique, described in the book by Rall [35, pp. 155 – 160], forms the basis of the program written in 1965 by Reiter [40] and later extended by Gray and Reiter [15] and Wertz [45, 46], the latter assisted by T. Ladner.

The principles are best illustrated by an example [35, pp. 155 – 156]. Suppose that one wishes to evaluate the function

$$f(x, y) = (xy + \sin x + 4)(3y^2 + 6) \tag{2.1}$$

* Research sponsored by the United States Army under Contract No. DAAG29-75-C-0024.

and some of its derivatives. The first step is to decode the formula

$$F = (X*Y + SIN(X) + 4)*(3*Y**2 + 6) \tag{2.2}$$

corresponding to (2.1) into a sequence of arithmetic operations and calls to subroutines. This gives the *code list*

$$\begin{aligned} T1 &= X*Y \\ T2 &= SIN(X) \\ T3 &= T1 + T2 \\ T4 &= T3 + 4 \\ T5 &= Y**2 \\ T6 &= 3*T5 \\ T7 &= T6 + 6 \\ F &= T4*T7. \end{aligned} \tag{2.3}$$

Note that the code list is itself a sequence of statements in the same language in which (2.2) is written, and hence can be translated by the same compiler into machine language. Now, to differentiate (2.3), the program refers to a dictionary containing the derivatives of the arithmetic operations and subroutines called. For example, the entry corresponding to

$$U2 = SIN(U1) \tag{2.4}$$

would be

$$\begin{aligned} V1 &= COS(U1) \\ DU2 &= V1*DU1 \end{aligned} \tag{2.5}$$

and so forth. Applying this procedure to (2.3), one obtains the code list for the differential of F,

$$\begin{aligned} V1 &= X*DY \\ V2 &= DX*Y \\ DT1 &= V1 + V2 \\ V3 &= COS(X) \\ DT2 &= V3*DX \\ DT3 &= DT1 + DT2 \\ DT4 &= DT3 \\ V4 &= Y**1 \\ V5 &= 2*V4 \\ DT5 &= V5*DY \\ DT6 &= 3*DT5 \\ DT7 &= DT6 \\ V6 &= T4*DT7 \\ V7 &= DT4* \\ DF &= V6 + V7 \end{aligned} \tag{2.6}$$

In order to obtain the code list for $\partial f/\partial x$, for example, one sets DX = 1, DY = 0 in (2.6) and obtains, after elimination of trivialities (multiplications by 1 or 0, exponentiation to the first power, addition of 0), the list

$$
\begin{aligned}
\text{DXT1} &= \text{Y} \\
\text{DXT2} &= \text{COS(X)} \\
\text{DXT4} &= \text{DXT1} + \text{DXT2} \\
\text{DXF} &= \text{DXT4} * \text{T7}
\end{aligned}
\tag{2.7}
$$

In this case, the list for the derivative DXF is simpler than the list for F, due to the polynomial terms in $f(x, y)$. For nonpolynomial functions, one would expect

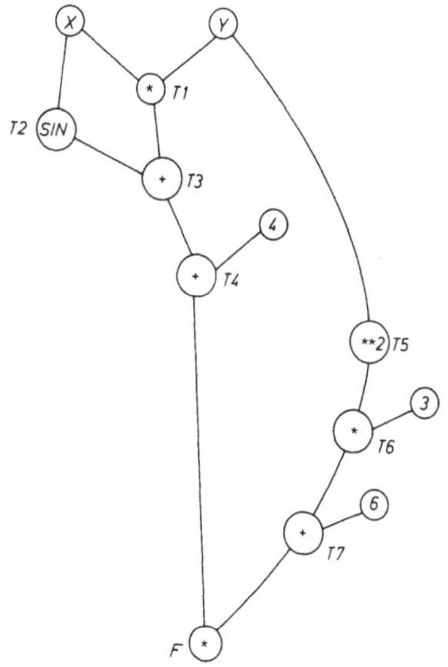

Fig. 2.1. The graph of the calculation of $f(x, y)$

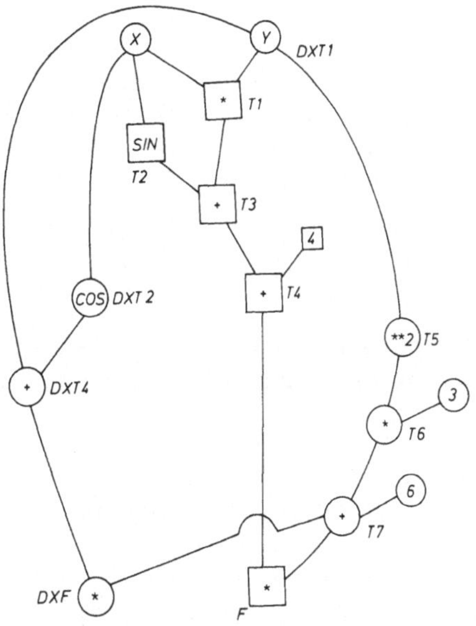

Fig. 2.2. The graph of the calculation of $f(x, y)$ and $f_x(x, y)$

differentiation to yield more complex lists; however, polynomials are not at all uncommon in the modelling and analysis of nonlinear problems.

The most important feature of differentiation by list processing is that the output (2.7) is a list of exactly the same form as the input list (2.3), and can itself be differentiated with respect to any variable entering into it, simply by invoking the same processor. Also, if one needs only some high derivative of a function, the lists for the intermediate derivatives can be discarded after they are processed, or at least it is not necessary to compile them into machine language. This can result in a considerable saving of computer time and storage locations.

It is instructive to view the procedure described above in terms of the Kantorovich graph [17] which is equivalent to the code list. Fig. 2.1 shows the graph corresponding to the list (2.3), Fig. 2.2 the graph for computing both F and its partial derivative DXF with respect to X. To obtain a graph for the computation of DXF only, the nodes indicated by squares in Fig. 2.2 are unnecessary, and can be eliminated.

3. Automatic Generation of Taylor Coefficients

This is the approach adopted by Moore et al. in 1964 [30] in connection with the solution of ordinary differential equations by series, and was implemented by Reiter in 1965 [38] and 1967 [41] as an independent program. Recursive generation of Taylor coefficients is also incorporated in the more general program differentiation software developed by Kedem [18], a system which is also capable of supervising the list processing approach. To illustrate this method, the notation

$$f_0 = f(x_0), \qquad f_\tau = \frac{1}{\tau!} f^{(\tau)}(x_0), \qquad \tau = 1, 2, \ldots, j, \tag{3.1}$$

will be used for the first $j + 1$ Taylor coefficients of the function $f(x)$ at $x = x_0$. (In case f depends on several variables, the derivatives in (3.1) are understood to be partial with respect to the variable of interest.) The recursion relations for generating the successive Taylor coefficients for the arithmetic operations and various functions are well known [27, pp. 107–118; 19]. For example,

$$(fg)_j = \sum_{\tau=0}^{j} f_\tau g_{j-\tau}, \qquad j = 0, 1, 2, \ldots . \tag{3.2}$$

These relationships can be used to process the code list for the evaluation of a function in the same way as the formulas for differentials were used in the previous section.

To interpret this method in terms of the Kantorovich graph, suppose that instead of simply values of terms, a vector consisting of the first $j + 1$ Taylor coefficients is received by each node along the edges coming into it from above. This vector is then processed at the node by invoking the corresponding recurrence relations, and the resulting vector is then transmitted to the lower nodes connected to it. Using the graph in Fig. 2.1, for example, one obtains the Taylor coefficients (3.1) by setting

$$X = (x_0, 1, 0, \ldots, 0),$$
$$Y = (y_0, 0, 0, \ldots, 0). \quad (3.3)$$

At the node labelled T3, one would compute

$$\text{T3(TAU)} = \text{T1(TAU)} + \text{T2(TAU)}, \text{ TAU} = 0(1)\text{J}, \quad (3.4)$$

and so on. The Taylor coefficients of f considered as a function of y may be obtained in the same way.

This method of differentiation has the advantage that only the evaluation list (2.3) is needed. It is not possible to discard intermediate values when only higher derivatives are wanted, and the list processing approach is better adaptable to the computation of mixed partial derivatives. However, if Taylor series expansions with respect to one variable or only first partial derivatives are required, then the method described in this section is highly effective.

It is one thing to describe methods for automatic differentiation in a general way as done here, but quite another to produce quality software for this purpose, as has been done by Reiter, Gray, Wertz, and others cited in this paper and elsewhere. Of course, although this software has performed satisfactorily for a number of years, there is always the possibility of improvement by making use of advances in computer science.

4. Some Traditional Applications of Automatic Differentiation

A number of possibilities for the use of software for automatic differentiation come to mind immediately. A few will be sketched here.

a) Solution of Differential Equations by Series Methods

The software developed by Moore et al. in 1964 [30] and Braun and Moore in 1967 [3] is based on the methods for Taylor series solution of ordinary differential equations described in the papers [25, 26] and books [27, 28] by Moore. Although these programs were designed to be used with interval arithmetic to provide automatic error estimates, they can be executed in ordinary real (floating-point) arithmetic to obtain results which are competitive with other methods of the same order. In particular, Moore [28] refutes the statement made in standard texts on numerical analysis [4, p. 214; 5, p. 330] that "the necessity of calculating the higher derivatives makes Taylor's algorithm completely unsuitable on high-speed computers for general integration purposes".

The related technique of the Lie-series method for the numerical solution of differential equations also makes use of the automatic generation of Taylor coefficients, as described by Knapp and Wanner [19]. The original program [20], written in 1968, required the user to write the sequence of subroutine calls corresponding to a code list of the form (2.3) for the integrand. In 1969, automatic generation of the code list was added by T. Szymanski, under the direction of Julia Gray, to obtain an improved program [21]. Once again, the result is apparently competitive with Runge-Kutta methods of the same order [19, 37].

b) Solution of Nonlinear Systems of Equations

A popular and effective method for the solution of systems of n nonlinear equations in n unknowns,

$$f_i(x_1, x_2, \ldots, x_n) = 0, \qquad i = 1, 2, \ldots, n, \tag{4.1}$$

is the Newton-Kantorovich algorithm [16, 35]. However, to implement this method in such a way as to obtain quadratic convergence, one must calculate the n^2 partial derivatives

$$f'_{ij} = \frac{\partial f_i}{\partial x_j}, \qquad i, j = 1, 2, \ldots, n, \tag{4.2}$$

to obtain the $n \times n$ Jacobian matrix

$$J = f'(x) = (f'_{ij}). \tag{4.3}$$

This is done by automatic differentiation in the 1967 program of Gray and Rall [10] and the 1972 version prepared by Kuba and Rall [23] and described in [11, 35]. Higher order methods, such as Chebyshev's method and the inversion of power series would require computation of the *Hessian operator*

$$H = f''(x) = \left(\frac{\partial^2 f_i}{\partial x_j \, \partial x_k}\right) \tag{4.4}$$

and perhaps higher derivatives [35]. The automatic formation of (4.4) is implemented in the programs [10, 23] in connection with verification of existence of solutions of (4.1) and error estimation.

Another approach to the solution of the system (4.1) using derivatives is a continuation or homotopy method based on the conversion of (4.1) into a system of differential equations for a curve connecting an initial approximation x_0 for which $f(x_0) = y$ to a solution x^*. Such a system is, using vector notation

$$J \frac{dx}{dt} = -y, \tag{4.5}$$

with the initial conditions $x(0) = x_0$. Integrating (4.5) by some curve-following technique [24] from $t = 0$ to $t = 1$ will give an approximation to $x(1) = x^*$. The system (4.5) can be constructed using software for differentiation.

c) Optimization of Nonlinear Functionals

A problem arising frequently in the applications of mathematics is to *optimize* (maximize or minimize) the value of a nonlinear functional

$$g = g(x_1, x_2, \ldots, x_n). \tag{4.6}$$

With sufficient smoothness, this can be converted into the problem of solving the system (4.1) by taking the *gradient*

$$\nabla g = \left(\frac{\partial g}{\partial x_1}, \frac{\partial g}{\partial x_2}, \ldots, \frac{\partial g}{\partial x_n}\right) \tag{4.7}$$

and setting $\nabla g = 0$, which is (4.1) with $f_i(x_1, x_2, \ldots, x_n) = \partial g/\partial x_i$, $i = 1, 2, \ldots, n$. If one wants to solve this system by a method of Newton type, then the Hessian matrix $(\partial^2 g/\partial x_i\, \partial x_j)$ is required, which is simply the Jacobian (4.3). In problems of this type, which are called *unconstrained* optimization problems, the formation of the gradient and the Hessian can be automated by the use of software for differentiation.

Another type of optimization problem requires the satisfaction of conditions

$$h_\tau(x_1, x_2, \ldots, x_n) = 0, \qquad \tau = 1, 2, \ldots, k, \tag{4.8}$$

which are called *constraints*. Given enough smoothness, the *constrained* optimization problem can be reduced to the solution of a system of equations of the form (4.1) by the introduction of new unknowns $\lambda_1, \lambda_2, \ldots, \lambda_k$, called *Lagrange multipliers*. One obtains the nonlinear equations

$$\frac{\partial g}{\partial x_i} + \sum_{\tau=1}^{k} \lambda_\tau \frac{\partial h_\tau}{\partial x_i} = 0, \qquad i = 1, 2, \ldots, n, \tag{4.9}$$

which, together with (4.8), form a system of order $n + k$ to solve. The equations (4.9) can be constructed by automatic differentiation as in the unconstrained case, and, if desired, the entire system can itself be differentiated to implement its solution by a method of Newton type.

5. Analysis of Roundoff Error Using Differentiation

A nontraditional application of software for automatic differentiation in numerical analysis would be the implementation of the method of Bauer [1] for the study of the propagation of roundoff error. This technique is based on the use of *relative* (logarithmic) *differentials*,

$$\rho f = df/f, \qquad \rho x = dx/x, \tag{5.1}$$

and the corresponding *relative derivatives*

$$\rho f/\rho x = xf'/f, \tag{5.2}$$

and the code list or Kantorovich graph for f. The quantities needed for the analysis of roundoff error in the calculation of the function defined by (2.2) are available immediately from the code lists (2.3) and (2.4), for example,

$$\mathrm{RHOT5} = \mathrm{DT5/T5}. \tag{5.3}$$

Setting DX = X*RHOX, DY = Y*RHOY would lead to a code list expressed entirely in terms of relative differentials. According to Bauer [1], a rounding error in the calculation of say, T4, is *harmless* if the *input condition number*

$$|\mathrm{RHOT4/RHOX}| \geqslant 1, \tag{5.4}$$

or the *output condition number*

$$|\mathrm{RHOF/RHOT4}| \leqslant 1. \tag{5.5}$$

Here, F is being considered to be a function of X only, with Y fixed. Similar conditions can be formulated for functions of several variables [1]. The significance of (5.4) is that a roundoff error in T4 does not exceed the result of a corresponding error in X, while (5.5) means that an error in T4 will not affect the final value of F by more than the same amount. An application of these ideas would be to examine final or intermediate condition numbers of several alternative graphs (or lists) for calculating the same function in order to pick one in which roundoff error is the least malignant. In order to automate this process, one can use the existing software or, perhaps better, write a list processor which produces relative rather than ordinary differentials. For example, for

$$U3 = U1*U2, \tag{5.6}$$

the dictionary entry for the relative differential would be

$$RHOU3 = RHOU1 + RHOU2, \tag{5.7}$$

and so on.

Another direct application of automatic differentiation to roundoff error analysis would be to automate the linearization technique due to Stummel [44].

6. Computational Verification of Existence of Solutions of Systems of Equations Using Interval Analysis and Automatic Differentiation

Interval analysis [27, 28] provides a method for obtaining bounds for the ranges of computable functions, including perturbations due to roundoff error in the actual computations. Software for computing with interval arithmetic was developed side-by-side with differentiation software by Reiter [39, 42], starting in 1965. A modern package has been prepared by Yohe [47]. Unlike the programs for differentiation, which can be written in a high level language, interval software was originally highly machine dependent, including input and output [2]. Using advances in computer science, Crary and Ladner [9], Crary [6, 7, 8], and Yohe [48, 49] have been able to reduce this dependence to a minimum, at the cost of speed.

For the present purposes, it is adequate to know that one can extend numerical computations from ordinary numbers x to closed intervals $X = [a, b]$ (which include numbers if $a = b$). Execution in interval arithmetic of a program for the calculation of a function $f(x)$ will result in an *interval extension* $F(X)$ of $f(x)$ such that

$$f(x) \in F(X), \qquad x \in X. \tag{6.1}$$

Thus, for $|X| = |[a, b]| = \max\{|a|, |b|\}$, one will have

$$|f(x)| \leqslant |F(X)|, \qquad x \in X. \tag{6.2}$$

These results extend to finite dimensional spaces, using the maximum absolute component norm for vectors, and the corresponding norm for matrices. In R^n_∞, the closed ball with center y and radius ρ can be represented as an interval,

$$\bar{U}(y, \rho) = \{x : \|x - y\|_\infty \leqslant \rho\} = [y - \rho e, y + \rho e], \tag{6.3}$$

where $e = (1, 1, \ldots, 1)$.

An immediate application of these techniques is to automate some theorems from classical and interval analysis on the existence of solutions of systems of equations (4.1) and to obtain error bounds.

a) The Contraction Mapping Theorem

Here, one transforms the system (4.1), written in vector form as $f(x) = 0$, as a fixed point problem

$$x = \phi(x) \tag{6.4}$$

for some operator ϕ, for example, to use an iteration method

$$x_{n+1} = \phi(x_n), \qquad n = 0, 1, 2, \ldots, \tag{6.5}$$

starting from some initial approximation $x_0 = y$. If ϕ is Lipschitz continuous in some sufficiently large ball $\bar{U}(y, \rho)$, that is,

$$\|\phi(x) - \phi(z)\| \leqslant \alpha\|x - z\|, \qquad x, z \in \bar{U}(y, \rho), \tag{6.6}$$

then it follows from the theorem of Banach [35, pp. 64–74] that the sequence generated by (6.5) converges to a solution $x^* \in \bar{U}(y, \rho)$, provided

$$\rho \geqslant \|x_1 - x_0\|/(1 - \alpha). \tag{6.7}$$

In order to obtain the Lipschitz constant α, automatic differentiation may be used to compile a program to evaluate the $n \times n$ matrix $\phi'(x) = (\partial\phi_i/\partial x_j)$. Using interval arithmetic, the same program will compute the extension $\Phi'(X)$ on the interval $X = \bar{U}(y, \rho)$ defined by (6.3). If then

$$\alpha = |\Phi'(X)| < 1 \tag{6.8}$$

and (6.7) holds, then the hypotheses of the contraction mapping theorem are satisfied. This computation, if successful, also yields the error bound $\|y - x^*\| \leqslant \rho$ for y as an approximate solution.

b) The Kantorovich Theorem on the Convergence of Newton's Method

Newton's method, as applied to the system (4.1), may be written

$$x_{n+1} = x_n - [f'(x_n)]^{-1}f(x_n), \qquad n = 0, 1, 2, \ldots, \qquad x_0 = y, \tag{6.9}$$

in vector form. The theorem of Kantorovich [16] requires the numbers $B_0 \geqslant \|[f'(y)]^{-1}\|$, $\eta_0 \geqslant \|x_1 - x_0\|$, both obtainable rigorously by interval computation, and a Lipschitz constant K for f' on some sufficiently large ball $X = \bar{U}(y, \rho)$. The programs described in [10, 11, 23] use automatic differentiation to obtain the bilinear operator $f''(x)$ defined by (4.4). Evaluation by interval arithmetic gives the extension $F''(X)$ and the desired Lipschitz constant

$$K = |F''(X)|. \tag{6.10}$$

Now if

$$h_0 = B_0\eta_0 K \leqslant \frac{1}{2}, \qquad \rho \geqslant \frac{1 - \sqrt{1 - 2h_0}}{h_0}\eta_0, \tag{6.11}$$

then the Kantorovich theorem guarantees the existence of x^* such that $f(x^*) = 0$ in X, the convergence of the sequence generated by (6.9) to x^*, and provides the error bound $\|y - x^*\| \leqslant \rho$.

c) *An Interval Newton's Method*

One may obtain an interval version of Newton's method by setting

$$X_{n+1} = X_n \cap \{m(X_n) - [F'(X_n)]^{-1}F(m(X_n))\}, \tag{6.12}$$

$n = 0, 1, 2, \ldots$, where $m(X_n)$ is the midpoint of the interval X_n, as described by Nickel [31] and Moore [27]. Here, X_0 can be an arbitrary interval, not necessarily a ball, and the condition $X_{n+1} \subset X_n$ implies the existence of a solution $x^* \in X_n$. Once again, automatic differentiation and interval arithmetic are used to obtain $F'(X_n)$, which is then inverted by interval methods. Inversion of interval matrices is a tedious process, however, so (6.12) is of little practical importance. This computation is available as an option in the program of Kuba and Rall [23].

d) *Moore's Theorem*

This is an interval theorem, based on the *Krawczyk transformation*

$$K(X) = y - Yf(y) + \{I - YF'(X)\}(X - y) \tag{6.13}$$

of an arbitrary interval X [22]. In (6.13), $y \in X$ and Y is an arbitrary, nonsingular real (not interval) matrix. Once again, $K(X)$ is computed by automatic differentiation and interval arithmetic, and has been implemented as an option in the program [23] by Julia Gray. If $K(X) \subset X$, then Moore's theorem [29] guarantees the existence of a solution $x^* \in X$, which, as in the previous method, also yields error bounds for $x^* - m(X)$ or any other approximate solution. Rall [36] has shown that the Kantorovich theorem has little theoretical advantage over Moore's theorem, while the latter is much less costly to apply. Hence, the elaborate implementation of the calculation of (6.10) in the programs [11, 23] can be discarded.

It should be pointed out that the purpose of the methods and programs described in this section is not to solve equations, but rather to guarantee existence of solutions and give rigorous error bounds for approximate solutions. This type of computation is expensive, but may be justifiable in certain applications, such as the design of critical components of aircraft.

7. Automation of Estimation of Error of Numerical Integration and Other Techniques of Classical Numerical Analysis

An ordinary applications of numerical analysis, error estimation is a tedious chore which should be mechanized as much as possible. As errors in data, roundoff error, and truncation error contaminate almost every actual computation, some analysis of their effects is required in order to judge the reliability of the results obtained. Interval arithmetic is suitable for keeping track of data and roundoff errors automatically. If expressions for the truncation error are known in terms of derivatives, as in classical numerical analysis, then automatic differentiation can be used in connection with interval arithmetic to obtain bounds for truncation error. All of this is illustrated aptly by the work of Moore on ordinary differential equations [3, 25, 26, 27, 28, 30], and by the following examples from ordinary numerical analysis.

The processes of interpolation and numerical integration and differentiation can be regarded as coming from some linear functional Λ applied to the function f under consideration. Thus, one writes

$$\Lambda f = \sum_{i=1}^{n} \alpha_i f(x_i) + \left(\frac{1}{n}\right)^p t(\xi), \qquad \xi \in [a, b], \tag{7.1}$$

where the *remainder term* $t(\xi)$ is some linear combination of derivatives of f at ξ, for example,

$$t(\xi) = C_p f^{(p-1)}(\xi), \tag{7.2}$$

where the constant C_p is known. The linear combination of values of f,

$$Rf = \sum_{i=1}^{n} \alpha_i f(x_i) \tag{7.3}$$

is usually called the *rule* of numerical interpolation, integration, or differentiation, etc. Taking an interval extension of (7.1) gives, for $X = [a, b]$,

$$\Lambda f \in \sum_{i=1}^{n} A_i F(x_i) + \overline{\left(\frac{1}{n}\right)^p} T(X), \tag{7.4}$$

where the interval on the right can be computed automatically by use of the software described in this paper. Thus, this furnishes a complete error analysis of the use of the rule Rf given by (7.3) in order to calculate the functional Λf. This type of analysis is carried out in the programs of Gray and Rall [12, 13, 14] for bounding the error of various rules for the numerical integration of functions of a single variable. For example, for Simpson's rule, the form of (7.4) is

$$\int_a^b f(x)\,dx \in \frac{w[a,b]}{6}\left[Y(A) + 4Y\left(\frac{A+B}{2}\right) + Y(B)\right] - \frac{(w[a,b])^5}{2880} Y^{iv}([a,b]) \tag{7.5}$$

where $w[a, b] = b - a$ is the width of the interval $[a, b]$.

Although the speed of the computer blurs the distinction between *a priori* and *a posteriori* error estimations, an expression of the form can furnish either. For example, (7.4) can be computed for a small value of n, and then its width can be extrapolated to larger values of n, as the width of $T(X)$ from the original computation will always be an upper bound for $T(X_n)$, $X_n \subset X$. This method for optimization of time or accuracy in numerical integration is implemented in the program [13]. *A posteriori* error estimates, of course, are always available directly from the computation of (7.4). Once it has been established that $\Lambda f \in [c, d]$, one may take as an estimate for Λf one of the numbers

$$m[a,b] = \frac{a+b}{2}, \qquad h[a,b] = \frac{2ab}{a+b}, \qquad g[a,b] = \sqrt{ab}, \tag{7.6}$$

which minimize the absolute error, relative error, or maximizes the relative precision in the sense of Olver [32], respectively. Interval results can be intersected, if more than one is available, to obtain a possible improvement [12, 13].

As an example of the application of this program, the region to the left and below of the staircase in Fig. 7.1 shows where 5-place accuracy for the calculation of the elliptic integral of first kind

$$F(k, \theta) = \int_0^{\varphi} (1 - k^2 \sin^2\theta)^{-1/2}\, d\theta \tag{7.7}$$

can be guaranteed, using Simpson's rule with 7 nodes (three times on $[0, \varphi]$). This guarantee includes roundoff on the UNIVAC 1110 in single precision (to about 9 decimal digits), and should be valid for calculators carrying more places. A result of this kind might be useful in the design of software for programmable hand calculators, however, the important point is that the results shown in Fig. 7.1 were obtained interactively at a computer terminal. The user supplied the integrand of (7.7), and then various values of k and φ. The detailed error analysis was carried out by the computer, using the program [13].

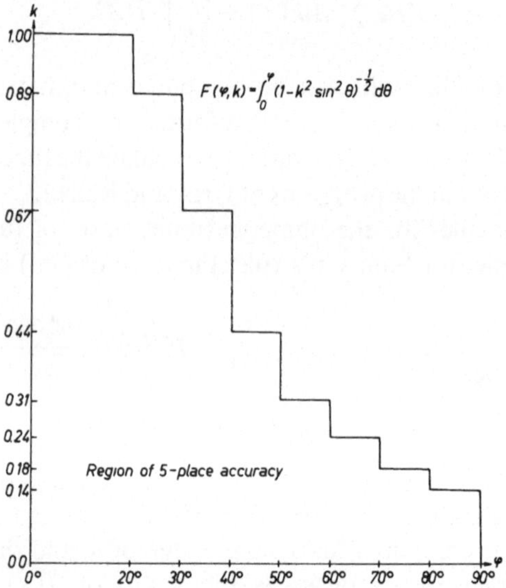

Fig. 7.1. Region of guaranteed 5-place accuracy for calculation of $F(\varphi, k)$ by Simpson's rule with 7 nodes

The error of numerical differentiation can be analyzed in the same way. It may seem strange to use an analytic differentiator for this purpose; however, it may turn out to be cheaper or more suitable to the input data to calculate derivatives as linear combinations of function values than to execute the differentiation subroutine, once the rule for numerical differentiation has been established to be of sufficient accuracy for the functions considered. Here again, the elaborate differentiation and interval software is used for error analysis in the design of a program for production computation, not for the computation itself.

8. Conclusions and Predictions

The above illustrations give an idea of the power and usefulness of software for automatic differentiation, however, in spite of its portability, it is not widely accepted at the present time by numerical analysts. One hears objections such as: (i) It won't work; (ii) it won't work well; or (iii) a problem is known to which it does not apply. Objections (i) and (ii), coming from people who use formula translators routinely, are demolished by the facts that differentiators are based on the same principles as translators and have been used with success at the Mathematics Research Center and elsewhere for more than a decade. Objection (iii) is simply a negative, nonconstructive remark which applies to any method or tool. A more positive approach would be to think of problems to which this software does apply, as the examples given above indicate.

There are, of course, alternatives to the use of automatic differentiators. One is to derive and code all needed derivatives by hand. This is a dull, routine task, highly prone to error. Thus, to make as much use of the speed and accuracy of the computer as possible, such mechanical jobs should be turned over to it. The other alternative is to avoid the use of derivatives and Taylor series altogether. This has been done with great success in a number of areas, and the resulting finite difference methods have enriched both the theory and practice of numerical analysis. However, there is a long tradition of the use of series and derivatives in numerical analysis, dating back at least to the 1730 book by Stirling [43]. Rather than turn one's back on this long history of development, it would seem to be better to make as much use of the power of mathematical analysis as possible, and employ derivatives and series where appropriate.

Another point brought up in this paper is the close relationship in applications between automatic differentiation and interval arithmetic, even though they are quite different in concept and implementation. The reason for this is that in floating-point (real) arithmetic, the derivative can often be successfully approximated by the difference quotient

$$f'(x) \sim \frac{f(x+h) - f(x)}{h}. \tag{8.1}$$

Numerical analysts working with finite derivatives, however, know that h cannot be taken too small in (8.1), as the significant digits in $f(x+h)$ and $f(x)$ will cancel out, and the resulting roundoff error will be multiplied by the huge number $1/h$ to give a result of no significance whatever. It is known that if $f(x)$ can be calculated with absolute precision η, then $h = \sqrt{\eta}$ is as small as h should be taken. Thus, a somewhat inaccurate value of the derivative is the price paid to keep the computation meaningful. However, this type of inaccuracy is of no consequence in many applications. In interval arithmetic, however, one has

$$[0,1] - [0,1] = [-1,1], \tag{8.2}$$

so cancellation is replaced by a widening of intervals, which is only aggravated by multiplying by large numbers. This "painful honesty" of interval arithmetic makes it more suitable for use with derivatives than difference quotient approximations such as (8.1).

From a historical standpoint, formula translators did not come into general use until hardware for floating-point arithmetic become widely available. Then, one could declare a variable to be "real", and forget the scaling, etc. which is better done in machine or assembly language for fixed-point computations. Because of the intimate relationship between interval analysis and derivatives, one may have to wait for hardware for interval arithmetic before differentiation by software is widely accepted.

Finally, the present breed of automatic differentiators, as all human products, can be improved. At present, this seems to be a task which is too exotic for numerical analysts and too mundane for computer scientists. However, as with formula translators, one can expect improvements to come with use and coupled with advances in hardware and computer science.

References

[1] Bauer, F. L.: Computational graphs and rounding error. SIAM J. Numer. Anal. **11**, 87–96 (1974).

[2] Binstock, W., Hawkes, J., Hsu, N.-T.: An interval input/output package for the UNIVAC 1108. Tech. Sum. Rpt. No. 1212, Math. Res. Center, Univ. Wisconsin-Madison, September 1973.

[3] Braun, J. A., Moore, R. E.: A program for the solution of differential equations using interval arithmetic (DIFEQ) for the CDC 3600 and 1604. Tech. Sum. Rpt. No. 907, Math. Res. Center, Univ. Wisconsin-Madison, June 1968.

[4] Conte, S. D.: Elementary Numerical Analysis: An Algorithmic Approach. New York: McGraw-Hill 1965.

[5] Conte, S. D., de Boor, C.: Elementary Numerical Analysis: An Algorithmic Approach, 2nd ed. New York: McGraw-Hill 1972.

[6] Crary, F. D.: Language extensions and precompilers. Tech. Sum. Rpt. No. 1319, Math. Res. Center, Univ. Wisconsin-Madison, February 1973.

[7] Crary, F. D.: The AUGMENT precompiler. I. User information. Tech. Sum. Rpt. No. 1469, Math. Res. Center, Univ. Wisconsin-Madison, December 1974, revised April 1976.

[8] Crary, F. D.: The AUGMENT precompiler. II. Technical documentation. Tech. Sum. Rpt. No. 1470, Math. Res. Center, Univ. Wisconsin-Madison, October 1975.

[9] Crary, F. D., Ladner, T. D.: A simple method of adding a new data type to FORTRAN. Tech. Sum Rpt. No. 1605, Math. Res. Center, Univ. Wisconsin-Madison, May 1970.

[10] Gray, Julia H., Rall, L. B.: NEWTON: A general purpose program for solving nonlinear systems. Tech. Sum. Rpt. No. 790, Math. Res. Center, Univ. Wisconsin-Madison, September 1967.

[11] Gray, Julia H., Rall, L. B.: NEWTON: A general purpose program for solving nonlinear systems. Proc. 1967 Army Numer. Anal. Conf., pp. 11–59. Durham, N. C.: Army Res. Office 1967.

[12] Gray, Julia H., Rall, L. B.: A computational system for numerical integration with rigorous error estimation. Proc. 1974 Army Numer. Anal. Conf., pp. 341–355. Durham, N. C.: Army Res. Office 1974.

[13] Gray, Julia H., Rall, L. B.: INTE: A UNIVAC 1108/1110 program for numerical integration with rigorous error estimation. Tech. Sum. Rpt. No. 1428, Math. Res. Center, Univ. Wisconsin-Madison, October 1975.

[14] Gray, Julia H., Rall, L. B.: Automatic Euler-Maclaurin integration. Proc. 1976 Army Numer. Anal. and Comp. Conf., pp. 431–444. Research Triangle Park, N. C.: Army Res. Office 1976.

[15] Gray, Julia H., Reiter, A.: Compiler of differentiable expressions (CODEX) for the CDC 3600. Tech. Sum. Rpt. No. 791, Math. Res. Center, Univ. Wisconsin-Madison, December 1967.

[16] Kantorovich, L. V.: Functional analysis and applied mathematics. Uspehi Mat. Nauk **3**, 89–185 (1948). (In Russian.) Translated by C. D. Benster, Natl. Bur. Std. Rpt. No. 1509, Washington, 1952.

[17] Kantorovich, L. V.: On a mathematical symbolism convenient for performing machine calculations. Dokl. Acad. Nauk SSR **113**, 738–741 (1957). (In Russian.)

[18] Kedem, G.: Automatic differentiation of computer programs. Tech. Sum. Rpt. No. 1697, Math. Res. Center, Univ. Wisconsin-Madison, November 1976.
[19] Knapp, H., Wanner, G.: Numerical solution of ordinary differential equations by Groebner's method of Lie-series, Tech. Sum. Rpt. No. 880, Math. Res. Center, Univ. Wisconsin-Madison, June 1968.
[20] Knapp, H., Wanner, G.: LIESE: A program for ordinary differential equations using Lie-series, Tech. Sum. Rpt. No. 881, Math. Res. Center, Univ. Wisconsin-Madison, June 1968.
[21] Knapp, H., Wanner, G.: LIESE II: A program for ordinary differential equations using Lie-series, Tech. Sum. Rpt. No. 1008, Math. Res. Center, Univ. Wisconsin-Madison, August 1969.
[22] Krawczyk, R.: Newton-Algorithmen zur Bestimmung von Nullstellen mit Fehlerschranken, Computing **4**, 187 – 201 (1969).
[23] Kuba, D., Rall, L. B.: A UNIVAC 1108 program for obtaining rigorous error estimates for approximate solutions of systems of equations, Tech. Sum. Rpt. No. 1168, Math. Res. Center, Univ. Wisconsin-Madison, January 1972.
[24] Li, T. Y., Yorke, J. A.: A simple, reliable numerical algorithm for following homotopy paths, Proc. MRC Symp. on Analysis and Computation of Fixed Points. New York: Academic Press (to appear 1979).
[25] Moore, R. E.: The automatic analysis and control of error in digital computation based on the use of interval numbers. [33], pp. 61 – 130. 1965.
[26] Moore, R. E.: Automatic local coordinate transformations to reduce the growth of error bounds in interval computation of solutions of ordinary differential equations. [34], pp. 103 – 140. 1965.
[27] Moore, R. E.: Interval Analysis. Englewood Cliffs, N. J.: Prentice-Hall 1966.
[28] Moore, R. E.: Methods and Applications of Interval Analysis. Philadelphia: SIAM Publications 1979.
[29] Moore, R. E.: A test for existence of solutions to nonlinear systems. SIAM J. Numer. Anal. **14**, 611 – 615 (1977).
[30] Moore, R. E., Davison, J. A., Jaschke, H. R., Shayer, S.: DIFEQ integration routine – user's manual. Tech. Rpt. LMSC 6-90-64-6, Lockheed Missiles and Space Co., Palo Alto, Calif., 1964.
[31] Nickel, K.: On the Newton method in interval analysis. Tech. Sum. Rpt. No. 1136, Math. Res. Center, Univ. Wisconsin-Madison, December 1971.
[32] Olver, F. W. J.: A new approach to error arithmetic. SIAM J. Numer. Anal. **15**, 368 – 393 (1978).
[33] Rall, L. B. (ed.): Error in Digital Computation, Vol. 1. New York: J. Wiley 1965.
[34] Rall, L. B. (ed.): Error in Digital Computation, Vol. 2. New York: J. Wiley 1965.
[35] Rall, L. B.: Computational Solution of Nonlinear Operator Equations. New York: J. Wiley 1969.
[36] Rall, L. B.: A comparison of the existence theorems of Kantorovich and Moore. SIAM J. Numer. Anal. **17** (1980, to appear).
[37] Rall, L. B., Wanner, G.: Experience with Lie series. Meth. und Verfahren der Math. Physik **5**, 29 – 42 (1971).
[38] Reiter, A.: Automatic generation of Taylor coefficients (TAYLOR). Prog. No. 3, Math. Res. Center, Univ. Wisconsin-Madison, July 1965.
[39] Reiter, A.: Interval arithmetic package (INTERVAL). Prog. No. 2, Math. Res. Center, Univ. Wisconsin-Madison, June 1965.
[40] Reiter, A.: Compiler of differential expressions (CODEX). Prog. No. 1, Math. Res. Center, Univ. Wisconsin-Madison, August 1965.
[41] Reiter, A.: Automatic generation of Taylor coefficients (TAYLOR) for the CDC 1604. Tech. Sum. Rpt. No. 830, Math. Res. Center, Univ. Wisconsin-Madison, November 1967.
[42] Reiter, A.: Interval arithmetic package (INTERVAL) for the CDC 1604 and CDC 3600. Tech. Sum. Rpt. No. 794, Math. Res. Center, Univ. Wisconsin-Madison, January 1979.
[43] Stirling, J.: Methodus Differentialis: sive Tractatus de Summatione et Interpolatione Serierum Infinitarum. London: Typis, Gul. Bowyer, Impensis, G. Strahan 1730.
[44] Stummel, F.: Rounding error analysis of numerical algorithms. (This volume.)
[45] Wertz, H. J.: SUPER-CODEX (Supervisor plus a compiler of differentiable expressions. Math. Res. Center, Univ. Wisconsin-Madison, June 1968.
[46] Wertz, H. J.: SUPER-CODEX: Analytic differentiation of FORTRAN statements. Rpt. No. TOR-0172(9320)-12, Aerospace Corporation, El Segundo, Calif., April 1972.
[47] Yohe, J. M.: The interval arithmetic package. Tech. Sum. Rpt. No. 1755, Math. Res. Center, Univ. Wisconsin-Madison, June 1977.

[48] Yohe, J. M.: Implementing nonstandard arithmetics. SIAM Rev. **21**, 34–56 (1979).
[49] Yohe, J. M.: Portable software for interval arithmetic. (This volume.)

Prof. L. B. Rall
Mathematics Research Center
University of Wisconsin-Madison
610 Walnut Street
Madison, WI 53706, U.S.A.

Computing, Suppl. 2, 157 – 164 (1980)

Small Bounds for the Solution of Systems of Linear Equations

S. M. Rump and **E. Kaucher**, Karlsruhe

Abstract

An algorithm is presented to solve a system of linear equations $Ax = b$ of high order. There are no restrictions for A; A may be a floating-point or interval matrix. The algorithm leads to small, guaranteed bounds for the solution even for ill-conditioned matrices. It takes about six times the computing time needs for the usual floating-point Gaussian algorithm with comparable accuracy.

0. Introduction

Here as throughout the paper "solving" always means giving guaranteed, provable bounds for the exact solution. The approximation of the exact solution may have a big relative error in case that the condition number is large (which in fact is not known in general). If there is no (provable) error estimation one is not able to decide whether an approximation is good or bad. Therefore an algorithm not yielding guaranteed bounds for the exact solution may cause a lot of damage (e.g. repetition of expensive experiments etc.) when a poor approximation is regarded as a good one. With respect to this and other reasons we set a high value on provable bounds to be computed by the algorithm.

Up to now most of those algorithms are either calculating bounds for the solution in a brute-force way (naive interval arithmetic) or depending on a first inclusion of the solution or even for the inverse of the matrix. We are looking for general algorithms working with single precision but giving a solution of high accuracy and moreover we wish to avoid the following lacks as well as possible:

computing with only single precision or without computing residual corrections produces intervals of a large width

to start an inclusion of the inverse or of the solution is needed

the algorithm converges slowly or cannot even start for more bad conditioned problems

there are certain restrictions for the matrix to be satisfied

the algorithm needs a lot of time and (or) space

the algorithm is suitable working only for lower degrees, at most for degrees less than 50

in general the algorithm cannot use the accuracy of classical methods

the iteration functions have to be inclusion isotone.

In the following an algorithm is presented satisfying the properties mentioned above. Furthermore several improvements have been introduced, especially

computing bounds for a "residual equation",

going in contrast "from the inner to the outer", that means starting with a certain interval (not necessarily including the solution) and by blowing it up receiving an inclusion of the solution

using good floating-point approximations as well as possible.

The algorithm is working for floating-point as well as for interval systems and has been implemented on the UNIVAC 1108 of the University of Karlsruhe. The algorithm is written in FORTRAN and therefore portable.

Some computational results are given at the end.

1. Theoretical Background

In the paper [4] we presented very general and widely usable theorems concerning Schauder's Fixpoint Theorem. We now present a special form of the cited theorem.

Theorem 1. *Let* $f\colon \mathbb{R}^n \to \mathbb{R}^n$ *be a continuous mapping. Let further* $F\colon \mathrm{I}\mathbb{R}^n \to \mathrm{I}\mathbb{R}^n$ *be a given arbitrary function with*

$$\bigwedge_{I \in \mathrm{I}\mathbb{R}^n} x \in I \Rightarrow f(x) \in F(I), \tag{1}$$

where $\mathrm{I}\mathbb{R}^n$ *denotes the set of n-dimensional interval-vectors over* $\mathbb{R}$. *If for an* $\Omega \in \mathrm{I}\mathbb{R}^n$ *holds*

$$F(\Omega) \subseteq \Omega \tag{2}$$

then there exists a fixpoint $\hat{x}$ *of* f *with*

$$\hat{x} \in F(\Omega) \tag{3}$$

and furthermore

$$\hat{x} \in \bigcap_{i=0}^{\infty} F^i(\Omega). \tag{4}$$

Note the very weak assumptions and the strong results of the theorem, especially that the interval function F is not assumed to be inclusion monotone or convergent in any sense.

Now let a system of linear equations $Ax = b$ be given with an $n \times n$-matrix A and an n-dimensional vector b. We first assume A and b to be real, i.e. $A \in \mathbb{R}^{n \times n}$ and $b \in \mathbb{R}^n$. To apply the theorem above we first have to look for a real function f and then for an interval function F satisfying (1). The first idea for f may be the well-known residual iteration (for invertible A)

$$f^*(y) = y + A^{-1}(b - Ay).$$

Of course the inverse matrix A^{-1} is in general not known; so one tries to replace it by a floating point approximation $R \approx A^{-1}$:

$$\bar{f}(y) = y + R(b - Ay). \tag{5}$$

For a fixpoint $\hat{y}$ of $\bar{f}$ we have

$$\hat{y} = \bar{f}(\hat{y}) = \hat{y} + R(b - A\hat{y}) \equiv R(b - A\hat{y}) = 0.$$

If R is not singular, then $\hat{y}$ is a solution of $Ax = b$. To find an interval function F we write (5) in the following way

$$\bar{f}(y) = x + Rb - RAx + y - x - RAy + RAx$$
$$= x + R(b - Ax) + (E - RA)(y - x)$$

where E denotes the $n \times n$ unit matrix. Now let F be defined as

$$F(Y) = x + R(b - Ax) + (E - RA) * (Y - x), \tag{6}$$

then (1) follows immediately. The function (6) occurs for instance in [5]. To get bounds for the solution of $Ax = b$ we need an interval-vector Ω with $F(\Omega) \subseteq \Omega$. To find it we start with the interval Y^0 consisting only of the point $x + R(b - Ax)$ for a certain x. Then we iterate

$$Y^{k+1} := F(Y^k) \quad \text{until} \quad Y^{k+1} \subseteq Y^k. \tag{7}$$

x may be any vector; however, we prefer a floating-point approximation $\tilde{x}$ for the solution $\hat{x}$ to get sharp bounds.

If the spectral radius of $E - RA$ is smaller than one then surely A and R are non-singular. So in this case the last Y^{k+1} of (7) contains the unique solution $\hat{x}$ of $Ax = b$.

2. Improvements

First we give a sample for our algorithm; an improved version will follow later.

1) Compute a floating-point approximation R of A^{-1}
2) Compute $B := E \ominus R \odot A$ by interval arithmetic; **if** $\|B\| \geqslant 1$ **then** stop
3) Let $\tilde{x} := R \cdot b$ be a floating-point approximation of $\hat{x}$
4) $Y^0 := Z := \tilde{x} \oplus R \odot (b \ominus A \odot \tilde{x})$ by interval arithmetic
5) **repeat** $Y^{k+1} := Z \oplus B \odot (Y^k \ominus \tilde{x})$ **until** $Y^{k+1} \subseteq Y^k$.

The operations in a circle always denote an interval operation.

One fundamental improvement is to construct an inclusion for the solution of the residual equation

$$Ay = b - A\tilde{x} \tag{*}$$

instead of the original equation $Ax = b$, where $\tilde{x}$ is a floating-point approximation due to step 3). If y is the solution of the residual equation (*) then $\tilde{x} + y$ is equal to $\hat{x}$, and if an interval Y contains y then $\hat{x}$ is contained in $\tilde{x} \oplus Y$. If furthermore $\tilde{y}$ is a floating-point approximation to y and z the solution of $Az = b - A\tilde{x} - A\tilde{y}$, then $\hat{x} = \tilde{x} + \tilde{y} + z$. If Z is an interval containing z, then $\tilde{x} \oplus \tilde{y} \oplus Z$ contains $\hat{x}$.

This method can be continued in an obvious way. However, these computations make only sense, if the residuals $b - A\tilde{x} - A\tilde{y} - \cdots$ are computed with higher accuracy.

For this purpose for instance the algorithm of Bohlender [2] can be used or a long accumulator as described in [3].

In step 3) of the sketched algorithm it is superior not to take $R \cdot b$ as an approximation of $\hat{x}$ but to iterate with the residual iteration

$$x^0 = R \cdot b;\ x^{k+1} = x^k + R(b - Ax^k). \tag{8}$$

In the case of a floating-point computation the problem arises when to stop the iteration (8). In our algorithm there is no essential dependence on this question. Here a simple stopping criterion is used which, roughly spoken, guarantees a "win of at least one decimal digit" in two iteration steps. So we proceed as follows: First (8) is executed, where the last iterative is noted as $\tilde{x}$. Then

$$y^0 := R(b - A\tilde{x}); \qquad y^{k+1} := y^k + R(b - A\tilde{x} - Ay^k) \tag{9}$$

is executed, where $\tilde{y}$ denotes the last iterative.

With

$$Y^0 := \tilde{y} \oplus R \odot (b \ominus A \odot \tilde{x} \ominus A \odot \tilde{y}) \tag{10}$$

we proceed in step (5) yielding an interval vector Y, thus $\hat{x} \in \tilde{x} \oplus Y$. The residual $b \ominus A \odot \tilde{x} \ominus A \odot \tilde{y}$ in (10) has to be computed in double precision interval arithmetic.

When proceeding in step 5) it turns out that it is better to use a "Einzelschrittverfahren", that means to calculate componentwise and using the computed components at once.

Furthermore an ε-expansion will be introduced. We define

$$I \circ \varepsilon := I + d(I) \cdot \varepsilon * [-1, 1] \tag{11}$$

It follows

$$d(I \circ \varepsilon) = d(I) + 2\varepsilon \cdot d(I) = (1 + 2\varepsilon) \cdot d(I),$$

so the width of the interval I is relatively enlarged by 2ε and can be interpreted as an "artificial rounding". We write step 5) now as follows

$$\begin{aligned}
&Y^0 := Z; \qquad k := -1;\\
&\textbf{repeat}\ k := k+1; \qquad Y^{k+1} := Y^k := Y^k \circ \varepsilon;\\
&\qquad \textbf{for}\ i := 1\ \textbf{to}\ n\ \textbf{do}\ Y_i^{k+1} := Z_i \oplus B_i \odot (Y^k \ominus \tilde{y})\\
&\textbf{until}\ \ Y^{k+1} \subseteq Y^k;
\end{aligned} \tag{12}$$

Here Y_i and Z_i are the ith component of Y and Z and B_i the ith row of B, resp. In practice it turns out that $\varepsilon = 0.1$ is a good value. It should be noted explicitely that with the ε-expansion we achieve an improvement of the speed of convergence. Regarding (3) one might proceed with

$$Y^{k+1} := \{Z \oplus B \odot (Y^k \ominus \tilde{y})\} \cap Y^k \tag{13}$$

until $Y^{k+1} = Y^k$. However, even computing with double precision interval arithmetic in (13) gives no significant improvement of the bounds.

Obviously, if the matrix A is satisfying special conditions then special improvements can be introduced. If, for instance, the matrix A is strong diagonal dominant, then we can use D^{-1} instead of R as an approximation of A^{-1}. The inverse D^{-1} of the diagonal of A is very easy to compute.

We should mention that if a bound σ of the spectral radius of $E - RA$ is known and $\sigma \ll 1$ holds, one might use the formula

$$\|x^0 - \hat{x}\| \leqslant \frac{1}{1 - \sigma} \cdot \|x^0 - x^1\| \tag{14}$$

to give a bound for the solution $\hat{x}$ as is proposed in [1], [6]. But from (14) we get an inclusion "sphere" for $\hat{x}$ independent of the location of its components. Up to now there is not general equilibration method known. Therefore all components of relatively small modulus are overestimated proportional to the maximum difference of the exponents of $\hat{x}$ and so (14) yields in general poor bounds. We avoid this disadvantage for it is more convenient to estimate bounds for $\hat{x}$ using all information (e.g. the whole matrix A) instead of using the only number σ. Furthermore the execution of (12) is of very low cost compared with whole computing time and the gained quality of the bounds.

Moreover for σ nearly 1 or $\sigma \geqslant 1$ or unknown σ the method (14) does not work. Remember that $\sigma < 1$ in our algorithm is only needed to prove R to be nonsingular. This can be done e.g. externally or by proving $R \cdot A$ to be of property M or strong diagonal dominant or by some other methods. For the reasons mentioned it is not worthwhile to include (14) in the algorithm, especially not to complicate the program unnecessarily.

3. The Algorithm

With the improvements introduced above we can write down the final version of the algorithm:

---

1) Compute $R \approx A^{-1}$ by floating-point arithmetic.
2) Compute $C = A \odot R$ and $B = I \ominus C$ by interval arithmetic. **If** C is not a matrix of property M (or strong diagonal dominant) and $\|B\| \geqslant 1$ **then** stop.
3) Set $x^0 = R \cdot b$; $k := -1$;
   repeat $k := k + 1$; $x^{k+1} = x^k + R \cdot (b - Ax^k)$
   until $|x^{k+1} - x^k|/|x^k| \geqslant 10^{-k/2}$ or
   $|x^{k+1} - x^k|/|x^k| < 10^{1-t}$;
   the final x^{k+1} is named $\tilde{x}$ and $r := k + 1$.
   Set $y^0 = R(b - A\tilde{x})$; $k := -1$,
   repeat $k := k + 1$; $y^{k+1} = y^k + R(b - A\tilde{x} - Ay^k)$
   until $|y^{k+1} - y^k|/|y^k| \geqslant 10^{-k/2}$ or
   $|y^{k+1} - y^k|/|y^k| < 10^{2-t}$;
   the final y^{k+1} is named $\tilde{y}$ and $r := r + k + 1$.
4) Compute $Y^0 := Z := \tilde{y} \oplus R \odot (b \ominus A \odot \tilde{x} \ominus A \odot \tilde{y})$; $k := -1$ by interval arithmetic.
5) **repeat** $k := k + 1$; $Y^{k+1} := Y^k := Y^k \circ \varepsilon$,
   for $i := 1$ **to** n **do** $Y_i^{k+1} := Z_i \oplus B_i \odot (Y^k \ominus \tilde{y})$
   until $Y^{k+1} \subseteq Y^k$ or $\{k > 2.25 \cdot r - 1\} =:$ bool;
   if bool **then** stop.
6) The final Y^{k+1} is named Y and $\hat{x} \in \tilde{x} \oplus Y$.

Remarks. In step 1) the Gauss-Jordan algorithm with pivoting is used. It turned out to be better than the Gaussian elimination method. In step 3) and 4) only the residuals are to be computed in double precision. t denotes the number of digits of the mantissa of the computer. The exit stop means that the algorithm fails to compute bounds and may be repeated with higher accuracy.

It is very easy to extend the algorithm to interval equations. Just replace every interval matrix A or interval vector b by the midpoint $m(A)$ or $m(b)$ where A or b is occurring in a floating-point computation, resp.

4. Complexity

Let α_1 be the cost to compute a floating-point approximation for $\hat{x}$ and α_2 be the cost to compute bounds for $\hat{x}$ with the presented algorithm, both with comparable accuracy. As a measure for the additional costs to compute bounds we take the ratio

$$\rho = \frac{\alpha_2}{\alpha_1}.$$

W.l.o.g. we use the Gaussian elimination method with r residual iterations (step 3)) to compute a floating-point approximation. Then (neglecting linear terms in n) we get $\alpha_1 \geqslant n^3/3 + 2rn^2$.

Adding the additional computing times for the steps 1) to 5), where step 5) was executed s times, yields (neglecting linear terms)

$$\begin{aligned}\alpha_2 &\leqslant 2n^3 + n^2(3r + 4s + 4)\\ &\leqslant 6 \cdot \alpha_1 + n^2(-9r + 4s + 4).\end{aligned}$$

So

$$\rho \leqslant 6 + \frac{-9r + 4s + 4}{n/3 + 2r}$$

can be estimated as follows: If $s \leqslant (9r - 4)/4$ then $\rho \leqslant 6$, otherwise $\rho \leqslant 6 \cdot \{1 + (2s - 3.5r)/n\}$. In fact we have $\rho = 6 + 0(1/n)$; in the algorithm $\rho \leqslant 6$ is satisfied. If the algorithm fails (via exit: stop) and is repeated with doubled accuracy etc., the finally ρ is in the most pessimistic case estimated by $\rho \leqslant 6 \cdot \sum_{i=0}^{\infty} 4^{-i} = 6 \cdot \frac{4}{3} = 8$. It is remarkable that ρ is independent of n and the condition number of A.

5. Empirical Results

The empirical results given were computed on the UNIVAC 1108 of the University of Karlsruhe with $t = 8\frac{1}{2}$ decimal digit accuracy for single precision. All computing were done in single precision arithmetic except the residuals, which were computed in double precision but stored in single precision. First we give an example to show what happens when only few iterations are executed in step 3). We take the Hilbert 7×7-matrix, where the coefficients are rational integers and $b = (1, \ldots, 1)$. The condition number is $4.8 \cdot 10^8$ and σ less than 0.55. Due to 3) we have $r = 6$ iterations and we get for the first component of x^k: 11.658, 5.336, 7.594, 6.788 and

of y^k: 0.2884, 0.1853, 0.2221, 0.209.

With the final values for $\tilde{x}$ and $\tilde{y}$ we get

$$\tilde{x} + \tilde{y} = 6.997,$$

far away from the exact value 7.0. With $Y^0 = [0.2136253_{00}^{11}]$ we get after 9 iterations of step 5)

$$\hat{x} \in [6.999999471, 7.000000294]$$

with an accuracy of 7 correct decimal positions. By the way, using the estimation (14) we would get the bad inclusion:

$$\hat{x} \in [-21.96, 35.95].$$

If we would have 18 iterations in step 3) the relative error bound of the final inclusion would have been bounded by $5 \cdot 10^{-11}$. We see that in special cases it might be better to iterate a bit longer in step 3); however, it is not necessary to achieve good results.

Finally, we give a sample of some bad-conditioned matrices of higher degree.

degree(A) $n =$	cond(A) $\lesssim$ $\|A\| \cdot \|R\| =$	at least guaranteed decimal positions in each component $[\log(d(Y)/\|Y\|)]$	r	s	ρ
25	$3.1 \cdot 10^7$	12	9	2	3.4
50	$9 \cdot 10^7$	11	10	3	4.0
100	$2.6 \cdot 10^8$	12	16	4	4.1
150	$8 \cdot 10^7$	13	8	2	5.1
200	$6 \cdot 10^7$	13	3	1	5.7
200	$7 \cdot 10^7$	13	4	2	5.7
200	$5 \cdot 10^7$	14	4	1	5.6

The maximum degree 200 is caused by the limited storage of the UNIVAC 1108 and is no bound for the algorithm.

6. Conclusion

A fast direct method to solve an arbitrary system of linear equations is the well-known Gaussian elimination. However, we have seen that the result may be arbitrary false without taking any precautions (see Hilbert$_7$-matrix). Even if the residual iterations "seem" to converge, one is not sure of the true value of the solution. To achieve guaranteed bounds for the solution one needs a special algorithm (using a specified rounding). We presented such an algorithm with the following advantages:

any floating-point algorithm to achieve a good approximation of the solution is usable there is no restriction on the matrix or the right-hand side

it is working for interval-matrices, too

no inclusion for the solution or the inverse matrix is needed

interval arithmetic is used very late

the algorithm is very fast.

After finishing the algorithm with step 6) it is proved, that

A is not singular

$Ax = b$ is uniquely solvable

exact bounds for the solution are given.

When the residuals are computed with double precision, we got the following experimental results:

matrices with a condition number up to 10^8 can be trieved when single precision has an 8 decimal digit mantissa

in step 4) usually 4, in very bad conditioned cases up to 16 iterations are necessary

in step 5) one iteration is necessary, in very bad conditioned cases up to four

with an 8 decimal digit mantissa at least 10 decimal digits accuracy is achieved in all components of the solution.

The computing time on the UNIVAC 1108 at the University of Karlsruhe were 2 minutes for degree 100 and 12 minutes for degree 200; the computing time for the Gaussian elimination were always $\frac{1}{6}$, independent of the degree. The algorithms were programmed in ALGOL and FORTRAN and are available.

References

[1] Alefeld, G., Apostolatos, N.: Praktische Anwendung von Abschätzungsformeln bei Iterationsverfahren, Bericht des Instituts für Angewandte Mathematik und Rechenzentrum der Universität Karlsruhe, Januar 1968, 9 p.

[2] Bohlender, G.: Floating-point computation of functions with maximum accuracy. IEEE Computer Society, Symposium on Computer Arithmetic, Dallas 1975, pp. 14–23.

[3] Kaucher, E., Klatte, R., Rump, S. M.: Der dynamische Intervallrechner, Bericht des Instituts für Angewandte Mathematik der Universität Karlsruhe, September 1978, 15 p.

[4] Kaucher, E., Rump, S. M.: Generalized iteration methods for bounds of the solution of fixed point operator equations. (This volume.)

[5] Krawczyk, R.: Newton-Algorithmen zur Bestimmung von Nullstellen mit Fehlerschranken. Computing **4**, 182–201 (1969).

[6] Kulisch, U.: Grundzüge der Intervallrechnung. Überblicke Mathematik 2 (Laugwitz, D., ed.), pp. 51–98. Mannheim: Bibliographisches Institut 1969.

Dipl.-Math. S. M. Rump
Dr. E. Kaucher
Institut für Angewandte Mathematik
Universität Karlsruhe
Kaiserstrasse 12
D-7500 Karlsruhe
Federal Republic of Germany

Computing Suppl. 2, 165 – 168 (1980)

Shorthand Notation for Rounding Errors

R. Scherer and **K. Zeller,** Tübingen

Abstract

The first modern papers dealing with rounding errors were rather long. Gradually the presentation became more concise. Here we discuss a shorthand notation. It leads almost immediately to basic results concerning polynomials, triangular systems, elimination, and other problems.

1. Introduction

The first modern papers dealing with rounding errors were rather long: von Neumann and Goldstine [4], Turing [9], Givens [2], Wilkinson [10]. Gradually the presentation became more concise, see e.g. Wilkinson [11], [12], Jörgens [3], Ehlich and Zeller [1], Stewart and Jensen [7], Reinsch [6]. Here we discuss a shorthand notation for rounding errors (cf. [13], [14]).

This notation (comparable with "O") dispenses with certain subtleties. In return it leads almost immediately to basic results and estimates. We explain this in three cases: polynomials, triangular systems, elimination. Some hints concerning refinements and further applications are given.

We assume that the machine arithmetic (fixed resp. floating point) obeys the rules

$$fi\,(ab) = ab + \sigma, \qquad \text{with} \qquad |\sigma| \leqslant s^*,$$

$$fl\,(a + b) = (a + b)\rho, \qquad \text{with} \qquad |\log \rho| \leqslant r^*,$$

where s^* and r^* are certain (small) machine constants; and corresponding rules for a/b resp. $a - b$, ab, a/b. Hence we work with the absolute resp. relative error.

Of course σ and ρ depend on the parameters, but we avoid to denote this dependence. So at different occurrences σ or ρ will in general denote different numbers (each satisfying the inequality). Consistently we write

$$2\sigma,\ 3\sigma,\ \ldots\ ;\ \rho^2,\ \rho^3,\ \ldots$$

for a sum (product) of two or more numbers of this type. The following calculations will show the use of this abbreviated notation. We shall always assume that no overflow occurs.

2. Polynomials

We evaluate a polynomial, using Horner's scheme and floating point arithmetic. Our notation yields

$$((\cdots((a_n x\rho + a_{n-1})\rho x\rho + a_{n-2})\rho x\rho + \cdots + a_1)\rho x\rho + a_0)\rho.$$

We insert an additional factor ρ for $k = n$ (similarly we proceed in other cases) and get

$$\sum_{k=0}^{n} (a_k \rho^{2k+1}) x^k.$$

Hence we have evaluated a modified polynomial whose coefficients are located in ranges described by the factors. More information is available since some of the ρ coincide; this can be utilized by employing the Horner sums, cf. Reimer [5]. On the other hand, in general some of the ρ are different; hence it is correct to associate the factors with the coefficients and not with the variable.

A similar consideration in fixed point arithmetic leads to

$$\sum_{k=0}^{n} (a_k + \sigma) x^k.$$

Again the σ for $k = n$ is superfluous.

3. Triangular Systems

Next we consider a system

$$d_{k1}x_1^* + d_{k2}x_2^* + \cdots + d_{kk}x_k^* = d_k \qquad (\text{with } d_{kk} \neq 0; k = 1, \ldots, n).$$

Using floating point arithmetic and a certain arrangement of the operations, we define the numerical values x_k recursively by

$$(\cdots((d_k - d_{k1}x_1\rho)\rho - d_{k2}x_2\rho)\rho - \cdots - d_{k,k-1}x_{k-1}\rho)\rho/(d_{kk}\rho) =: x_k.$$

We note that the factor ρ can also be applied in the denominator, as done in the last term. In a direct way we now get

$$-d_k\rho^{k-1} + d_{k1}\rho^k x_1 + d_{k2}\rho^{k-1}x_2 + \cdots + d_{k,k-1}\rho^2 x_{k-1} + d_{kk}\rho^1 x_k = 0.$$

But it is also possible to shift the factors in a certain way, by performing suitable divisions successively in the first error expression. One of these modifications reads as follows:

$$\sum_{j=1}^{k} d_{kj}\rho^j x_j = d_k \qquad (k = 1, 2, \ldots, n).$$

Using fixed point arithmetic we arrive at

$$\sum_{j=1}^{k} d_{kj}x_j = d_k + k\sigma \qquad (k = 1, 2, \ldots, n).$$

All the numerical processes amount to the exact solution of systems with slightly modified coefficients.

4. Elimination

We consider a matrix A of type (n, n) and perform an elimination process (assuming that the pivots are admissible). Using fixed point arithmetic we determine the factors by

$$f_{k1} := a_{k1}/a_{11} + \sigma = (a_{k1} + \sigma)/a_{11} \qquad (k = 2, \ldots, n).$$

The manipulation (and change) of σ is justified by $|a_{11}| \leqslant 1$ (assuming fixed point with numbers in this range). Then we compute the elements of the new matrix B:

$$b_{kj} := a_{kj} - f_{k1}a_{1j} + \sigma = (a_{kj} + \sigma) - f_{k1}a_{1j} \qquad (j, k = 2, \ldots, n).$$

Our numerical process can be interpreted as an exact elimination process, applied to a matrix A, most of whose coefficients are modified by terms of type σ. Next we apply the same consideration to the new matrix B (the previous factors f_{k1} are not affected). Finally we arrive at a triangular matrix D. This matrix D belongs (considering exact operations) to

$$A + \begin{pmatrix} 0 & . & . & . & . & . & 0 \\ 1 & . & . & . & . & . & 1 \\ . & 2 & . & . & . & . & 2 \\ . & . & 3 & . & . & . & 3 \\ . & . & . & . & . & . & . \\ 1 & 2 & 3 & . & . & . & n-1 \end{pmatrix} \sigma$$

hence to a slightly modified matrix.

Now we employ floating point. Here it is useful to pass to the absolute error in each step. Assuming all |operands| $\leqslant 1$ for instance we arrive at the error representation above

with σ replaced by $2(\rho - 1)$.

Again modifications and refinements are possible.

5. Remarks

The considerations concerning elimination and triangular systems can be combined to give a backward analysis for the Gauß method for solving linear equations. It is useful to compute an inverse and then derive error bounds. The Crout reduction can be treated similarly. See e.g. Wilkinson [11], [12], Jörgens [3], Ehlich and Zeller [1], Stewart and Jensen [7].

Our shorthand notation dispenses with certain subtleties. Also the backward analysis combined with forward estimates of the norm type is in certain cases not very sharp. Hence one will – if necessary – look for better inequalities. One such case has been indicated in connection with the Horner sums. More generally the perturbation theory given by Stummel [8] seems very useful.

Of course one can consider more applications like Gauß-Jordan and simplex, expressions and recursions, approximation and stability, handling of various kinds machine arithmetic. Finally we call attention to H. Kneser's motto "notio et notatio": Notion and Notation are both important.

References

[1] Ehlich, H., Zeller, K.: Rundefehler bei der Lösung linearer Gleichungen. ZAMM **41**, T69 – T71 (1961).

[2] Givens, W.: Numerical computation of the characteristic values of a real symmetric matrix. Oak Ridge Nat. Lab. ORNL – 1574 (1954).
[3] Jörgens, K.: Fehlerabschätzung für die numerische Inversion von Matrizen. ZAMM **40**, T15 (1960).
[4] Von Neumann, J., Goldstine H. H.: Numerical inverting of matrices of high order. Bull. Amer. Math. Soc. **53**, 1021 – 1099 (1947).
[5] Reimer, M.: Bounds for the Horner sums. SIAM J. Numer. Anal. **5**, 461 – 469 (1968).
[6] Reinsch, C.: Die Behandlung von Rundungsfehlern in der numerischen Analysis. Jahrbuch Überblicke Mathematik 1979, pp. 43 – 62. Mannheim-Wien-Zürich: Bibliographisches Institut 1979.
[7] Stewart, N. F., Jensen F.: Solution numérique des problèmes matriciels. Les Presses de l'Université de Montreal: 1975.
[8] Stummel, F.: Rounding error analysis of elementary numerical algorithms. (This volume.)
[9] Turing, A. M.: Rounding-off errors in matrix processes. Quart. J. Mech. **1**, 287 – 308 (1948).
[10] Wilkinson, J. H.: Error analysis of direct methods of matrix inversion. J. Assoc. Comp. Mach. **8**, 281 – 330 (1961).
[11] Wilkinson, J. H.: Rundungsfehler. Berlin-Heidelberg-New York: Springer 1969.
[12] Wilkinson, J. H.: Modern error analysis. SIAM Rev. **13**, 548 – 568 (1971).
[13] van der Sluis, A.: Syllabus toegepaste analyse. Rijks-Universiteit, Utrecht, 1969.
[14] van Veldhuizen, M.: A note on partial pivoting and Gaussian elimination. Numer. Math. **29**, 1 – 10 (1977).
([13] and [14] added in proof.)

Priv.-Doz. Dr. R. Scherer
Prof. Dr. K. Zeller
Mathematisches Institut
Universität Tübingen
Auf der Morgenstelle 10
D-7400 Tübingen
Federal Republic of Germany

Computing, Suppl. 2, 169 – 195 (1980)

Rounding Error Analysis of Elementary Numerical Algorithms

F. Stummel, Frankfurt am Main

Abstract

The paper presents a new rounding error analysis of product and summation algorithms, Horner's scheme, evaluations of finite continued fractions, computations of determinants of tridiagonal systems, of determinants of second order and a 'fast' complex multiplication. The error analysis uses the linearization method and new condition numbers constituting optimal bounds in appraisals of the possible errors. The new error estimates are tested by numerous numerical examples.

Introduction

Our error analysis is based on the well-known linearization method (see McCracken-Dorn [12], Bauer [6]). In a series of textbooks and papers this method is explained and applied to simple examples. But it seems that hitherto there had not been given a theoretical foundation to this method, in particular, associated error estimates had been missing. These questions have now been thoroughly studied and answered in our perturbation theory for evaluation algorithms of arithmetic expressions (see Stummel [18]). The theoretical results are applied in this paper to a series of elementary but important algorithms and, at the same time, tested numerically. The results show that our new condition numbers yield much more detailed informations on possible errors than Wilkinson's backward analysis.

Chapter 1 analyzes the computation of sequences of partial products and partial sums. The computation of partial products is always a well-conditioned algorithm. In addition, using the linear error analysis, it is easily seen in which cases rounding errors cancel and in which cases this does not happen. The error analysis of the common summation algorithm yields, among other results, that in evaluating sums of both positive and negative terms, in contrast to Wilkinson [22, I. 25], not the ordering by increasing absolute values of the terms but with respect to a smallest sum of the absolute values of the partial sums is best. In § 1.3 it will be shown that the method of cascade summation, also called 'binary summation' or 'Linz's method', is easily and concisely analyzed by our means.

In Chapter 2 the class of elementary one-step algorithms is introduced, collecting most of the algorithms considered here. In particular, the associated absolute and relative a priori and a posteriori condition numbers are established. The error analysis of Horner's scheme in § 2.2 shows that our a posteriori condition numbers, in essence, coincide with error bounds described by Adams [2] and Tsao [21]. Further a numerical example of Mesztenyi-Witzgall [10] is analyzed exhibiting the differing numerical stability of evaluating the Taylor and the Newton form of a polynomial. Another important member of the class of elementary one-step algorithms is the Horner-like algorithm for evaluating finite continued fractions. A first numerical example deals with rounding errors in the evaluation of partial fractions for $z/\tan z$. Next the recursive computation of determinants of tridiagonal matrices is analyzed and the error estimate tested numerically with success by a matrix of 100 rows. In comparison with the stability constants of Babuška [4] for solving tridiagonal linear systems, our condition numbers constitute explicit measures for the accumulated errors.

Finally, Chapter 3 deals with the evaluation of determinants of second order. As an application the numerical stability of the common complex multiplication and a 'fast' complex multiplication are analyzed and compared. Let us remark in this context, that a detailed rounding error analysis of numerically solving two linear equations in two unknowns has already been established in [17]. An error analysis of Gaussian elimination for general linear systems is in preparation. The rounding error analysis of difference and extrapolation schemes by our methods is found in Stummel [16], [19].

The study of the numerical algorithms proceeds in the following way. First the sequence $F = (F_0, \ldots, F_n)$ of input and arithmetic operations is determined, specifying the algorithm. Then the graph of the functional dependences, the paths of error propagation, and the associated weights are described. The building blocks of these graphs and of the associated linear error equations are found in Fig. 2.3 of our paper [18], in the following quoted by I. The system of linear error equations can easily be read from the graph of the algorithm. Its solutions yield approximations of the absolute and relative a priori and a posteriori errors, respectively. Excepting the 'fast' complex multiplication, the graphs of all algorithms in this paper are trees. It has been shown in I – § 2.3 that in this case the condition numbers can easily be determined by recursion formulae.

Let $u = (u_0, \ldots, u_n)$, $v = (v_0, \ldots, v_n)$ denote the sequences of input data, intermediate and final results and their numerical approximations. By $\Delta u_t = v_t - u_t$ are meant the absolute and by $Pu_t = \Delta u_t/u_t$ the relative a priori errors of the approximations v_t of u_t. The fundamental a priori error estimates then read

$$|\Delta u_t| \leqslant \sigma_t \eta + O_t(\eta^2), \qquad |Pu_t| \leqslant \rho_t \eta + O_t(\eta^2), \qquad t = 0, \ldots, n, \tag{1}$$

where σ_t, ρ_t are the absolute and relative a priori condition numbers and η is the floating point accuracy constant. The numerical examples in the paper are computed in the decimal floating point arithmetic of the calculators HP 97 and TI 59. The symmetric rounding function rd_N is performed by rounding the 10- or 13-digit results to N decimal places. N-digit decimal floating point arithmetic is realized by rounding to N places after each operation. The constant η in the numerical examples is thus specified by

$$\eta = 5 \times 10^{-N}. \tag{2}$$

In general, $\eta \ll 1$ such that the remainder terms $O_t(\eta^2)$ in (1) can be neglected against the first order terms. In applying the error estimates (1) to numerical examples, it should be noticed that condition numbers yield optimal bounds of the possible errors. The actual errors, as a rule, are significantly smaller than indicated by $\sigma_t\eta$, $\rho_t\eta$. By variation of the parameters, however, we have, in most cases, found examples where the actual maximal errors are overestimated at most by a factor 5 to 10 such that the magnitude of the error is described correctly. The sign $\doteq$ indicates that the numerical result has been computed in higher precision than the given digits show. A priori condition numbers and a priori errors have always been computed with the highest precision of 10 or 13 decimal places.

1. Computation of Products and Sums

The methods of our error analysis differ essentially from Wilkinson's rounding error analysis (see Wilkinson [22]). Therefore, the usual product and summation algorithms are discussed thoroughly in this chapter. In particular, the condition numbers for perturbations by rounding errors in the arithmetic operations and for data perturbations will be determined and the general error estimates (1) be tested by selected numerical examples. The error analysis of cascade summation in § 1.3 shows that also in more complex algorithms the linear error equations and associated condition numbers are readily obtained by use of the graph-theoretic tools established in I – § 2.3.

1.1 Sequences of Partial Products

A finite or infinite product is computed by means of the sequence of partial products

$$u_0 = b_0, \qquad u_t = b_t u_{t-1}, \qquad t = 1, 2, \ldots, n, \tag{1}$$

such that

$$u_t = \prod_{j=0}^{t} b_j. \tag{2}$$

Introducing the intermediate steps $u_{t-1/2} = b_t$, this algorithm is defined, analogously to I – 1.2.(A), by the sequence of functions

$$F_0(x) = b_0, \qquad F_{t-1/2}(x) = b_t, \qquad F_t(x) = x_{t-1/2}x_{t-1}, \tag{3}$$

for $t = 1, \ldots, n$. The associated linear relative a priori error equations read

$$r_0 = e_0^b, \qquad r_{t-1/2} = e_t^b, \qquad r_t = r_{t-1/2} + r_{t-1} + e_t^\times,$$

or

$$r_0 = e_0^b, \qquad r_t = r_{t-1} + e_t^b + e_t^\times, \qquad t = 1, \ldots, n. \tag{4}$$

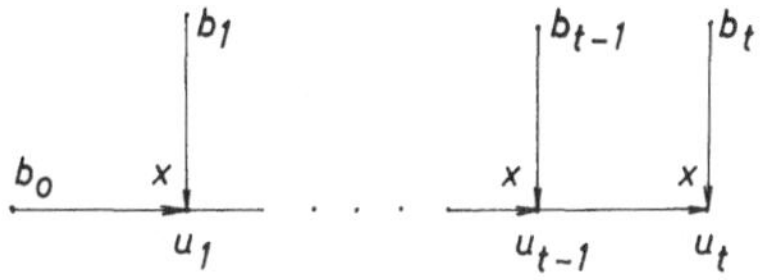

Fig. 1.1. Graph of the linear relative a priori error equations for sequences of partial products

To simplify notation, as here so in the following by $e_t^b, e_t^c, \ldots$ are meant the relative errors $Pb_t, Pc_t, \ldots$ and by $e_t^\times, e_t^+, e_t^/, \ldots$ the relative rounding errors of the floating point operations $\times, +, /, \ldots$. The graph of this algorithm is a tree, all nonzero weights b_{tk}^{rel} are equal to one (see Fig. 1.1).

Obviously, the linear system (4) has the solution

$$r_t = e_0^b + \sum_{j=0}^{t} (e_j^b + e_j^\times), \qquad t = 1, \ldots, n. \tag{5}$$

From this representation one immediately derives the associated relative a priori condition numbers ρ_t^D of data perturbations only ($e_j^\times = 0, |e_j^b| \leqslant \eta$), and ρ_t^R of perturbations by rounding errors in the arithmetic operations only ($e_j^b = 0$, $|e_j^\times| \leqslant \eta$):

$$\rho_t^D = t + 1, \quad \rho_t^R = t, \quad \rho_t = \rho_t^D + \rho_t^R = 2t + 1. \tag{6}$$

Consequently,

$$\rho_t^R = \rho_t^D - 1, \qquad t = 1, \ldots, n, \tag{7}$$

so that the computation of partial products is a well-conditioned algorithm.

Numerical Example 1.

$$u_t = (\exp .01)^t, \qquad t = 1, 2, \ldots. \tag{8}$$

This sequence of powers is established by the above algorithm for

$$b_0 = 1, \qquad b_t = b = \exp .01 = 1.010050\ldots.$$

We compute the sequence numerically in 5-digit decimal floating point. The basis b, rounded to five digits, becomes $b' = 1.0101$ having the relative error $e_t^b = e^b = 4.93-05$. In this case, (5) yields the representation

$$r_t = t \times 4.93-05 + \sum_{j=1}^{t} e_j^\times. \tag{9}$$

Table 1.1.1. *Relative errors in the computation of* $(\exp .01)^t$

t	$\frac{10^5}{t} Pu_t$	t	$\frac{10^5}{t} Pu_t$	t	$\frac{10^5}{t} Pu_t$
10	3.88	110	4.85	210	4.88
20	4.49	120	4.87	220	4.88
30	4.55	130	4.89	230	4.87
40	4.65	140	4.88	240	4.87
50	4.71	150	4.88	250	4.88
60	4.83	160	4.88	260	4.89
70	4.79	170	4.87	270	4.91
80	4.75	180	4.87	280	4.92
90	4.83	190	4.88	290	4.91
100	4.86	200	4.88	300	4.94

Tab. 1.1.1 contains the relative errors Pu_t divided by t for $t = 10, 20, \ldots, 300$. It turns out that Pu_t increases, in essence, linearly with t as te^b. The relative errors $e_t^\times$ seem not to participate in the growth of the error. Tab. 1.1.1 shows that the mean values over $e_j^\times$ can be estimated by

$$\left|\frac{1}{t}\sum_{j=1}^{t} e_j^\times\right| \doteq \left|\frac{1}{t}Pu_t - 4.93-05\right| \leqslant 5.-07 \tag{10}$$

for $t = 120, \ldots, 300$. Thus it is seen that the mean values of the sequence $(e_j^\times)$ are by a factor of $1.-02$ less than the floating point accuracy constant $\eta = 5.-05$. That is, randomly distributed rounding errors $e_t^\times$ cancel to a considerable extent. □

Numerical Example 2.

$$\frac{\sin \pi x}{\pi x} = \prod_{k=1}^{\infty}\left(1 - \frac{x^2}{k^2}\right) \tag{11}$$

(see Abramowitz-Stegun [1, p. 75]). The associated sequence of partial products is computed for

$$x = \frac{13}{6}, \qquad \frac{\sin \pi x}{\pi x} = \frac{3}{13\pi} = 7.34561-02$$

in the form

$$v_0 = 1, \qquad v_t = rd_N(rd_N(b_t)v_{t-1}), \qquad b_t = 1 - \frac{x^2}{t^2}, \qquad t = 1, 2, \ldots, \tag{12}$$

in 6-digit decimal floating point such that $\eta = 5.-06$. The exact partial products u_t are computed approximately in 10-digit decimal floating point. Tab. 1.1.2 lists, blockwise, the index t, the numerical approximation v_t, the associated relative error $Pu_t = (v_t - u_t)/u_t$ and the error sums

Table 1.1.2. *Computation of partial products for* $\sin \pi x/\pi x$, $x = 13/6$

100	200	300	400	500
7.69685−02	7.51955−02	7.46116−02	7.43205−02	7.41460−02
−6.21250−06	−1.10424−05	−1.45763−05	−2.30239−05	−3.17491−05
−3.48184−06	−1.61624−06	−4.07357−06	−3.42983−06	−2.48271−06
2.38333−05	5.05023−05	7.71145−05	1.02476−04	1.26851−04
−2.72941−06	−9.42562−06	−1.05023−05	−1.95940−05	−2.92661−05
5.37617−05	8.58803−05	1.20189−04	1.53491−04	1.85734−04
600	**700**	**800**	**900**	**1000**
7.40306−02	7.39488−02	7.38854−02	7.38405−02	7.38005−02
−2.73902−05	−1.69475−05	−3.74896−05	5.86592−06	−1.49325−05
−5.34780−07	−2.53890−06	−1.36779−06	8.68927−07	−1.07688−06
1.52375−04	1.76783−04	2.03862−04	2.24900−04	2.49224−04
−2.68550−05	−1.44085−05	−3.61227−05	4.99673−06	−1.38575−05
2.23253−04	2.56284−04	2.84328−04	3.25447−04	3.71101−04

$$\sum_{j=1}^{t} e_j^b, \quad \sum_{j=1}^{t} |e_j^b|, \quad \sum_{j=1}^{t} e_j^\times, \quad \sum_{j=1}^{t} |e_j^\times|. \tag{13}$$

The numerical results show that the absolute error sums of the sequences (e_j^b), $(e_j^\times)$ increase, in essence, linearly with t. In the error sums, however, the errors e_j^b, $e_j^\times$ again cancel to a large extent. Accordingly, the relative errors Pu_t do not grow systematically but remain bounded in modulus by about $8\eta = 4.-05$. In addition, it is easily verified using the listed results that the linear error approximations r_t in (5) approximate the relative errors Pu_t very accurately. □

1.2 The Summation Algorithm

The algorithm

$$u_0 = c_0, \qquad u_t = u_{t-1} + c_t, \qquad t = 1, \ldots, n, \tag{1}$$

yields the sequence of partial sums

$$u_t = \sum_{j=0}^{t} c_j. \tag{2}$$

Specifying the input of coefficients by intermediate steps $u_{t-1/2} = c_t$, the algorithm is defined by the sequence of functions

$$F_0(x) = c_0, \qquad F_{t-1/2}(x) = c_t, \qquad F_t(x) = x_{t-1} + x_{t-1/2}, \qquad t = 1, \ldots, n. \tag{3}$$

The linear absolute a priori error equations read

$$s_0 = c_0 e_0^c, \qquad s_{t-1/2} = c_t e_t^c, \qquad s_t = s_{t-1} + s_{t-1/2} + u_t e_t^+,$$

or

$$s_0 = c_0 e_0^c, \qquad s_t = s_{t-1} + c_t e_t^c + u_t e_t^+, \qquad t = 1, \ldots, n. \tag{4}$$

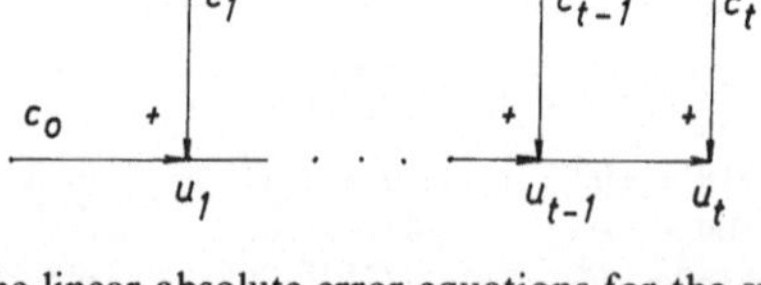

Fig. 1.2. Graph of the linear absolute error equations for the summation algorithm

The associated graph (see Fig. 1.2) is a tree, all nonvanishing weights b_{tk}^{abs} are equal to one. The absolute a priori condition numbers σ_t are determined according to I−2.3.(23) with the weights

$$\alpha_{t-1/2} = |c_t| \gamma_t^c, \qquad \alpha_t = |u_t| \gamma_t^+,$$

recursively by

$$\sigma_0 = |c_0| \gamma_0^c, \qquad \sigma_t = \sigma_{t-1} + |c_t| \gamma_t^c + |u_t| \gamma_t^+, \qquad t = 1, \ldots, n. \tag{5}$$

These condition numbers have the explicit representation

$$\sigma_t = |c_0|\gamma_0^c + \sum_{j=1}^{t} (|c_j|\gamma_j^c + |u_j|\gamma_j^+). \tag{6}$$

When the errors in the coefficients c_j have the same order of magnitude as rounding errors in the arithmetic operations, the partial condition numbers σ_t^D, σ_t^R of the summation algorithm under data perturbations only, assuming exact arithmetic operations ($\gamma_j^c = 1$, $\gamma_j^+ = 0$), and under rounding errors in the arithmetic operations, assuming exact data ($\gamma_j^c = 0$, $\gamma_j^+ = 1$), become

$$\sigma_t^D = \sum_{j=0}^{t} |c_j|, \qquad \sigma_t^R = \sum_{j=1}^{t} |u_j|. \tag{7}$$

The associated relative a priori condition numbers are

$$\rho_t^D = \frac{\sum_{j=0}^{t} |c_j|}{\left|\sum_{j=0}^{t} c_j\right|}, \qquad \rho_t^R = \frac{\sum_{j=1}^{t} |u_j|}{\left|\sum_{j=0}^{t} c_j\right|}, \tag{8}$$

such that

$$\rho_t = \frac{\sigma_t}{|u_t|} = \rho_t^D + \rho_t^R. \tag{9}$$

Let us first study the behaviour of the summation algorithm under data perturbations but for exact arithmetic operations ($\gamma_t^+ = 0$). If all terms c_j are positive, evidently,

$$\rho_t^D = 1, \qquad t = 1, 2, \ldots, n. \tag{10}$$

The situation is completely different in the case of alternating signs. As an example, consider the partial sums of the exponential series

$$\exp x = 1 + \frac{x}{1!} + \frac{x^2}{2!} + \cdots = \sum_{j=0}^{\infty} c_j.$$

Then

$$\sum_{j=0}^{\infty} |c_j| = \exp|x|$$

and for sufficiently large n one has

$$\sigma_n^D \doteq \exp|x|, \qquad \rho_n^D \doteq \frac{\exp|x|}{\exp x} = \begin{cases} 1, & x \geqslant 0, \\ \exp 2|x|, & x < 0. \end{cases} \tag{11}$$

Hence for negative x the condition numbers increase exponentially with increasing $|x|$. Forsythe [8] and Stegun-Abramowitz [15] discuss the example of computing $\exp(-5.5) = 4.0868-03$ where the terms c_j of the series are computed exactly in floating point to five decimal places such that $|Pc_j| \leqslant \eta = 5.-05$. The summation is extended over so many terms until the first five digits of the partial sums remain unchanged, that is, $rd_5(v_n) = rd_5(v_{n+1})$. Thus $n = 25$, $v_n = 2.6363-03$ and the associated condition number $\sigma_n^D \doteq \exp 5.5 \doteq 2.45+02$. From the error estimate (1) in the Introduction we obtain that

$$|\Delta u_n| \mathrel{\dot{\leqslant}} \sigma_n^D \eta \doteq 1.22-02 \tag{12}$$

whereas the actual absolute error $|\Delta u_n| \doteq 1.45-03$.

Analogous conditions are met in computing partial sums of the series for $\sin x$, $\cos x$, $\ln x$, $J_\nu(x)$, and similar alternating series.

Numerical Example 1. In the same way as above, partial sums of the Bessel functions

$$J_0(x) = \sum_{j=0}^{\infty} (-1)^j \frac{(x^2/4)^j}{(j!)^2} \tag{13}$$

and simultaneously, the absolute a priori condition numbers

$$\sigma_n^D = \sum_{j=0}^{n} \frac{(x^2/4)^j}{(j!)^2} \doteq I_0(x) \tag{14}$$

are computed. Tab. 1.2.1 shows some values of $J_0(x)$, the computed approximations $J_0(x)'$ and the absolute errors $\Delta J_0(x) = J_0(x)' - J_0(x)$ in the neighbourhood of the fourth zero 11.79153.... The function I_0 increases from 1.5351+04 to 1.6104+04 in this interval. Hence $1.6+04 \times 5.-05 = .8$ is the approximate a priori bound for the absolute errors $\Delta J_0(x)$. This bound is six times the largest error at $x = 11.82$. □

Table 1.2.1. *Numerical computation of $J_0(x)$*

x	$J_0(x)$	$J_0(x)'$	$\Delta J_0(x)$
11.78	− 2.6821 − 03	− 5.2142 − 02	− 4.95 − 02
11.79	− 3.5590 − 04	7.5159 − 02	7.52 − 02
11.80	1.9670 − 03	− 9.1710 − 02	− 9.37 − 02
11.81	4.2887 − 03	− 2.4652 − 02	− 2.89 − 02
11.82	6.6079 − 03	1.3835 − 01	1.32 − 01
11.83	8.9250 − 03	3.7195 − 02	2.83 − 02

When the terms c_j of the sum (2) are positive we have

$$\rho_t^D = 1, \qquad \rho_t^R \leqslant t, \tag{15}$$

and thus

$$\rho_t^R \leqslant t\rho_t^D, \qquad t = 1, \ldots, n. \tag{16}$$

Hence the computation of sums with positive terms is always quasi-well-conditioned. In contrast to the condition numbers ρ_t^D, the condition numbers ρ_t^R are not independent of the ordering of the terms $c_0, \ldots, c_t$. In computing the sums (2) in floating point arithmetic of a computer, the ordering of the terms should be so chosen that the condition numbers σ_n^R, ρ_n^R and thus the bounds in the error estimates (1) become as small as possible. For a sum with positive terms this is achieved if the terms c_j constitute an increasing sequence. Note that the ordering by increasing absolute values is commonly recommended without any restrictions. However, if both positive and negative terms c_j occur in the sum, this is, in general, no longer true as the following example shows. In this case the terms have to be ordered with respect to a smallest sum of the absolute values of the partial sums.

Numerical Example 2.

$$s = 1025 + (-912.3) + (-96.63) + (-9.315) \tag{17}$$

(see Wilkinson [22, I.25]). Adding $c_0 = 1025, \ldots, c_3 = -9.315$, one obtains the partial sums and, by (7), (9), the condition numbers

$$u_1 = 112.7, \quad u_2 = 16.07, \quad u_3 = 6.755; \quad \sigma_3^R \doteq 135.5, \quad \rho_3^R \doteq 20.06. \tag{18}$$

In 4-digit decimal floating point the summation is performed without rounding errors and gives the result $v_3 = u_3 = 6.755$. In converse order $c_0 = -9.315$, $\ldots, c_3 = 1025$, the terms are arranged with respect to increasing absolute values. Then

$$u_1 = -105.945, \quad u_2 = -1018.245, \quad u_3 = 6.755; \quad \sigma_3^R \doteq 1130, \quad \rho_3^R \doteq 167.4. \tag{19}$$

These condition numbers are more than eight times larger than the above condition numbers (18). In 4-digit decimal floating point now the approximate sum $v_3 = 7.000$ is computed having the relative error $Pu_3 \doteq 3.6-02$ and the approximate error bound $\rho_3^R\eta \doteq 8.4-02$. □

Numerical Example 3.

$$u_n = \sum_{k=0}^{n} \frac{1}{(k+1)^2}. \tag{20}$$

We compute this partial sum of the infinite series of $\pi^2/6 = 1.644934\ldots$ for $n = 1023$ in 6-digit decimal floating point. The exact partial sum rounded to six places is $u_n = 1.64396$. The associated absolute a priori condition numbers (5) are computed recursively from the equation

$$\sigma_0 = |c_0|, \qquad \sigma_t = \sigma_{t-1} + |u_t| + |c_t|, \qquad t = 1, \ldots, n. \tag{21}$$

The relative a priori condition number is finally determined by $\rho_n = \sigma_n/|u_n|$. The numerical summation in natural ordering yields partial sums which are constant for $t \geqslant m = 446$,

$$v_m = 1.64308, \qquad \rho_m \doteq 445, \qquad Pu_m \doteq 2.29-04. \tag{22}$$

The error bound $\rho_m\eta \doteq 2.23-03$ overestimates the error Pu_m by a factor of about 10. In converse order of increasing terms the algorithm computes

$$v_n = 1.64396, \qquad \rho_n \doteq 5.57. \tag{23}$$

This value coincides with the above rounded value of u_n in all decimal places. We observe a considerable difference in the magnitude of the condition numbers (22), (23). In this context, let us refer already to the result of § 1.3 dealing with cascade summation. There we shall compute

$$v_n = 1.64395, \qquad \rho_n = 11. \tag{24}$$

This approximation differs by one unit of the last decimal place from the rounded u_n. The last condition number (24) lies between the two condition numbers above such that the second summation procedure is best for this series. □

1.3 Cascade Summation

In the literature this summation algorithm is called 'binary summation' or 'Linz summation'. However, the method is discussed for the case $n = 4$ already by McCracken-Dorn [12, p. 62] and mentioned in Babuška [3] as 'process L_{1B}'. Linz [9] gives an error bound with $\log_2 n$. It seems that our error bounds for this algorithm, obtained by condition numbers, have not been known up to now. Let us remark in passing that the method discussed below and the associated error analysis can be applied in an analogous multiplicative form to the computation of products and the study of relative errors.

The preceding § 1.2 has dealt with the usual recursive computation of sums

$$y = \sum_{k=1}^{n} c_k = c_1 + c_2 + \cdots + c_n. \tag{1}$$

The *cascade summation* computes from the initial values

$$u_k^0 = c_k, \qquad k = 1, \ldots, n, \tag{2i}$$

recursively the partial sums

$$u_k^i = u_{2k-1}^{i-1} + u_{2k}^{i-1}, \qquad k = 1, \ldots, n_i, \qquad i = 1, \ldots, m, \tag{2ii}$$

where

$$n_i = \operatorname{int}\left(\frac{n}{2^i}\right), \qquad m = \operatorname{int}(\log_2 n).$$

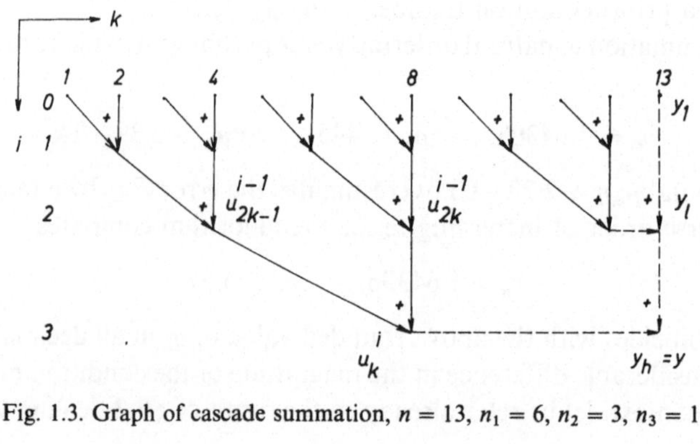

Fig. 1.3. Graph of cascade summation, $n = 13$, $n_1 = 6$, $n_2 = 3$, $n_3 = 1$

The graph of this algorithm is illustrated in Fig. 1.3. At each point (i, k) of the graph, u_k^i is the sum of the two preceding partial sums u_{2k-1}^{i-1}, u_{2k}^{i-1} at the two points $(i-1, 2k-1), (i-1, 2k)$.

We shall now analyze the error propagation in this algorithm. The floating point implementation of the algorithm is specified by

$$v_k^i = \mathrm{rd}(v_{2k-1}^{i-1} + v_{2k}^{i-1}), \qquad k = 1, \ldots, n_i, \qquad i = 1, \ldots, m, \tag{3}$$

where the initial values v_k^0 are appropriate approximations of the terms $c_k = u_k^0$. The associated linear absolute a priori error equations have the form

$$\begin{aligned} s_k^0 &= u_k^0 e_k^0, \\ s_k^i &= s_{2k-1}^{i-1} + s_{2k}^{i-1} + u_k^i e_k^i, \qquad k = 1, \ldots, n_i, \qquad i = 1, \ldots, m. \end{aligned} \tag{4}$$

The solutions s_k^i of this linear system are approximations of the a priori errors $\Delta u_k^i = v_k^i - u_k^i$, e_k^0 are the relative errors of the approximation v_k^0 of u_k^0, and e_k^i for $i \geqslant 1$ are the relative errors of the floating point additions in (3).

The coefficient matrix of the linear system (4) consists only of zeros and ones and thus is a binary matrix. As suggested by Fig. 1.3, the graph of this linear system is a union of trees. By virtue of I – 2.3.(23) with the weights $\alpha_k^i = \gamma_k^i |u_k^i|$, the absolute a priori condition numbers σ_k^i can be determined recursively from the equations

$$\begin{aligned} \sigma_k^0 &= \gamma_k^0 |u_k^0| \\ \sigma_k^i &= \sigma_{2k-1}^{i-1} + \sigma_{2k}^{i-1} + \gamma_k^i |u_k^i|, \qquad k = 1, \ldots, n_i, \qquad i = 1, \ldots, m. \end{aligned} \tag{5}$$

The weights γ_k^i are bounds for e_k^i in the appraisals

$$|e_k^i| \leqslant \gamma_k^i \eta. \tag{6}$$

By replacing u_k^i by v_k^i in (5), the corresponding recursion of the absolute a posteriori condition numbers is obtained. Note that these condition numbers can easily be computed together with v_k^i. In this way, valuable informations are computed concerning the possible errors of the computed partial sums.

Let us now consider the case that $n = 2^m$. Then $n_i = 2^{m-i}$ and the graph of the algorithm is a tree. It follows immediately from (2) that

$$y = \sum_{k=1}^{2^m} c_k = \sum_{k=1}^{2^{m-i}} u_k^i = u_1^m, \qquad i = 0, \ldots, m. \tag{7}$$

The solutions of the linear system (5) are in this case, according to I – 2.3.(25), simply sums of the weights $\alpha_k^i = \gamma_k^i |u_k^i|$. In this way, we conclude the representation

$$\sigma_1^m = \sum_{i=0}^{m} \sum_{k=1}^{2^{m-i}} \gamma_k^i |u_k^i|, \qquad \rho_1^m = \frac{1}{|y|} \sigma_1^m, \tag{8}$$

of the absolute and relative a priori condition numbers of cascade summation for computing the sum (1). When the sum has positive terms c_k and the weights γ_k^i are 1, (7), (8) yield the simple expressions

$$\sigma_1^m = (m+1)y, \qquad \rho_1^m = m + 1 = \log_2 n + 1. \tag{9}$$

If the exact terms c_k are used as initial values v_k^0, then we can put $e_k^0 = 0$ and $\gamma_k^0 = 0$. In this case, it is seen from (7), (8) with $\gamma_k^i = 1$, $i \geqslant 1$, that

$$\sigma_1^m = my, \qquad \rho_1^m = m = \log_2 n. \tag{10}$$

If n is not a power of 2, the graph consists of so many trees as there are ones in the binary representation of

$$n = 2^m + a_{m-1}2^{m-1} + \cdots + a_0. \tag{11}$$

To each $a_i \neq 0$, the point (i, n_i) of the graph is the root of a tree. The associated partial sums yield the total sum y in the form

$$y = \sum_{k=1}^{m} c_k = \sum_{\substack{i=0\\ a_i \neq 0}}^{m} u_{n_i}^i. \tag{12}$$

An error estimate for the numerically computed total sum is easily infered from the above results by extending the summation trivially over $c_k = 0$ to the next power of two, 2^{m+1}. The condition numbers σ_1^{m+1}, ρ_1^{m+1} of the extended summation are then upper bounds for condition numbers of the given summation. Finer estimates are obtained from (8) by choosing $\gamma_k^i = 0$ at all points where vanishing terms are input or added, that is, where $e_k^i = 0$.

Note that the computation of the partial sums from (2) can be organized columnwise (see Caprani [7]). In this ordering it suffices to store the last column of partial sums such that only $m + 1 = \log_2 n + 1$ storage places are needed.

Numerical Example.

$$y = \sum_{k=1}^{n} \frac{1}{k^2}.$$

By cascade summation in 6-digit decimal floating point for $n = 1024$, the approximation

$$y' = 1.64395, \qquad \rho = 11, \tag{13}$$

is computed. This result differs by one unit of the last decimal place from the exact sum y rounded to 6 places. A comparison with forward and backward recursive summation has been given in § 1.2. □

2. Elementary One-Step Algorithms

This class collects a series of related algorithms having the same basic structure: the product and summation algorithms of §§ 1.1, 1.2, Horner's scheme for Taylor and Newton polynomials, algorithms for solving bidiagonal linear systems and inhomogeneous linear recursions of first order, as well as the Horner-like evaluation of continued fractions. The graph of the general algorithm, defining this class, is a tree so that condition numbers can recursively be determined a priori and be computed a posteriori. The condition numbers are solutions of inhomogeneous linear recursions of first order. Thus the associated evaluation algorithms belong themselves to the class of elementary one-step algorithms. In addition, the recursions have positive coefficients so that the numerical computation of condition numbers is always quasi-stable.

2.1 Condition Numbers and Stability Properties

Given two sequences of real numbers $a, b_1, b_2, \ldots, c_1, c_2, \ldots$, the elementary one-step algorithm computes the sequence of numbers

$$u_0 = a, \qquad u_t = b_t \circ_t u_{t-1} + c_t, \qquad t = 1, \ldots, n, \tag{1i}$$

where the operations $\circ_t$ are, for the present, specified only by

$$\circ_t \in \{\times, /, \backslash\}. \tag{1ii}$$

That is, $y \circ_t z$ is the product yz or the quotient y/z or z/y. This algorithm consists of the following single steps

$$\begin{aligned} u_0 = a, \qquad u_{t-3/4} &= b_t, \qquad u_{t-1/2} = b_t \circ_t u_{t-1}, \\ u_{t-1/4} &= c_t, \qquad u_t = u_{t-1/2} + u_{t-1/4}, \end{aligned} \tag{2}$$

for $t = 1, \ldots, n$. By use of the functions

$$\begin{aligned} F_0(x) = a, \qquad F_{t-3/4}(x) &= b_t, \qquad F_{t-1/2}(x) = x_{t-3/4} \circ_t x_{t-1}, \\ F_{t-1/4}(x) &= c_t, \qquad F_t(x) = x_{t-1/2} + x_{t-1/4}, \end{aligned} \tag{3}$$

the algorithm has a form corresponding to I−1.2.(A), only the numbering $t = 0, \ldots, n$ is replaced by $t = 0, \frac{1}{4}, \frac{2}{4}, \ldots, n$. Fig. 2.1 shows the graph of this

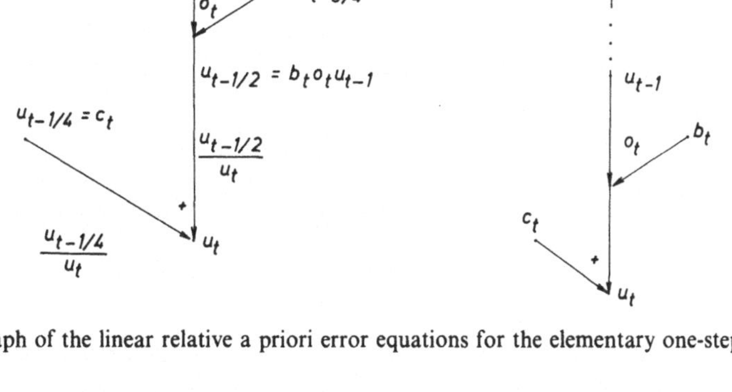

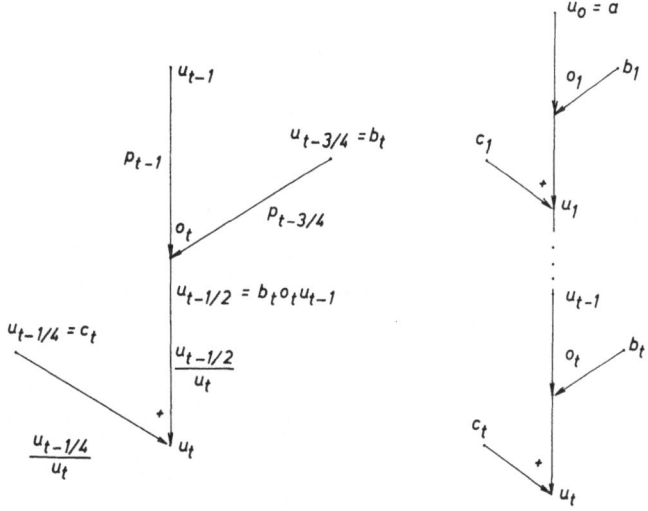

Fig. 2.1. Graph of the linear relative a priori error equations for the elementary one-step algorithm

algorithm and Tab. 2.1 the weights $p_{t-1}, p_{t-3/4}$ for the operations $\circ_t$. The following linear relative a priori error equations are read from the graph,

$$r_0 = e_0^a, \qquad r_{t-3/4} = e_t^b, \qquad r_{t-1/2} = p_{t-1} r_{t-1} + p_{t-3/4} r_{t-3/4} + e_t^0,$$

$$r_{t-1/4} = e_t^c, \qquad r_t = \frac{u_{t-1/4}}{u_t} r_{t-1/4} + \frac{u_{t-1/2}}{u_t} r_{t-1/2} + e_t^+, \tag{4}$$

where the notation of the local errors has been adapted to the present algorithm. Consequently, the approximations r_t of the relative a priori errors Pu_t suffice the recursion

$$r_0 = e_0^a,$$

$$r_t = \frac{b_t \circ_t u_{t-1}}{u_t} (p_{t-1} r_{t-1} + p_{t-3/4} e_t^b + e_t^0) + \frac{c_t}{u_t} e_t^c + e_t^+, \qquad t = 1, \ldots, n. \tag{5}$$

Table 2.1. *Weights for the operations* $\circ_t$

$\circ_t$	p_{t-1}	$p_{t-3/4}$
×	1	1
/	1	-1
\	-1	1

The graph of the algorithm is a tree as it is readily seen from Fig. 2.1. Hence the associated condition numbers can be determined recursively according to I – 2.3.(23). The relative a priori condition numbers ρ_t for the computation of u_t are thus obtained by

$$\rho_0 = \gamma_0^a, \qquad \rho_{t-3/4} = \gamma_t^b, \qquad \rho_{t-1/2} = \rho_{t-1} + \rho_{t-3/4} + \gamma_t^0,$$

$$\rho_{t-1/4} = \gamma_t^c, \qquad \rho_t = \left|\frac{u_{t-1/4}}{u_t}\right| \rho_{t-1/4} + \left|\frac{u_{t-1/2}}{u_t}\right| \rho_{t-1/2} + \gamma_t^+, \tag{6}$$

or simpler

$$\rho_0 = \gamma_0^a, \qquad \rho_t = \left|\frac{b_t \circ_t u_{t-1}}{u_t}\right| (\rho_{t-1} + \gamma_t^b + \gamma_t^0) + \left|\frac{c_t}{u_t}\right| \gamma_t^c + \gamma_t^+ \tag{7}$$

for $t = 1, \ldots, n$. By γ_0^a, γ_t^b, γ_t^c are meant the weights of the input errors of the coefficients a, b_t, c_t such that

$$|e_0^a| \leqslant \gamma_0^a \eta, \qquad |e_t^b| \leqslant \gamma_t^b \eta, \qquad |e_t^c| \leqslant \gamma_t^c \eta, \tag{8}$$

and by γ_t^0, γ_t^+ the weights of the rounding errors in the arithmetic operations such that

$$|e_t^0| \leqslant \gamma_t^0 \eta, \qquad |e_t^+| \leqslant \gamma_t^+ \eta. \tag{9}$$

The absolute a priori condition numbers $\sigma_t = |u_t| \rho_t$ are determined correspondingly by the recursion formulae

$$\sigma_0 = |a| \gamma_0^a, \qquad \sigma_t = |b_t \circ_t u_{t-1}| \left(\frac{\sigma_{t-1}}{|u_{t-1}|} + \gamma_t^b + \gamma_t^0\right) + |c_t| \gamma_t^c + |u_t| \gamma_t^+ \tag{10}$$

for $t = 1, \ldots, n$. Finally, the relative a posteriori condition numbers are specified by recursions analogous to (6) where $v_t, v_{t-1/2}, v_{t-1/4}$ replace $u_t, u_{t-1/2}, u_{t-1/4}$.

Elementary one-step algorithms possess favorable numerical stability properties with respect to relative errors under the condition

$$|u_t| = |b_t \circ_t u_{t-1}| + |c_t|, \qquad t = 1, 2, \ldots . \tag{11}$$

This condition holds trivially when $c_t = 0$ for all t as in the computation of sequences of partial products. Further (11) is evidently valid in case the coefficients a, b_t, c_t are nonnegative for all t. This happens in computations of sequences of partial sums with nonnegative terms, in Horner's scheme with nonnegative coefficients c_t and argument $b_t = z$, and in evaluations of continued fractions with nonnegative coefficients. Under the assumption (11),

$$\left|\frac{b_t \circ_t u_{t-1}}{u_t}\right| \leqslant 1, \qquad \left|\frac{c_t}{u_t}\right| \leqslant 1, \qquad t = 1, 2, \ldots, n,$$

thus

$$\rho_t \leqslant \rho_{t-1} + \max(\gamma_t^b + \gamma_t^0, \gamma_t^c) + \gamma_t^+,$$

that is,

$$\rho_t \leqslant \gamma_0^a + \sum_{j=1}^{t} \{\max(\gamma_j^b + \gamma_j^0, \gamma_j^c) + \gamma_j^+\}, \qquad t = 1, \ldots, n. \tag{12}$$

For computations of sequences of partial products ($\gamma_0^a = \gamma_t^b = \gamma_t^0 = 1, \gamma_t^c = \gamma_t^+ = 0$) and partial sums ($\gamma_t^b = \gamma_t^0 = 0, \gamma_0^a = \gamma_t^c = \gamma_t^+ = 1$) one concludes from (12) that

$$\rho_t \leqslant 2t + 1, \qquad t = 1, \ldots, n. \tag{13}$$

If a condition of this form holds, we call the algorithm quasi-stable.

The algorithm has favorable numerical stability with respect to absolute errors under the condition

$$\left|\frac{b_t \circ_t u_{t-1}}{u_{t-1}}\right| \leqslant q \leqslant 1. \tag{14}$$

For, in this case we infer from (10) the appraisal

$$\sigma_t \leqslant q\sigma_{t-1} + q|u_{t-1}|(\gamma_t^b + \gamma_t^0) + |c_t|\gamma_t^c + |u_t|\gamma_t^+$$

for $t = 1, \ldots, n$. Thus

$$\sigma_t \leqslant q^t|a|\gamma_0^a + \sum_{j=1}^{t} q^{t-j}\{q|u_{j-1}|(\gamma_j^b + \gamma_j^0) + |c_j|\gamma_j^c + |u_j|\gamma_j^+\}. \tag{15}$$

If $q < 1$, a damping of preceding error effects is observed. The smaller q and the earlier an error originated the stronger is the damping. Let μ be some constant such that

$$|a|\gamma_0^a \leqslant \mu, \qquad q|u_{t-1}|(\gamma_t^b + \gamma_t^0) + |c_t|\gamma_t^c + |u_t|\gamma_t^+ \leqslant \mu \tag{16}$$

for all t, then we conclude from (15) the estimate

$$\sigma_t \leqslant \frac{\mu}{1-q}, \qquad t = 1, \ldots, n. \tag{17}$$

In case $\mu/(1-q)$ is not much greater than 1 the elementary one-step algorithm is said to be strongly stable with respect to absolute errors. In Horner's scheme, for instance, $b_t = z$, $\circ_t = \times$ such that the above condition (14) holds for $q = |z| < 1$. Another example will be obtained by the computation of determinants for tridiagonal matrices.

2.2 *Horner's Scheme*

The well-known Horner's scheme has the form

$$u_0 = c_0, \qquad u_t = zu_{t-1} + c_t, \qquad t = 1, \ldots, n. \tag{1}$$

It computes the sequence of polynomials

$$u_t = p_t(z) = c_t + c_{t-1}z + \cdots + c_0 z^t, \qquad t = 1, \ldots, n. \tag{2}$$

This algorithm is a special case of the elementary one-step algorithm with the coefficients and operations

$$a = c_0, \qquad b_t = z, \qquad \circ_t = \times\,. \tag{3}$$

The perturbations of function values $p(z)$, due to perturbations of the argument z, have been described already by I – 1.1.(10). Hence we limit the further study to the case of unperturbed arguments, realized in practice by the input of machine numbers as arguments z.

The general error analysis of the elementary one-step algorithm is first applied to perturbations of the coefficients c_t. On choosing

$$\gamma_0^a = \gamma_t^c = 1, \qquad \gamma_t^b = \gamma_t^0 = \gamma_t^+ = 0, \tag{4}$$

in § 2.1, the associated relative a priori condition numbers ρ_t^D are specified by the recursion

$$\rho_0^D = 1, \qquad \rho_t^D = \left|\frac{zu_{t-1}}{u_t}\right| \rho_{t-1}^D + \left|\frac{c_t}{u_t}\right|, \tag{5}$$

and the absolute a priori condition numbers $\sigma_t^D = |u_t|\rho_t^D$ by

$$\sigma_0^D = |c_0|, \qquad \sigma_t^D = |z|\sigma_{t-1}^D + |c_t|, \qquad t = 1, \ldots, n. \tag{6}$$

Evidently, these recursion formulae have the solutions

$$\sigma_t^D = \sum_{j=0}^{t} |c_j z^{t-j}|, \qquad \rho_t^D = \frac{\sum_{j=0}^{t} |c_j z^{t-j}|}{\left|\sum_{j=0}^{t} c_j z^{t-j}\right|}, \qquad t = 1, \ldots, n. \tag{7}$$

When the coefficients c_j and arguments z are positive,

$$\sigma_t^D = p_t(z), \qquad \rho_t^D = 1. \tag{8}$$

In the case of alternating signs, however, the condition numbers may become large. The Tschebyscheff polynomials constitute a particularly unfavourable case (see Reimer-Zeller [14], Reimer [13]). For instance

$$T_{10}(x) = 512x^{10} - 1280x^8 + 1120x^6 - 400x^4 + 50x^2 - 1$$

is a polynomial of fifth degree in $z = x^2$. In the interval $-1 \leqslant x \leqslant 1$, we have $0 \leqslant z \leqslant 1$, $|T_{10}(x)| \leqslant 1$ and

$$\sigma_5^D = 512z^5 + 1280z^4 + 1120z^3 + 400z^2 + 50z + 1 \leqslant 3363.$$

Obviously, there is a neighbourhood of $x = \pm 1$ or $z = 1$ where the relative a priori condition number $\rho_5^D = \sigma_5^D/|T_{10}(x)| \geqslant 3000$.

If the coefficients c_t are floating point numbers, but the arithmetic operations are perturbed, the condition numbers σ_t^R, ρ_t^R of Horner's scheme are of interest. Choosing, in § 2.1, the weights

$$\gamma_0^a = \gamma_t^b = \gamma_t^c = 0, \qquad \gamma_t^0 = \gamma_t^+ = 1, \tag{9}$$

we obtain

$$\rho_0^R = 0, \qquad \rho_t^R = \left|\frac{zu_{t-1}}{u_t}\right|(\rho_{t-1}^R + 1) + 1, \qquad t = 1, \ldots, n. \tag{10}$$

The associated absolute a priori condition numbers $\sigma_t^R = |u_t|\rho_t^R$ thus suffice the recursion

$$\sigma_0^R = 0, \qquad \sigma_t^R = |z|(\sigma_{t-1}^R + |u_{t-1}|) + |u_t|, \qquad t = 1, \ldots, n, \tag{11}$$

having the solution

$$\sigma_t^R = \sum_{j=1}^{t} |z|^{t-j}(|u_j| + |zu_{j-1}|). \tag{12}$$

The corresponding recursion formulae of the relative a posteriori condition numbers have the form

$$\rho_0^R = 0, \qquad \rho_t^R = \left|\frac{v_{t-1/2}}{v_t}\right|(\rho_{t-1}^R + 1) + 1, \qquad t = 1, \ldots, n. \tag{13}$$

Now $v_{t-1/2} = (1 + O(\eta))zv_{t-1}$. Hence the associated a posteriori condition numbers $\sigma_t^R = |v_t|\rho_t^R$ are, save for a factor $1 + O(\eta)$, equal to

$$\sum_{j=1}^{t} |z|^{t-j}(|v_j| + |v_{j-1/2}|). \tag{14}$$

This bound for the absolute a posteriori error $|\Delta v_t|/\eta$ has been stated by Tsao [21, (5.5)] who also proposes the recursive computation of the bound. A similar result is found in Adams [2] where it is attributed to Kahan.

Numerical Example 1. The polynomial

$$p(x) = 5x^3 - 1515x^2 + 153010x - 5151000 \tag{15}$$

has the zeros 100, 101, 102. The polynomial is evaluated in 5-digit decimal floating point such that $\eta = 5.-05$. The computed polynomial values are denoted by $v_3(z)$. The absolute errors $\Delta p(z) = v_3(z) - p(z)$, and the absolute a priori condition numbers $\sigma_3(z)$ have been computed to 10 decimal places. The numerical results in Tab. 2.2.1 show the absolute error $\Delta p(z) = -770.27$ at $z = 99.005$ and the condition number $\sigma_3(z) = 3.5150+07$. That is, the error bound $\sigma_3(z)\eta = 1.7575+03$ is only by a factor of 2.28 greater than the absolute value of the error. □

Table 2.2.1. *Argument $z, p(z)$, approximation $v_3(z)$ of $p(z)$, absolute a priori error $\Delta p(z)$ and associated absolute a priori condition number $\sigma_3(z)$*

9.9001+01	9.9002+01	9.9003+01	9.9004+01	9.9005+01
−2.9946+01	−2.9892+01	−2.9840+01	−2.9778+01	−2.9728+01
0.0000+00	1.0000+02	1.0000+02	2.0000+02	−8.0000+02
2.9946+01	1.2989+02	1.2984+02	2.2978+02	−7.7027+02
3.5148+07	3.5148+07	3.5149+07	3.5149+07	3.5150+07
9.9006+01	9.9007+01	9.9008+01	9.9009+01	9.9010+01
−2.9669+01	−2.9621+01	−2.9564+01	−2.9508+01	−2.9453+01
−7.0000+02	−7.0000+02	−6.0000+02	−6.0000+02	−5.0000+02
−6.7033+02	−6.7038+02	−5.7044+02	−5.7049+02	−4.7055+02
3.5150+07	3.5151+07	3.5151+07	3.5152+07	3.5152+07

Polynomials in Newton's representation are evaluated by the generalized Horner's scheme

$$u_0 = c_0, \qquad u_t = (z - z_{t-1})u_{t-1} + c_t, \qquad t = 1, \dots, n, \tag{16}$$

computing the sequence of polynomials

$$u_t = p_t(z) = c_t + c_{t-1}(z - z_0) + \cdots + c_0(z - z_0) \cdots (z - z_{t-1}). \tag{17}$$

Numerical Example 2. The paper of Mesztenyi-Witzgall [10] gives an example for the differing stability properties of evaluating a polynomial p first in Taylor form and then in a particularly suitable Newton form. The Taylor representation of the polynomial p is

$$p^T(x) = c_5^T + c_4^T x + c_3^T x^2 + c_2^T x^3 + c_1^T x^4 + x^5 \tag{18}$$

with the coefficients

$$c_0^T = 1, \qquad c_3^T = 8.4247503796,$$
$$c_1^T = -3.0593781606, \qquad c_4^T = -11.291738407,$$
$$c_2^T = 0.921133131886, \qquad c_5^T = 4.1007470240.$$

The Newton representation of p reads

$$p^N(x) = c_5^N + c_3^N(x - x_0)^2 + c_2^N(x - x_0)^2(x - 1) + c_1^N(x - x_0)^2(x - 1)^2 + (x - x_0)^2(x - 1)x \tag{19}$$

where

$$c_0^N = 1, \quad c_4^N = 0, \qquad c_3^N = 3.4126984127,$$

$$c_1^N = \quad 0.60784\,31372\,5, \qquad c_5^N = 1.03199\,1744-03,$$

$$c_2^N = -\,1.87912\,08791, \qquad x_0 = 0.83361\,06489\,2.$$

The polynomial values $p(x)$ are positive and possess at the point $x = x_0$ their minimum in the interval $[0, 1]$. At this point $p(x_0) = c_5^N$ is very small. This is the reason why Horner's scheme for evaluating p in Taylor form is ill-conditioned in the neighbourhood of x_0. We compute $p^N(x)$, $p^T(x)$ by Horner's scheme in 10-digit decimal floating point. The coefficient c_0 is 1 and the arguments x are exactly floating point numbers such that $e_0^a = e_t^b = 0$, for $t = 1, \ldots, 5$ in evaluating p^T. Instead of x_0, we use its approximation to 10 places in computing $p^N(x)$. Since

$$b_1, \ldots, b_5 = x, x - 1, x - 1, x - x_0, x - x_0,$$

also for the Newton form $b_1, \ldots, b_5$ are exactly floating point numbers. The associated relative a posteriori condition numbers are obtained, according to 2.1.(6), by

$$\rho_0 = 0, \qquad \rho_t = \left|\frac{v_{t-1/2}}{v_t}\right|(\rho_{t-1} + 1) + \left|\frac{c_t'}{v_t}\right| + 1, \qquad t = 1, \ldots, 5. \tag{20}$$

Table 2.2.2. *Evaluation of Taylor's and Newton's form p^N, p^T of a polynomial p, associated relative condition numbers and errors*

x	$p^N(x)$	ρ_5^N	$p^T(x)$	ρ_5^T	$\frac{p^N - p^T}{p^N}(x) \times 10^{10}$
.2	2.1821 83418	6.93	2.1821 83417	6.15+00	4.6
.4	9.2288 41614−01	6.75	9.2288 41600−01	2.19+01	1.5+01
.6	2.3884 34630−01	6.50	2.3884 34620−01	1.22+02	4.2+01
.8	5.3754 09972−03	5.44	5.3754 08000−03	7.70+03	3.7+03
.82	1.7353 22570−03	3.71	1.7354 21000−03	2.47+04	9.0+03
.83361	1.0319 91746−03	2.00	1.0319 90000−03	4.25+04	1.7+04
.84	1.1850 98272−03	2.54	1.1850 97000−03	3.74+04	1.1+04
.86	3.6118 28966−03	4.98	3.6118 27000−03	1.27+04	5.4+03
1.0	9.5513 96752−02	5.96	9.5513 96500−02	6.06+02	2.6+02

The condition numbers ρ_5^N in Tab. 2.2.2 show that the Newton polynomial p^N is evaluated in a stable and very accurate way for all x, the condition number is nearly constant in the interval. The evaluation of tne Taylor polynomial, however, is comparatively ill-conditioned in the neighbourhood of x_0, the loss of significant figures amounts up to four leading digits. We observe that the variation of the computed condition number ρ_5^T is in good agreement with that of the relative error $(p^N - p^T)/p^T$. □

2.3 Evaluation of Continued Fractions

Finite continued fractions

$$w = c_n + \frac{b_n}{c_{n-1} +} \cdots \frac{b_2}{c_1 +} \frac{b_1}{c_0} \tag{1}$$

are evaluated, analogously to Horner's scheme, by the algorithm

$$u_0 = c_0, \qquad u_t = c_t + \frac{b_t}{u_{t-1}}, \qquad t = 1, \ldots, n. \tag{2}$$

This is the elementary one-step algorithm with the coefficients and operations

$$a = c_0, \quad b_t, \quad \circ_t = /, \quad t = 1, \ldots, n. \tag{3}$$

On choosing the weights $p_{t-1} = 1, p_{t-3/4} = -1$, Fig. 2.1 shows the graph of this algorithm. The sequence of relative a priori condition numbers is obtained recursively from 2.1.(7). If the coefficients b_t, c_t are floating point numbers, the numerical evaluation of the continued fraction is perturbed only by rounding errors in the arithmetic operations. In this case, one puts

$$\gamma_0^a = \gamma_t^b = \gamma_t^c = 0, \qquad \gamma_t^/ = \gamma_t^+ = 1. \tag{4}$$

The sequence of relative a priori condition numbers then satisfies the recursion

$$\rho_0^R = 0, \qquad \rho_t^R = \left|\frac{b_t}{u_t u_{t-1}}\right|(\rho_{t-1}^R + 1) + 1, \qquad t = 1, \ldots, n, \tag{5}$$

and the absolute a priori condition numbers $\sigma_t^R = |u_t|\rho_t^R$ are obtained from

$$\sigma_0^R = 0, \qquad \sigma_t^R = \left|\frac{b_t}{u_{t-1}}\right|\left(\frac{\sigma_{t-1}^R}{|u_{t-1}|} + 1\right) + |u_t|, \qquad t = 1, \ldots, n. \tag{6}$$

Numerical Example 1. The continued fraction of $\tan z$ (see Abramowitz-Stegun [1, p. 75]) yields the approximations

$$\tan_n z = \frac{z}{1-}\,\frac{z^2}{3-} \cdots \frac{z^2}{2n+1}$$

Table 2.3.1. *Evaluation of the continued fraction for $z/\tan_n z$, associated relative a priori condition numbers ρ_t and errors Pu_t*

t	v_t	ρ_t^R	$10^4 Pu_t$	v_t	ρ_t^R	$10^4 Pu_t$	v_t	ρ_t^R	$10^4 Pu_t$
0	21.00	0	0	21.00	0	0	21.00	0	0
1	18.88	1.006	−1.311	18.53	1.025	−1.413	15.96	1.190	−1.790
2	16.87	1.016	.4169	16.47	1.065	−.2388	13.00	1.673	−3.441
3	14.85	1.020	−2.490	14.40	1.085	−2.834	10.09	2.299	−5.482
4	12.83	1.026	−2.997	12.32	1.115	1.140	6.679	4.119	−5.021
5	10.81	1.036	2.132	10.20	1.165	−3.173	1.451	34.57	−30.65
6	8.772	1.053	.4056	8.038	1.259	−.0627	−34.96	45.76	39.87
7	6.719	1.086	.5376	5.779	1.477	.1164	8.824	10.70	−8.656
8	4.633	1.165	.7260	3.302	2.274	1.243	−2.228	39.02	28.09
9	2.467	1.468	−1.058	.02800	359.0	298.1	31.63	37.24	−24.28
10	0	8243	−10000	−349.5	362.0	−291.3	−1.016	77.03	44.06

$\frac{z}{\tan z} = -2.995{-}04$, $z^2 = 2.468$; $\quad \frac{z}{\tan z} = -360.0$, $z^2 = 9.815$; $\quad \frac{z}{\tan z} = -1.062$, $z^2 = 63.78$

and, consequently,

$$\frac{z}{\tan_n z} = 1 + \frac{-z^2}{3+} \cdots \frac{-z^2}{2n+1} \qquad (\tan_n z \neq 0). \tag{7}$$

This continued fraction has the form (1) with the coefficients

$$b_t = -z^2, \quad c_t = 2(n-t)+1, \quad t = 0, \ldots, n.$$

Tab. 2.3.1 lists numerical results of evaluating (7) in 4-digit decimal floating point. The associated relative errors Pu_t and condition numbers ρ_t^R have been computed in 13-digit floating point. The error bound $\rho_t^R \times 5.-04$ overestimates $|Pu_t|$ to various extent. For $t = 4(5, 3)$ the estimates are closest and $5\rho_t^R/(10^4 Pu_t)$ is approximately 2. In most cases the error is overestimated by a factor between 5 and 10. □

The computation of determinants of tridiagonal matrices

$$A = \begin{pmatrix} a_0 & c_0 & & 0 \\ b_1 & a_1 & c_1 & \\ & \ddots & \ddots & \ddots \\ & b_{n-1} & a_{n-1} & c_{n-1} \\ 0 & & b_n & a_n \end{pmatrix} \tag{8}$$

by Gaussian elimination without pivoting yields another application of the continued fraction algorithm (2). It is well known that the determinant of A is obtained by the sequence

$$d_0 = a_0, \qquad d_t = a_t - \frac{b_t c_{t-1}}{d_{t-1}} = \frac{1}{d_{t-1}} \begin{vmatrix} d_{t-1} & c_{t-1} \\ b_t & a_t \end{vmatrix}, \tag{9}$$

for $t = 1, \ldots, n$, in the form

$$\det A = d_0 d_1 \cdots d_n. \tag{10}$$

The sequence (d_t) plays an important role also in solving associated tridiagonal linear systems. The recursion (9) is the continued fraction algorithm (2) with the coefficients $a_0, -b_t c_{t-1}, a_t$ instead of a, b_t, c_t. The recursion of the relative a priori condition numbers 2.1.(7) now takes on the form

$$\rho_0 = \gamma_0^a, \qquad \rho_t = \left|\frac{b_t c_{t-1}}{d_t d_{t-1}}\right| (\rho_{t-1} + \gamma_t^{bc} + \gamma_t') + \left|\frac{a_t}{d_t}\right| \gamma_t^a + \gamma_t^+, \tag{11}$$

and the associated absolute a priori condition numbers $\sigma_t = |d_t|\rho_t$ are obtained from

$$\sigma_0 = |a|\gamma_0^a, \qquad \sigma_t = \frac{|b_t c_{t-1}|}{d_{t-1}^2} (\sigma_{t-1} + |d_t|(\gamma_t^{bc} + \gamma_t')) + |a_t|\gamma_t^a + |d_t|\gamma_t^+ \tag{12}$$

for $t = 1, \ldots, n$.

When the weak row sum criterion (see Stummel-Hainer [20, § 6.3])

$$|b_t| + |c_t| \leqslant |a_t|, \quad |b_{t+1}| < |a_t|, \quad |c_t| < |a_t|, \quad t = 0, \ldots, n, \tag{13}$$

holds for the matrix A and its transposed A^T, where $b_0 = c_{-1} = 0$ and $b_{n+1} = c_n = 0$, then

$$d_t \neq 0, \qquad \left|\frac{c_t}{d_t}\right| < 1, \qquad \left|\frac{b_{t+1}}{d_t}\right| < 1, \qquad t = 0, \ldots, n. \tag{14}$$

In this case

$$|b_t c_{t-1}| < d_{t-1}^2 \tag{15}$$

so that the sequence of condition numbers σ_t is majorized by

$$\sigma_t \leqslant |a|\gamma_0^a + \sum_{j=1}^{t} \{|d_{j-1}|(\gamma_j^{bc} + \gamma_j') + |a_j|\gamma_j^a + |d_j|\gamma_j^+\} \tag{16}$$

for $t = 1, \ldots, n$. Simple numerical examples are specified by the coefficients

$$a_t = a, \qquad b_t = c_t = 1.$$

If a is a floating number, we can put $\gamma_t^a = \gamma_t^{bc} = 0$ and $\gamma_t' = \gamma_t^+ = 1$ in (12) so that

$$\sigma_0^R = 0, \qquad \sigma_t^R = \frac{1}{|d_{t-1}|}\left(\frac{\sigma_{t-1}}{|d_{t-1}|} + 1\right) + |d_t|, \qquad t = 1, \ldots, n. \tag{17}$$

The weak row sum criterion is true in this case whenever $|a| \geqslant 2$. Then the following appraisal is inferred from (16),

$$\sigma_t^R \leqslant \sum_{j=1}^{t} \{|d_{j-1}| + |d_j|\}. \tag{18}$$

Numerical Example 2. The sequence

$$d_0 = a, \qquad d_t = -\frac{1}{d_{t-1}} + a, \qquad t = 1, \ldots, 100, \tag{19}$$

is computed in 4-digit decimal floating point. Tab. 2.3.2 shows the computed values d_t' for $a = 1.922$ and $t = 10, 20, \ldots, 100$ together with the relative a priori condition

Table 2.3.2. *Numerical solutions d_t' of (19) for $a = 1.922$, associated relative a priori condition numbers ρ_t^R, and errors Pu_t*

t	d_t'	ρ_t^R	$10^4 Pu_t$
0	1.922	0	0
10	−3.454	819.4	−640.2
20	.3140	465.5	294.2
30	.6610	107.0	57.02
40	.8150	65.94	30.15
50	.9210	56.35	28.99
60	1.016	60.36	27.73
70	1.126	80.68	39.28
80	1.294	142.3	62.56
90	1.714	403.4	180.5
100	23.20	12460	9952

numbers ρ_t^R and errors Pu_t. Here too, the bound $\rho_t^R \times 5.-04$ overestimates the error by a factor between about 5 and 10. □

3. Determinants of Second Order and 'Fast' Complex Multiplication

The computation of determinants of second order is a plain example in our error analysis. However the associated condition numbers play a crucial role in the error analysis of Cramer's rule for two linear equations in two unknowns, as has been shown in [17]. Let us remark in passing that similarly the linear error equations of the basic step in Gaussian elimination are derived (see Bauer [5]). For the sake of brevity, we can only discuss a simple but interesting application in which the numerical stability of the common complex multiplication is compared to that of a so-called 'fast' complex multiplication.

3.1 Determinants of Second Order

For evaluating determinants of second order

$$D = ad - bc = \begin{vmatrix} a & b \\ c & d \end{vmatrix}, \tag{1}$$

first the coefficients

$$u_0 = a, \qquad u_1 = b, \qquad u_2 = c, \qquad u_3 = d, \tag{2i}$$

are input, then the products, and finally the difference is computed

$$u_4 = ad, \qquad u_5 = bc, \qquad u_6 = u_4 - u_5. \tag{2ii}$$

This algorithm has the form I – 1.2.(A) with the functions

$$F_0(x) = a, \qquad F_1(x) = b, \qquad F_2(x) = c, \qquad F_3(x) = d,$$
$$F_4(x) = x_0x_3, \qquad F_5(x) = x_1x_2, \qquad F_6(x) = x_4 - x_5. \tag{3}$$

The graph of the linear relative a priori error equations is illustrated in Fig. 3.1.

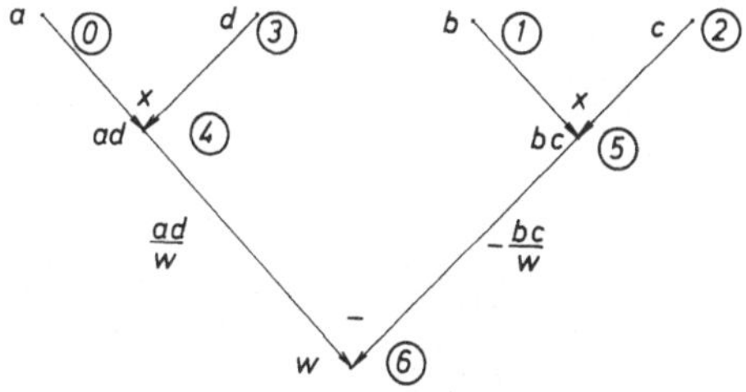

Fig. 3.1. Graph of the linear relative a priori error equations of evaluating $ad - bc$

The corresponding system of linear error equations thus reads

$$r_0 = e_0, \qquad r_1 = e_1, \qquad r_2 = e_2, \qquad r_3 = e_3,$$

$$r_4 = r_0 + r_3 + e_4, \qquad r_5 = r_1 + r_2 + e_5, \qquad r_6 = \frac{ad}{w}r_4 - \frac{bc}{w}r_5 + e_6. \tag{4}$$

The error approximation $r = r_6$ of PD has the explicit representation

$$r = \frac{ad}{D}(e_0 + e_3 + e_4) - \frac{bc}{D}(e_1 + e_2 + e_5) + e_6, \tag{5}$$

where $e_0, \ldots, e_3$ denote the relative errors of $a, \ldots, d$, e_4, e_5 the relative errors of the two multiplications, and e_6 of the final subtraction. The graph of the algorithm is a tree such that

$$\rho = 3\delta + 1, \qquad \delta = \frac{|ad| + |bc|}{|ad - bc|}, \tag{6}$$

is the relative a priori condition number of computing D.

If a, b, c, d are already floating point numbers, that is, $e_0 = \cdots = e_3 = 0$, the relative a priori condition number, with respect to rounding errors in the arithmetic operations only, reads

$$\rho^R = \delta + 1. \tag{7}$$

Similarly, the dependence of the algorithm on data perturbations, assuming exact arithmetic operations ($e_4 = e_5 = e_6 = 0$), has the relative a priori condition number

$$\rho^D = 2\delta. \tag{8}$$

Consequently,

$$\rho^R = \tfrac{1}{2}\rho^D + 1, \tag{9}$$

so that rounding errors in the arithmetic operations act, in essence, as data perturbations do and thus the evaluation of determinants of second order is well-conditioned.

The above condition numbers can be applied in a simple manner to Cramer's rule for solving two linear equations in two unknowns,

$$ax + by = f, \qquad cx + dy = g. \tag{10}$$

Using the determinants

$$D_0 = \begin{vmatrix} a & b \\ c & d \end{vmatrix}, \qquad D_1 = \begin{vmatrix} f & b \\ g & d \end{vmatrix}, \qquad D_2 = \begin{vmatrix} a & f \\ c & g \end{vmatrix}, \tag{11}$$

the unknowns x, y are determined by

$$x = \frac{D_1}{D_0}, \qquad y = \frac{D_2}{D_0}, \tag{12}$$

provided that $D_0 \neq 0$. Applying the simple error estimates of I – § 1.1 to the computation of x, y as quotients D_1/D_0, D_2/D_0 we obtain, assuming exact data $a, \ldots, g$ and putting $\rho_a = \delta_1 + 1, \delta_2 + 1, \rho_b = \delta_0 + 1$, $\gamma = 1$ in I – 1.1.(19), the following bounds of the relative a priori condition numbers ρ_x^R, ρ_y^R,

$$\tau_x^R = \delta_0 + \delta_1 + 3, \qquad \tau_y^R = \delta_0 + \delta_2 + 3. \tag{13}$$

Under data perturbations but exact arithmetic operations, putting $\rho_a = 2\delta_1, 2\delta_2,$

$\rho_b = 2\delta_0$, $\gamma = 0$ in I – 1.1.(19), the bounds

$$\tau_x^D = 2\delta_0 + 2\delta_1, \qquad \tau_y^D = 2\delta_0 + 2\delta_1, \tag{14}$$

are infered. These are condition numbers for data perturbations x, y in the case that the coefficients b, d and a, c are perturbed independently in the nominator and denominator of x, y in (12). The general condition numbers of computing x, y by Cramer's rule have been determined in [17]. It has been shown there that $\rho_x^R = \tau_x^R$, $\rho_y^R = \tau_y^R$ and that Cramer's rule is a well-conditioned algorithm. Note that [17] contains a series of numerical examples illustrating our error analysis of Cramer's rule and Gaussian elimination for two linear equations in two unknowns.

3.2 'Fast' Complex Multiplication

The real and imaginary parts of the product ab of two complex numbers $a = a_1 + ia_2$, $b = b_1 + ib_2$ have the form

$$\mathrm{Re}(ab) = a_1 b_1 - a_2 b_2, \qquad \mathrm{Im}(ab) = a_1 b_2 + a_2 b_1. \tag{1}$$

These are determinants of second order whose computation has been analyzed in the preceding section. As one readily verifies, the product ab is also obtained by use of the algorithm

$$\begin{aligned} &x_1 = a_1 + a_2, \qquad x_2 = b_1 + b_2, \qquad x_3 = b_1 - b_2,\\ &y_1 = x_1 b_1, \qquad y_2 = a_2 x_2, \qquad y_3 = a_1 x_3,\\ &\mathrm{Re}(ab) = y_1 - y_2, \qquad \mathrm{Im}(ab) = y_1 - y_3, \end{aligned} \tag{2}$$

(see Miller [11, p. 104]). To evaluate (1) four multiplications are needed whereas (2) requires three multiplications. In complexity theory, therefore, the second method is called fast. The second method, however, uses five additions or subtractions instead of only two in the common complex multiplication.

Fig. 3.2 shows the graph of the linear error equations of computing $w = \mathrm{Re}(ab)$ by 'fast' complex multiplication. This graph is not a tree because there are two paths

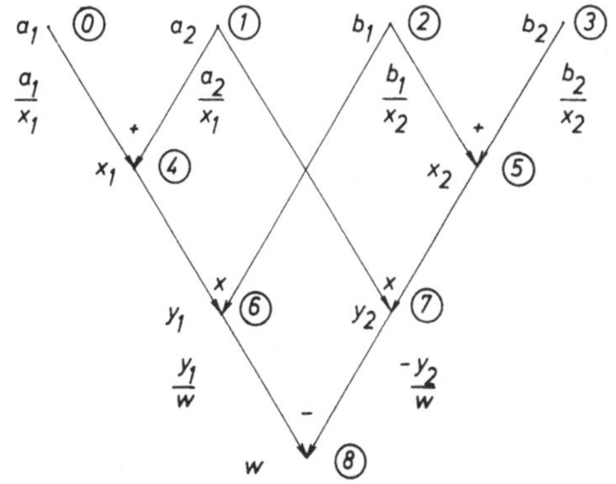

Fig. 3.2. Graph of linear relative a priori error equations of computing $w = \mathrm{Re}(ab)$ by 'fast' complex multiplication

from the nodes 1, 2 to the node 8. From the graph we immediately infer the linear error equation

$$r = \frac{y_1}{w}\left(\frac{a_1}{x_1}e_0 + e_4 + e_6\right) + \left(\frac{a_2}{x_1}\frac{y_1}{w} - \frac{y_2}{w}\right)e_1$$

$$-\frac{y_2}{w}\left(\frac{b_2}{x_2}e_3 + e_5 + e_7\right) + \left(\frac{y_1}{w} - \frac{b_1}{x_2}\frac{y_2}{w}\right)e_2 + e_8$$

for the approximation $r = r_8$ of the relative a priori error Pw of approximations w' of $w = \mathrm{Re}(ab)$. Using (2), the coefficients in the last equation can be simplified whereby

$$r = \frac{a_1 b_1}{w}(e_0 + e_2) + \frac{a_2 b_2}{w}(e_1 - e_3)$$

$$+ \frac{y_1}{w}(e_4 + e_6) - \frac{y_2}{w}(e_5 + e_7) + e_8. \tag{3}$$

Thus the associated relative a priori condition number reads

$$\rho = \frac{2}{|w|}(|a_1 b_1| + |a_2 b_2| + |y_1| + |y_2|) + 1. \tag{4}$$

The special condition numbers ρ^R of the accumulated rounding error for exact data ($e_0 = \cdots = e_3 = 0$), and ρ^D for data perturbations but exact arithmetic operations ($e_4 = \cdots = e_8 = 0$) have the form

$$\rho^R = 2\frac{|y_1| + |y_2|}{|w|} + 1, \qquad \rho^D = 2\delta, \qquad \delta = \frac{|a_1 b_1| + |a_2 b_2|}{|w|}. \tag{5}$$

In general, ρ^R cannot be majorized by ρ^D so that the 'fast' algorithm is not well-conditioned. The ill-conditioning of this algorithm, compared to the usual complex multiplication, is seen from the following

Numerical Example.

$$a_1 = 1.5-03, \qquad a_2 = 1.01; \qquad b_1 = 1.01, \qquad b_2 = 1.00-03; \tag{6}$$

gives $w = \mathrm{Re}(ab) = 5.05-04$. Computing w from (1) in 3-digit floating point yields

$$w' = 5.10-04, \qquad \rho_D^R = \delta + 1 = 6, \qquad Pw \doteq 1.-02, \qquad \rho_D^R\eta \doteq 3.-02,$$

whereas the 'fast' multiplication computes

$$w' = 0, \qquad \rho^R = 8091, \qquad Pw = -1, \qquad \rho^R\eta \doteq 40.$$ □

References

[1] Abramowitz, M., Stegun, I. A.: Handbook of mathematical functions. New York: Dover 1968.
[2] Adams, D. A.: A stopping criterion for polynomial root finding. Comm. ACM **10**, 655–658 (1967).
[3] Babuška, I.: Numerical stability in numerical analysis. Proc. IFIP Congress 1968, pp. 11–23. Amsterdam: North Holland 1969.

[4] Babuška, I.: Numerical stability in problems of linear algebra. SIAM J. Numer. Anal. **9**, 53–77 (1972).

[5] Bauer, F. L.: Genauigkeitsfragen bei der Lösung linearer Gleichungssysteme. Z. Angew. Math. Mech. **46**, 409–421 (1966).

[6] Bauer, F. L.: Computational graphs and rounding error. SIAM J. Numer. Anal. **11**, 87–96 (1974).

[7] Caprani, O.: Implementation of a low round-off summation method. BIT **11**, 271–275 (1971).

[8] Forsythe, G. E.: Pitfalls in computation, or why a math book isn't enough. Amer. Math. Monthly **77**, 931 – 956 (1970).

[9] Linz, P.: Accurate floating-point summation. Comm. ACM **13**, 361–362 (1970).

[10] Mesztenyi, C., Witzgall, C.: Stable evaluation of polynomials. J. Res. Nat. Bur. Standards **71B**, 11–17 (1967).

[11] Miller, W.: Computational complexity and numerical stability. SIAM J. Comput. **4**, 97–107 (1975).

[12] McCracken, D. D., Dorn, W. S.: Numerical methods and Fortran programming. New York-London: Wiley 1964.

[13] Reimer, M.: Bounds for the Horner sums. SIAM J. Numer. Anal. **5**, 461–469 (1968).

[14] Reimer, M., Zeller, K.: Abschätzung der Teilsummen reeller Polynome. Math. Zeitschr. **99**, 101–104 (1967).

[15] Stegun, I. A., Abramowitz, M.: Pitfalls in computation. J. Soc. Indust. Appl. Math. **4**, 207–219 (1956).

[16] Stummel, F.: Fehleranalyse numerischer Algorithmen. Vorlesungsskriptum WS 1977/78, 215 pp. Universität Frankfurt 1978.

[17] Stummel, F.: Rounding errors in numerical solutions of two linear equations in two unknowns. (Submitted to Numer. Math.)

[18] Stummel, F.: Perturbation theory for evaluation algorithms of arithmetic expressions. (To appear in Math. Comp.)

[19] Stummel, F.: Rounding error analysis of difference and extrapolation schemes. (To appear.)

[20] Stummel, F., Hainer, K.: Introduction to numerical analysis. Edinburgh-London: Scottish Academic Press 1977.

[21] Tsao, Nai-Kuan: Some a posteriori error bounds in floating point computations. J. ACM **21**, 6–17 (1974).

[22] Wilkinson, J. H.: Rounding errors in algebraic processes. Englewood Cliffs, N. J.: Prentice-Hall 1963.

Prof. Dr. F. Stummel
Fachbereich Mathematik
Johann-Wolfgang-Goethe-Universität
Robert-Mayer-Strasse 10
D-6000 Frankfurt am Main
Federal Republic of Germany

Computing, Suppl. 2, 197 – 209 (1980)

Iterative Methods in the Spaces of Rounded Computations

Ch. Ullrich, Karlsruhe

Abstract

Numerical algorithms are usually executed in a space over the finite set of floating-point numbers. Numerous properties are missing in such spaces in contrast to mathematical spaces in which we are used to study algorithms. For this reason numerical algorithms show another behaviour than we would expect based on theoretical investigations. This paper summarizes some results for iterative methods, which can be derived directly by algebraic and order properties of the spaces of rounded computations.

1. Introduction

In analysis there are used three components to describe a mathematical space: (a) algebraic structure, (b) order structure, (c) topological structure, where each component is connected to the other by certain compatibility properties. Respecting the spaces of rounded computations the structures (a) and (b) are known as well as the compatibility properties between them [7], [8], [10]. They are given by the structure of the weakly-ordered (or ordered) ringoid and the weakly-ordered (or ordered) vectoid, respectively. In the case of the topological structure the situation is different. Clearly, all spaces of rounded computations are metric spaces since they are subsets of the corresponding real or complex spaces. But the compatibility properties between the topological structure and the structures (a) and (b), which we are used to work with, are no longer valid (compare [3]). We consider the following

Example. We denote by $\mathbb{R}$ the field of real numbers and by $\hat{\mathbb{R}} := \mathbb{R}(b, l, e1, e2) \subseteq \mathbb{R}$ the following floating-point-system:

$$\begin{aligned}\mathbb{R}(b, l, e1, e2) = \{x \in \mathbb{R} \mid x = *0 \cdot d_1 d_2 \cdots d_l \cdot b^e, * \in \{+, -\},\\ d_i \in \{0, 1, \ldots, b-1\} \text{ for } i = 1, 2, \ldots, l,\ d_1 \neq 0,\ e \in Z,\\ e1 \leqslant e \leqslant e2\} \cup \{0\}.\end{aligned}$$

Then, $\{\hat{\mathbb{R}}, d\}$ with $d(a, b) := |a - b|$, for example, is a metric space. But usual properties like $d(a \boxplus b, a \boxplus c) = d(a, c)$ cannot be verified, where $\boxplus$ denotes the addition in $\hat{\mathbb{R}}$ introduced according to

$$\text{(RG)} \qquad \bigwedge_{a, b \in \hat{\mathbb{R}}} a \boxed{*}\, b := \square(a * b), \qquad * \in \{+, \cdot, /\},$$

and a rounding $\square\colon \mathbb{R} \to \hat{\mathbb{R}}$. So we compute in $\mathbb{R}(10, 2, e1, e2)$

$$0.95 = d(18, 0.85) = d(0.27 \boxplus 1.5, 0.27 \boxplus 0.58) = d(1.5, 0.58) = 0.92,$$

if $\square$ is the rounding to the nearest number in $\hat{\mathbb{R}}$.

As an immediate consequence iterative methods show, in general, another behaviour during their execution than we would expect according to theoretical investigations. To illustrate this fact we consider the following example [6]:

Example. Let $\mathscr{A}$ be the 2 × 2 real matrix

$$\mathscr{A} = \begin{pmatrix} 0.11 & 0.87 \\ 0.96 & 0.03 \end{pmatrix}$$

and b the vector

$$b = \begin{pmatrix} 3.4 \\ -3.6 \end{pmatrix}$$

and let us seek the solution of $x = \mathscr{A}x + b$.

The matrix $\mathscr{A}$ is non-negative with spectral radius $\rho(\mathscr{A})$, $0.98 < \rho(\mathscr{A}) < 0.99$. Therefore, the mapping $f(x) = \mathscr{A}x + b$ is a contraction and the iterative method

$$x^{(n+1)} = \mathscr{A}x^{(n)} + b, \qquad n = 0, 1, 2, \ldots$$

converges to the unique fixed point of $x = \mathscr{A}x + b$ for arbitrary $x^{(0)}$.

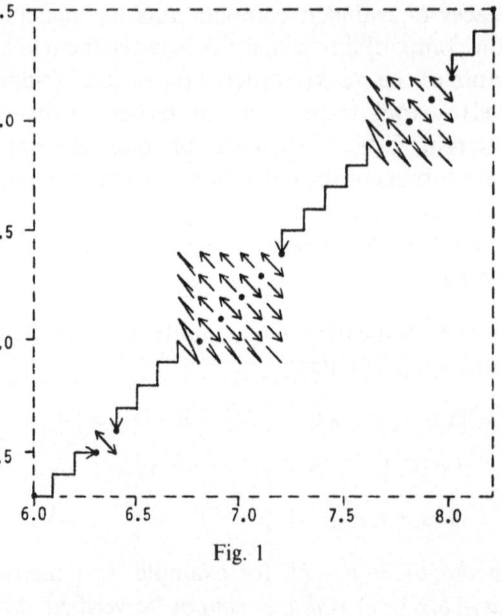

Fig. 1

We now execute this iterative method in $\hat{\mathbb{R}} = \mathbb{R}(10, 2, e1, e2)$ (see first example), i.e. we compute

$$x^{(n+1)} = \mathscr{A} \boxdot x^{(n)} \boxplus b, \qquad n\ 3\ 1, 2, \ldots.$$

Fig. 1 shows that in the interval

$$\left[\begin{pmatrix} 6 \\ 2.3 \end{pmatrix}, \begin{pmatrix} 8.2 \\ 4.5 \end{pmatrix}\right]$$

already exist 11 fixed points and 17 pairs of periodic points of the mapping

$$\mathscr{f}(x) := \mathscr{A} \boxdot x \boxplus \mathscr{b}.$$

Therefore, in the following sections we consider iterative methods directly in the spaces of rounded computations and ask for their properties. Because of the described situation we use only the algebraic and order structure of these spaces for our investigations. Hence the main results for the considered iterative methods naturally are monotonicity properties, which however give us interesting informations about fixed and periodic points of the iteration sequences.

2. Ordered Ringoids and Vectoids

We denote by $\{R, \leqslant\}$ an ordered set and by $\leqslant_I$ the order relation for elements $x = (x_i)$, $y = (y_i)$ of the n-dimensional vector space V_nR over R according to

$$x \leqslant_I y :\Leftrightarrow \left(\bigwedge_{i \in I} x_i \leqslant y_i \wedge \bigwedge_{i \in A} x_i \geqslant y_i \right),$$

where I is a subset of $N := \{1, 2, \ldots, n\}$ and $A := N \backslash I$ (see [4]). The order relation $\geqslant_I$ describes the order $d(\leqslant_I)$ dual to $\leqslant_I$. Further we consider the ordered ringoid $\{R, +, \cdot, \leqslant\}$ with the special elements $\{-e, 0, e\}$, the ordered R-vectoid $\{V_nR, R, \leqslant_I\}$ over R and the ordered M_nR-vectoid $\{V_nR, M_nR, \leqslant_I\}$ over R, where M_nR denotes the set of $n \times n$-matrices over R ordered by the following relation which is consistent with $\{V_nR, \leqslant_I\}$:

$$\mathscr{A} \leqslant_I \mathscr{B} :\Leftrightarrow \left(\bigwedge_{(i,j) \in I \times I \cup A \times A} a_{ij} \leqslant b_{ij} \wedge \bigwedge_{(i,j) \in I \times A \cup A \times I} a_{ij} \geqslant b_{ij} \right)$$

with $\mathscr{A} = (a_{ij})$, $\mathscr{B} = (b_{ij}) \in M_nR$.

The elements $\mathscr{A} \in M_nR$ with $\mathscr{A} \geqslant_I \mathscr{O}$ (resp. $\mathscr{A} \leqslant_I \mathscr{O}$) represent the isotone (resp. antitone) mappings from $\{V_nR, \leqslant_I\}$ to itself in M_nR ($\mathscr{O}$ denotes the null matrix). As usually we call a mapping isotone, if it is order preserving, antitone, if it is order reverting, and monotone, if the mapping is either isotone or antitone.

An $n \times n$ matrix $\mathscr{A} = (a_{ij}) \geqslant_I \mathscr{O}$ can be recognized by the corresponding sign distribution (see [4]):

(i) $\bigwedge_{i \in N} a_{ii} \geqslant 0$

(ii) $\bigwedge_{i,j \in N} (a_{ij} \geqslant 0 \Leftrightarrow a_{ji} \geqslant 0)$

(iii) lines and columns with index i, $i \in I$ (resp. $i \in A$) have the same sign distribution

(iiii) two lines (resp. columns) with indices i and j, $i \in I$ and $j \in A$, have dual sign distribution.

Note that in the case $a_{ij} = 0$ either positive or negative sign can be assumed. Therefore a sparse $n \times n$ matrix can be isotone or antitone with respect to different orderings $\leqslant_I$, $\leqslant_{I'}$ with $I, I' \subseteq N$. We illustrate this situation by the diagonal matrices $\mathscr{E}$ and $-\mathscr{E}$ respectively, which are isotone and antitone, respectively, with respect to each ordering $\leqslant_I$, $I \subseteq N$, and the matrix $\mathscr{A}$

$$\mathscr{E} = \begin{pmatrix} e & & 0 \\ & \ddots & \\ 0 & & e \end{pmatrix}, \quad -\mathscr{E} = \begin{pmatrix} -e & & 0 \\ & \ddots & \\ 0 & & -e \end{pmatrix},$$

$$\mathscr{A} = \left(\begin{array}{c|c|cccc} e & 0 & -e & \cdots & -e \\ \hline 0 & e & 0 & \cdots & 0 \\ \hline -e & 0 & e & \cdots & e \\ \vdots & \vdots & \vdots & & \vdots \\ -e & 0 & e & \cdots & e \end{array}\right)$$

with two corresponding sign distributions:

$$\left(\begin{array}{c|ccc} + & - & \cdots & - \\ \hline - & + & \cdots & + \\ \vdots & \vdots & & \vdots \\ - & + & \cdots & + \end{array}\right) \qquad \left(\begin{array}{cc|ccc} + & + & - & \cdots & - \\ + & + & - & \cdots & - \\ \hline - & - & + & \cdots & + \\ \vdots & \vdots & \vdots & & \vdots \\ - & - & + & \cdots & + \end{array}\right)$$

In the first case the ordering $\leqslant_I$ is given by $I = \{1\}$ and in the second case by $I = \{1, 2\}$. In a similar way we can consider isotone mappings from $\{V_n R, \leqslant_{I_n}\}$ into $\{V_M R, \leqslant_{I_m}\}$ (see [12]). Then an $m \times n$ matrix $\mathscr{A} = (a_{ij}) \in M_{mn} R$ is isotone, if for example the columns of $\mathscr{A}$ with indices in I_n (resp. in A_n) are not less o (resp. not greater o) with respect to the ordering $\{V_m R, \leqslant_{I_m}\}$. However, we note that an isotone mapping from $\{V_n R, \leqslant_{I_n}\}$ into $\{V_m R, \leqslant_{I_m}\}$ is an antitone mapping from $\{V_n R, \leqslant_{I_n}\}$ into $\{V_m R, \geqslant_{I_m}\}$ but an isotone mapping from $\{V_n R, d(\leqslant_{I_n})\}$ into $\{V_m R, d(\leqslant_{I_m})\}$. Therefore, for given orderings $\{V_n R, \leqslant_{I_n}\}$ and $\{V_m R, \leqslant_{I_m}\}$ we get the sign distribution of an antitone matrix replacing $+$ by $-$ and conversely $-$ by $+$ in the known sign distribution of an isotone matrix.

3. Fixed and Periodic Points

In the first section we have seen that during the execution of an iterative method the computed sequence of iterates does not converge to a fixed point in the usual way. However, because of the finiteness of the considered spaces it is clear that the sequence becomes periodical after a finite number of iteration steps. We describe this fact by the following definition (compare [2], [5]):

Definition 1. A sequence $\{x^{(k)}\}_{k=0}^{\infty}$, $x^{(k)} \in V_n R$ for $k = 0, 1, 2, \ldots$ is called "cyclic" if property (Z) holds:

(Z)
$$\bigvee_{\mu, k_0 \in \mathbb{N}} \bigwedge_{k \geqslant k_0} x^{(k+\mu)} = x^{(k)}.$$

The number $m := \min\{\mu \in \mathbb{N} \mid (Z) \text{ holds for } \mu\}$ is called "length of the cycle". The set $Z(x^{(0)}) := \{x^{(k_0)}, x^{(k_0+1)}, \ldots, x^{(k_0+m-1)}\} \subseteq V_nR$ is called "cycle of $\{x^{(k)}\}$" and we call the elements $x \in Z(x^{(0)})$ "periodic points" of $\{x^{(k)}\}$.

If the sequence $\{x^{(k)}\}$ is computed by an iterative method

$$\text{(IT)} \qquad x^{(k+1)} := f(x^{(k)}), \qquad k = 0, 1, 2, \ldots,$$

where f is a mapping from V_nR into V_nR, we obtain immediately for each periodic point $z \in Z(x^{(0)})$:

$$f^m(z) = z \quad \text{and} \quad f^i(z) \neq z \quad \text{for} \quad 1 \leqslant i < m.$$

Furthermore, for chosen element $z \in Z(x^{(0)})$ the remaining periodic points can be denoted by $f^i z$, $1 \leqslant i \leqslant m-1$, and for all $x, y \in Z(x^{(0)})$ the property $x \leqslant_I y$ does not hold, if f is an isotone mapping of the ordered set $\{V_nR, \leqslant_I\}$. Besides some interesting relations between fixed and periodic points of isotone mappings the following theorem represents a fundamental result [5]:

Theorem 2. *Let $\{V_nR, \leqslant_I\}$ be an ordered set and f: $V_nR \to V_nR$ a mapping with the property*

$$\text{(KI)} \qquad \bigwedge_{x, z \in V_nR} x \leqslant_I z \Rightarrow f^k(x) \leqslant_I f^k(y) \qquad (k\text{-isotone})$$

for a certain natural number k. Further let $\{x^{(k)}\}_{k=0}^{\infty}$ be a sequence computed according to (IT), *which satisfies property* (Z) *and contains two elements $x^{(i)}$, $x^{(j)}$ with $x^{(i)} \leqslant_I x^{(j)}$ and k divides $|i-j|$.*

Then, $\{x^{(k)}\}$ has m periodic points, where m divides $|i-j|$, and it is $Z(y^{(0)}) = Z(x^{(0)})$ for all sequences $\{y^{(k)}\}_{k=0}^{\infty}$ computed by (IT) *with*

$$y^{(0)} \in [\inf(x^{(i)}, x^{(j)}), \sup(x^{(i)}, x^{(j)})].$$

Evidently Theorem 2 can be applied immediately to isotone (resp. antitone) mappings, but also to mappings which satisfy property (KI) only for $k > 2$. For example, in [11], [12], the so-called weakly cyclic mappings are considered which are k-isotone under certain assumptions but not monotone.

Summarizing this section we can say that we are interested in property (KI) of the operator which is used in (IT) for computing iteration sequences. We want to investigate this question for several iterative methods in the following sections.

4. Splittings of a Mapping

Let $f(x)$: $V_nR \to V_nR$ be a mapping from V_nR into itself. Then there exist a lot of different iterative methods, which can be associated with $f(x)$. For description of these methods we define:

Definition 3. Let $f(x) = (f_i(x))$: $V_nR \to V_nR$ a mapping from V_nR to V_nR. A mapping $g(x, y)$ from $V_nR \times V_nR$ to V_nR with

$$\text{(S)} \qquad \bigwedge_{x \in V_nR} g(x, x) = f(x)$$

is called a "splitting" of $f(x)$. A splitting $g(x, y)$ is called "explicit", if the component functions $g_i(x, y)$ are depending at most on the first $i - 1$ components of x, i.e. $g_i(x, y) \equiv g_i((x_1, \ldots, x_{i-1}), y)$, $x = (x_i) \in V_n R$.

Now, let $\{V_n R, \leqslant\}$ be an ordered set. A splitting $g(x, y)$ of $f(x)$ is called "isotone (resp. antitone) in x", if (M1) holds:

$$\text{(M1)} \qquad \bigwedge_{x_1, x_2, y \in V_n R} (x_1 \leqslant x_2 \Rightarrow g(x_1, y) \leqslant (\text{resp.} \geqslant)\, g(x_2, y))$$

and "isotone (resp. antitone) in y", if (M2) holds:

$$\text{(M2)} \qquad \bigwedge_{x, y_1, y_2 \in V_n R} (y_1 \leqslant y_2 \Rightarrow g(x, y_1) \leqslant (\text{resp.} \geqslant)\, g(x, y_2)).$$

Remark. If $f(x) = (f_i(x))$ is an isotone mapping from the ordered set $\{V_n R, \leqslant_I\}$ to $\{V_n R, \leqslant_I\}$, we get a splitting $g(x, y)$ of $f(x)$ by the following manipulation: Define $g_i(x, y)$, $i = 1, \ldots, n$ replacing all y_j by x_j, $j \in N_i \subseteq \{1, \ldots, n\}$ in $f_i(y)$. Then, $g(x, y)$ is a splitting of $f(x)$, which is isotone in x and y. The splitting $g(x, y)$ is explicit, if $\max N_i < i$ for all $i \in \{1, \ldots, n\}$.

However, we may guarantee the existence of a splitting $g(x, y)$ of $f(x)$ with properties (M1) and (M2), if $f(x)$ is not isotone, but certain monotonicity assumptions hold for the component functions $f_i(x)$, $i = 1, \ldots, n$:

Theorem 4. *Let $f(x) = (f_i(x))$: $V_n R \to V_n R$ be a mapping of $\{V_n R, \leqslant_I\}$ to itself with*

$$\bigwedge_{i=1(1)n} \bigwedge_{j=1(1)n} (x_j \leqslant y_j \Rightarrow f_i(x_1, \ldots, x_n) \leqslant f_i(x_1, \ldots, x_{j-1}, y_j, x_{j+1}, \ldots, x_n) \vee$$
$$f_i(x_1, \ldots, x_n) \geqslant f_i(x_1, \ldots, x_{j-1}, y_j, x_{j+1}, \ldots, x_n)).$$

Then, there exist a splitting $g(x, y)$ of $f(x)$, which is isotone in x and antitone in y. This splitting is explicit, if the first unequality holds for all arguments with index $j \geqslant i$ of $f_i(x)$, $i = 1, \ldots, n$ and the second unequality otherwise.

Proof. We construct the splitting $g(x, y)$ by replacing x_j in $f_i(x)$ by y_j, if (i, j) is an element of $I \times I \cup A \times A$ and the second inequality holds or (i, j) is an element of $I \times A \cup A \times I$ and the first inequality holds. The completion of the proof is left to the reader.

As an immediate consequence of Theorem 4 we note the following corollary which is an extension of a well-known result in real vector spaces:

Corollary 5. *Let $f(x) = \mathcal{A} x + b$ a mapping from $\{V_n R, \leqslant_I\}$ to itself with $\mathcal{A} = (a_{ij}) \in M_n R$, $b = (b_i) \in V_n R$. Then, the mapping*

$$g(x, y) := \left(\sum_{j=1}^{n} (a_{ij}^{+} x_j - a_{ij}^{-} y_j) + b_i \right),$$

where $x = (x_i)$, $y = (y_i) \in V_n R$ and $\mathcal{A}^+ = (a_{ij}^+)$, $\mathcal{A}^- = (a_{ij}^-) \in M_n R$ are defined by

$$\bigwedge_{1 \leqslant i, j \leqslant n} a_{ij}^{+} := \begin{cases} a_{ij} & \text{if } (i, j) \in I \times I \cup A \times A \text{ and } a_{ij} \geqslant 0 \\ & \text{or } (i, j) \in I \times A \cup A \times I \text{ and } a_{ij} \leqslant 0 \\ 0 & \text{otherwise} \end{cases}$$

and

$$\bigwedge_{1\leqslant i,j\leqslant n} a_{ij}^- := \begin{cases} -a_{ij} & \text{if } (i,j)\in I\times I\cup A\times A \text{ and } a_{ij}\leqslant 0 \\ & \text{or } (i,j)\in I\times A\cup A\times I \text{ and } a_{ij}\geqslant 0 \\ 0 & \text{otherwise,}\end{cases}$$

is a splitting of $f(x)$ *which is isotone in* x *and antitone in* y *with respect to* $\{V_nR, \leqslant_I\}$.

Remark. However, the $n\times n$ matrices $\mathscr{A}^+$ and $\mathscr{A}^-$ are isotone mappings referring to $\{V_nR, \leqslant_I\}$. We show this statement by (OV1), (OV2), and (OV3) in $\{V_nR, M_nR, \leqslant_I\}$:

$$\bigwedge_{x_1,x_2,y\in V_nR} (x_1\leqslant_I x_2 \Rightarrow \mathscr{A}^+x_1 \leqslant_I \mathscr{A}^+x_2 \Rightarrow \mathscr{A}^+x_1 - \mathscr{A}^-y \leqslant_I \mathscr{A}^+x_2 - \mathscr{A}^-y$$
$$\Rightarrow (\mathscr{A}^+x_1 - \mathscr{A}^-y) + b \leqslant_I (\mathscr{A}^+x_2 - \mathscr{A}^-y) + b)$$

$$\bigwedge_{x,y_1,y_2\in V_nR} (y_1\leqslant_I y_2 \Rightarrow \mathscr{A}^-y_1 \leqslant_I \mathscr{A}^-y_2 \Rightarrow -\mathscr{A}^-y_1 \geqslant_I -\mathscr{A}^-y_2$$
$$\Rightarrow \mathscr{A}^+x - \mathscr{A}^-y_1 \geqslant_I \mathscr{A}^+x - \mathscr{A}^-y_2$$
$$\Rightarrow (\mathscr{A}^+x - \mathscr{A}^-y_1) + b \geqslant_I (\mathscr{A}^+x - \mathscr{A}^-y_2) + b).$$

But the mapping $(\mathscr{A}^+x - \mathscr{A}^-y) + b$ is not a splitting of $f(x)$, for the equality $\mathscr{A}x + b = (\mathscr{A}^+x - \mathscr{A}^-x) + b$ is not valid because of the missing associative law.

5. Iterative Methods Associated to Splittings

A typical result for iterative methods in the spaces of rounded computations we obtain in the following (compare [1], [9])

Theorem 6. *Let* $f(x) := (f_i(x))$ *be a mapping from* $\{V_nR, \leqslant_I\}$ *to itself and* $g(x, y)$ *a splitting of* $f(x)$, *which is isotone in* x *and antitone in* y. *With* $v^{(0)}, w^{(0)}\in V_nR$, $v^{(0)}\leqslant_I w^{(0)}$, *we define two sequences* $\{v^{(n)}\}_{n=0}^{\infty}$, $\{w^{(n)}\}_{n=0}^{\infty}$ *by*

$$v^{(n+1)} = g(v^{(n)}, w^{(n)}),$$
$$w^{(n+1)} = g(w^{(n)}, v^{(n)}), \qquad n\geqslant 0.$$

If $v^{(0)}\leqslant_I v^{(1)}\leqslant_I w^{(1)}\leqslant_I w^{(0)}$, *then* $v^{(n)}\leqslant_I v^{(n+1)}\leqslant_I w^{(n+1)}\leqslant_I w^{(n)}$ *for all* $n\geqslant 1$ *and the sequence* $\{u^{(n)}\}_{n=0}^{\infty}$, *generated by*

$$u^{(n+1)} := f(u^{(n)}), \qquad n\geqslant 0$$

starting with $u^{(0)}\in[v^{(0)}, w^{(0)}]$, *has the property*

$$u^{(n)}\in[v^{(n)}, w^{(n)}] \qquad \text{for all } n\leqslant 0.$$

Proof. Assuming $v^{(n)}\leqslant_I v^{(n+1)}\leqslant_I w^{(n+1)}\leqslant_I w^{(n)}$ we obtain

$$v^{(n+2)} = g(v^{(n+1)}, w^{(n+1)}) \geqslant_I g(v^{(n)}, w^{(n+1)}) \geqslant_I g(v^{(n)}, w^{(n)}) = v^{(n+1)},$$
$$w^{(n+2)} = g(w^{(n+1)}, v^{(n+1)}) \leqslant_I g(w^{(n)}, v^{(n+1)}) \leqslant_I g(w^{(n)}, v^{(n)}) = w^{(n+1)},$$
$$w^{(n+2)} = g(v^{(n+1)}, w^{(n+1)}) \leqslant_I g(w^{(n+1)}, w^{(n+1)}) \leqslant_I g(w^{(n+1)}, v^{(n+1)}) = w^{(n+2)}$$

i.e. $v^{(n+1)}\leqslant_I v^{(n+2)}\leqslant_I w^{(n+2)}\leqslant_I w^{(n+1)}$.

Consider now an element $u \in [v^{(n)}, w^{(n)}]$, i.e. $v^{(n)} \leqslant_I u \leqslant_I w^{(n)}$. Then

$$v^{(n+1)} = g(v^{(n)}, w^{(n)}) \leqslant_I g(u, w^{(n)}) \leqslant_I g(u, u) \leqslant_I f(u) \leqslant_I$$
$$\leqslant_I g(w^{(n)}, u) \leqslant_I g(w^{(n)}, v^{(n)}) = w^{(n+1)},$$

i.e. $f(u) \in [v^{(n+1)}, w^{(n+1)}]$.

Theorem 6 shows us that the behaviour of the sequence $\{u^{(n)}\}_{n=0}^{\infty}$ is determined by the behaviour of the two sequences $\{v^{(n)}\}_{n=0}^{\infty}$ and $\{w^{(n)}\}_{n=0}^{\infty}$, which are monotone increasing (resp. decreasing) and bounded by each other. Therefore, $\sup\{v^{(n)}\} =: v$ and $\inf\{w^{(n)}\} =: w$ exist and we obtain the

Corollary 7. *Let $f(x)$, $g(x, x)$, $\{v^{(n)}\}_{n=0}^{\infty}$ and $\{w^{(n)}\}_{n=0}^{\infty}$ be defined as in Theorem 6 and* $\sup\{v^{(n)}\} = \inf\{w^{(n)}\} =: x^*$. *Then, x^* is the unique fixed point of $x = f(x)$ in the interval $[v^{(0)}, w^{(0)}]$ and x^* can be computed by a finite number of steps of the iterative method*

$$u^{(n+1)} := f(u^{(n)}), \qquad n \geqslant 0$$

for arbitrary $u^{(0)} \in [v^{(0)}, w^{(0)}]$.

Proof. For all $n \geqslant 0$ follows

$$v^{(n+1)} = g(v^{(n)}, w^{(n)}) \leqslant_I g(v^{(n)}, \inf\{w^{(n)}\}) \leqslant_I g(\sup\{v^{(n)}\}, \inf\{w^{(n)}\})$$
$$= g(x^*, x^*) = g(\inf\{w^{(n)}\}, \sup\{v^{(n)}\}) \leqslant_I g(w^{(n)}, \sup\{v^{(n)}\})$$
$$\leqslant_I g(w^{(n)}, v^{(n)}) = w^{(n+1)},$$

i.e. $g(x^*, x^*) \in [v^{(n+1)}, w^{(n+1)}]$. With $\sup\{v^{(n)}\} = \inf\{w^{(n)}\} = x^*$ we get

$$x^* = g(x^*, x^*) = f(x^*).$$

Another class of iterative methods is associated to $f(x)$ in the following way:

Definition 8. Let $f(x)$: $V_nR \to V_nR$ be a mapping and $g(x, y)$ a splitting of $f(x)$. The operator T: $V_nR \to V_nR$ defined by $T(x) := g(Tx, x)$ is called "iterative operator associated to $g(x, y)$".

Theorem 9. *Let $\{V_nR, \leqslant_I\}$ be an ordered set, $f(x) = (f_i(x))$ a mapping from V_nR to V_nR, $g(x, y)$ an explicit splitting of $f(x)$ and T the corresponding iterative operator. Then,*

(a) T is an isotone mapping from $\{V_nR, \leqslant_I\}$ to $\{V_nR, \leqslant_I\}$, if $g(x, y)$ is isotone in x and y.

(b) T is an antitone mapping from $\{V_nR, \leqslant_I\}$ to $\{V_nR, \leqslant_I\}$, if $g(x, y)$ is isotone in x and antitone in y.

Proof. (a) For two elements $x_1, x_2 \in V_nR$ with $x_1 \leqslant_I x_2$ follows in the first component of Tx_1 and Tx_2

$$(Tx_1)_1 = (g(Tx_1, x_1))_1 = g_1(x_1) \leqslant_I^{1} g_1(x_2) = (g(Tx_2, x_2))_1 = (Tx_2)_1$$

---

[1] $\leqslant_I$ means here $\leqslant$, if $i \in I$, and $\geqslant$, otherwise.

Now, we assume for $k < n$

$$\bigwedge_{i=1(1)k} (T x_1)_i \leqslant_I{}^1 (T x_2)_i$$

and get in the components with index $k + 1$

$$\begin{aligned}(Tx_1)_{k+1} = (g(Tx_1, x_1))_{k+1} &= g_{k+1}(((Tx_1)_1, \ldots, (Tx_1)_k), x_1) \\ &\leqslant_I g_{k+1}(((Tx_2)_1, \ldots, (Tx_2)_k), x_1) \quad \text{by (M1)} \\ &\leqslant_I g_{k+1}(((Tx_2)_1, \ldots, (Tx_2)_k), x_2) \quad \text{by (M2)} \\ &= (g(Tx_2, x_2))_{k+1} = (Tx_2)_{k+1}.\end{aligned}$$

(b) can be proved analogously.

Theorem 9 permits us to formulate some formerly proved results in a simple way (see [4], [5]).

Theorem 10. *Let $\mathscr{A}$ be an $n \times n$ matrix of M_nR, $\mathscr{b}$ an element of V_nR and let $\mathscr{A}$ be expressed as the matrix sum $\mathscr{A} = \mathscr{L} + \mathscr{D} + \mathscr{R}$, where $\mathscr{D}$ is a diagonal matrix and $\mathscr{L}$ and $\mathscr{R}$ are strictly lower and upper triangular $n \times n$ matrices, respectively. Then*

(*a*) *The operator T defined by the total-step-method*

$$x^{(k+1)} = \mathscr{A} x^{(k)} + \mathscr{b}, \qquad k \geqslant 0$$

is isotone (respectively antitone) with respect to $\{V_nR, \leqslant_I\}$, if $\mathscr{A}$ is an isotone (respectively antitone) mapping from $\{V_nR, \leqslant_I\}$ to itself.

(*b*) *The operator T defined by the single-step-method*

$$x^{(k+1)} = \mathscr{L} x^{(k+1)} + (\mathscr{D} + \mathscr{R}) x^{(k)} + \mathscr{b}, \qquad k \geqslant 0$$

is isotone (respectively antitone) with respect to $\{V_nR, \leqslant_I\}$, if $\mathscr{L}$ is an isotone and $\mathscr{D} + \mathscr{R}$ an isotone (respectively antitone) mapping from $\{V_nR, \leqslant_I\}$ to itself.

(*c*) *The operator T defined by the overrelaxation-method*

$$x^{(k+1)} = (\omega \mathscr{L}) x^{(k+1)} + ((e - \omega)\mathscr{E} + \omega(\mathscr{D} + \mathscr{R})) x^{(k)} + \omega \mathscr{b}, \qquad k \geqslant 0$$

is isotone in the case $0 \leqslant \omega \leqslant e$ (respectively antitone in the case $\omega \geqslant e$) with respect to $\{V_nR, \leqslant_I\}$, if $\mathscr{L}$ is an isotone and $\mathscr{D} + \mathscr{R}$ an isotone (respectively antitone) mapping from $\{V_nR, \leqslant_I\}$ to itself.

Proof. The statement (a) (resp. (b)) is an immediate consequence of the properties (OV1) and (OV2) in $\{V_nR, M_nR, \leqslant_I\}$ if we consider the splitting

$$\bigwedge_{x, y \in V_nR} g(x, y) := \mathscr{A} y + \mathscr{b}$$

$$\text{(resp.} \bigwedge_{x, y \in V_nR} g(x, y) := \mathscr{L} x + (\mathscr{D} + \mathscr{R}) y + \mathscr{b}).$$

(c): The mapping

$$\bigwedge_{x, y \in V_n R} g(x, y) := (\omega \mathscr{L})x + ((e - \omega)\mathscr{E} + \omega(\mathscr{D} + \mathscr{R}))y + \omega b$$

satisfies the assumptions of Theorem 9 (except the unnecessary property of a splitting), for $\omega \mathscr{L}$ is isotone with $\mathscr{L}$ isotone and $\omega \geqslant 0$ and $(e - \omega)\mathscr{E} + \omega(\mathscr{D} + \mathscr{R})$ monotone in the following way:

$$\left.\begin{array}{r} 0 \leqslant \omega \leqslant e : e - \omega \geqslant 0 \Rightarrow (e - \omega)\mathscr{E} \text{ isotone} \\ \mathscr{D} + \mathscr{R} \text{ isotone} \Rightarrow \omega(\mathscr{D} + \mathscr{R}) \text{ isotone} \end{array}\right\} \underset{\text{(OV1)}}{\Rightarrow} (e - \omega)\mathscr{E} + \omega(\mathscr{D} + \mathscr{R}) \text{ isotone}$$

$$\left.\begin{array}{r} \omega \geqslant e : e - \omega \leqslant 0 \Rightarrow (e - \omega)\mathscr{E} \text{ antitone} \\ \mathscr{D} + \mathscr{R} \text{ antitone} \Rightarrow \omega(\mathscr{D} + \mathscr{R}) \text{ antitone} \end{array}\right\} \Rightarrow (e - \omega)\mathscr{E} + \omega(\mathscr{D} + \mathscr{R}) \text{ antitone}$$

An extension of Theorem 10 is given in the following theorem:

Theorem 11. *Let $\mathscr{A} \in M_n R$ be an $n \times n$-matrix expressed as the matrix sum $\mathscr{A} = \mathscr{L} + \mathscr{D} + \mathscr{R}$, $\mathscr{D} + \mathscr{R}$ antitone with respect to $\{V_n R, \leqslant_I\}$ and $b \in V_n R$. Then, there exists a splitting $g(x, y)$: $V_n R \times V_n R \to V_n R$ of $f(x) = \mathscr{A}x + b$ such that the iterative operator T associated to $g(x, y)$ is antitone.*

Proof. The mapping g is given by

$$g(x, y) := \mathscr{L}x + (\mathscr{D} + \mathscr{R})y + b,$$

if $\mathscr{L}$ is isotone with respect to $\{V_n R, \leqslant_I\}$ (compare Theorem 10(b)). Otherwise we define $n \times n$ matrices $\mathscr{A}' = (a'_{ij})$, $\mathscr{A}'' = (a''_{ij})$ according to

$$\bigwedge_{1 \leqslant i, j \leqslant n} a''_{ij} := \begin{cases} a_{ij} & \text{if } i \leqslant j \text{ or } a_{ij} < 0 < a_{ji} \text{ or } a_{ij} > 0 > a_{ji} \\ 0 & \text{otherwise} \end{cases}$$

and

$$\bigwedge_{1 \leqslant i, j \leqslant n} a'_{ij} := \begin{cases} a_{ij} & \text{if } a_{ij} \neq a''_{ij} \\ 0 & \text{otherwise.} \end{cases}$$

Evidently $\mathscr{A}'$ is isotone and $\mathscr{A}''$ antitone with respect to $\{V_n R, \leqslant_I\}$. Hence the mapping

$$g(x, y) := \left(\sum_{j=1}^{n} (a'_{ij} x_j + a''_{ij} y_j) + b_i \right),$$

$x = (x_i)$, $y = (y_i) \in V_n R$, satisfies the properties (M1), (M2) and by Theorem 10(b) the associated iterative operator is antitone.

Note that the sum in each component of g is to compute from left to right. Otherwise property (S) is not guaranteed for $g(x, y)$ because of the missing associative law. (Nevertheless the mapping $\mathscr{A}'x + \mathscr{A}''x + b$ also satisfies the properties (M1) and (M2)).

We illustrate the last theorems by the following

Examples. We consider a 6×6 matrix with the given sign distribution:

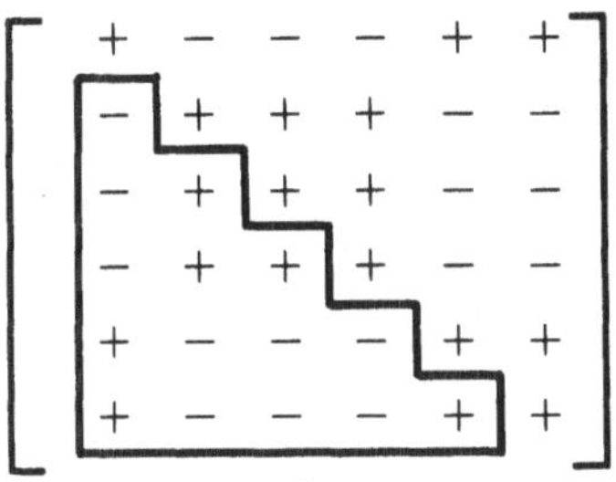

$\mathscr{A}_1$

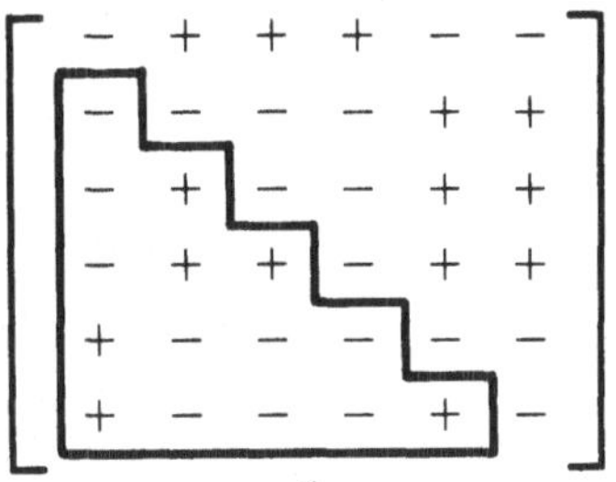

$\mathscr{A}_2$

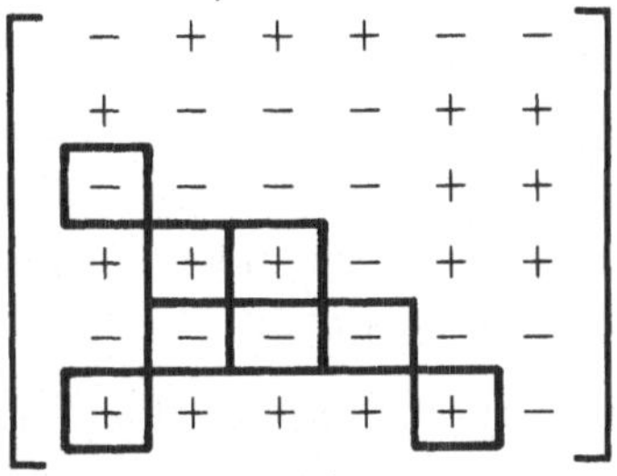

$\mathscr{A}_3$

$\mathscr{A}_1$ is isotone with respect to the ordering $\leqslant_I$ introduced by $I := \{1, 5, 6\}$. According to Theorem 10(a) and (b) total-step-method and single-step-method are represented by isotone operators. The second sign distribution for a matrix $\mathscr{A}_2$ expresses that $\mathscr{A}_2$ itself is not monotone but $\mathscr{L}_2$ is isotone and $\mathscr{D}_2 + \mathscr{R}_2$ is antitone with respect to $\{V_nR, \leqslant_I\}$. In matrix $\mathscr{A}_3$ $\mathscr{D}_3 + \mathscr{R}_3$ is antitone again but $\mathscr{L}_3$ is neither isotone nor antitone. By Theorem 11 we obtain the splitting into the isotone part $\mathscr{A}_3'$ (framed sign distribution) and the antitone part $\mathscr{A}_3''$ and hence an antitone associated iterative operator.

Finally we want to prove a statement which gives us a relation between the fixed points of an isotone mapping $\not{f}(x)$ and associated iterative operators.

Lemma 12. *Let $\not{f}(x) = (f_i(x))$ be an isotone mapping from $\{V_nR, \leqslant_I\}$ to itself, $g(x, y)$ an explicit splitting of $\not{f}(x)$ and T: $V_nR \to V_nR$ the iterative operator associated to $g(x, y)$. Then*

$$\bigwedge_{x \in V_nR} (\not{f}(x) \leqslant_I x \Rightarrow Tx \leqslant_I x).$$

Proof. Clearly, the statement is valid for the first component $(Tx)_1$ of Tx:

$$(Tx)_1 = (g(Tx, x))_1 = g_1(x, x) = f_1(x) \leqslant_I (x)_1.$$

Now we assume for arbitrary $k < n$

$$\bigwedge_{i=1(1)k} (Tx)_i \leqslant_I (x)_i,$$

and get the inequality

$$(Tx)_{k+1} = (g(Tx, x))_{k+1} = g_{k+1}(((Tx)_1, \ldots, (Tx)_k, x_{k+1}, \ldots, x_n), x)$$
$$\leqslant g_{k+1}((x_1, \ldots, x_k, x_{k+1}, \ldots, x_n), x) = f_{k+1}(x) \leqslant (x)_{k+1}.$$

Theorem 13. *Let $f(x) = (f_i(x))$ be an isotone mapping from $\{V_nR, \leqslant_I\}$ to itself, $g(x, y)$ an explicit splitting of $f(x)$ and T: $V_nR \to V_nR$ the iterative operator associated to $g(x, y)$.*

Suppose further that the sequence $\{x^{(n)}\}_{n=0}^{\infty}$ generated by

$$x^{(n+1)} := f(x^{(n)}), \qquad n \geqslant 0,$$

for an element $x^{(0)} \in V_nR$ has property (Z) and the inequality $f(x^{(i0)}) \leqslant_I x^{(i0)}$ holds for one index $i0$.

Then, the sequence $\{x^{(n)}\}_{n=0}^{\infty}$ has a fixed point x^* and each sequence $\{y^{(n)}\}_{n=0}^{\infty}$ with $x^{(i0+1)} \leqslant_I y^{(0)} \leqslant_I x^{(i0)}$ computed by

$$y^{(n+1)} := Ty^{(n)}, \qquad n = 1, 2, \ldots$$

has the same fixed point x^*.

Proof. Under the given propositions the monotonicity of $f(x)$ implies that the sequence $\{x^{(n)}\}_{n=0}^{\infty}$ is ending in a fixed point x^*.

Without loss of generality we write $x^{(0)}$ instead of $x^{(i0)}$ and choose an element $y^{(0)}$ with $f(x^{(0)}) \leqslant_I y^{(0)} \leqslant_I x^{(0)}$.

(a) We show at first: $y^{(k)} \leqslant_I x^{(k)}$, $k = 0, 1, 2, \ldots$: From T isotone follows $y^{(k)} = T(y^{(k-1)}) \leqslant_I T(x^{(k-1)}) = g(T(x^{(k-1)}, x^{(k-1)}))$ and then, since $Tx^{(k-1)} \leqslant_I x^{(k-1)}$ according to Lemma 12, $y^{(k)} \leqslant_I g(x^{(k-1)}, x^{(k-1)}) = f(x^{(k-1)}) = x^{(k)}$.

(b) In a second step we prove the statement: $x^{(nk+1)} \leqslant_I y^{(k)}$, $k = 0, 1, 2, \ldots$: This is valid in the case $k = 0$. Now we assume $x^{(n(k-1)+1)} \leqslant_I y^{(k-1)}$ and obtain

$$x^{(nk+1)} = f^n(x^{(n(k-1)+1)} \underset{(*)}{\leqslant_I} Tx^{(n(k-1)+1)} \leqslant_I Ty^{(k-1)} = y^{(k)}$$

applying property $(*)$ which will be proved at the end.

Summarizing (a) and (b) we get for $k = 0, 1, 2, \ldots$: $x^{(nk+1)} \leqslant_I y^{(k)} \leqslant_I x^{(k)}$, i.e. x^* is fixed point of $\{y^{(n)}\}_{n=0}^{\infty}$.

Finally, we show

$$(*) \qquad \bigwedge_{i=1(1)n} f^i(x^{(t)})_i \leqslant_I (Tx^{(t)})_i$$

(abbreviating $n(k-1)+1$ by t): It is

$$f_1(x^{(t)}) = f_1(x^{(t)}, x^{(t)}) = g_1(x^{(t)}, x^{(t)}) = g_1(Tx^{(t)}, x^{(t)}) = (Tx^{(t)})_1.$$

Now we assume for $k < n$:

$$\bigwedge_{i=1(1)k} f^i(x^{(t)}))_i \leqslant_I (Tx^{(t)})_i$$

Then, using the proposition $f^i(x^{(t)}) \leqslant_I x^{(t)}$, it follows

$$(f^{i+1}(x^{(t)}))_{i+1} = f_{i+1}(f^i(x^{(t)})) = g_{i+1}(f^i(x^{(t)}), f^i(x^{(t)}))$$
$$\leqslant_I g_{i+1}(f^i(x^{(t)}), x^{(t)})$$

$$\leqslant_I g_{i+1}(((f(x^{(t)}))_1, (f^2(x^{(t)}))_2, \ldots, (f^i(x^{(t)}))_i, x_{i+1}^{(t)}, \ldots, x_n^{(t)}), x^{(t)})$$
$$\leqslant_I g_{i+1}(((Tx^{(t)})_1, \ldots, (Tx^{(t)})_i, x_{i+1}^{(t)}, \ldots, x_n^{(t)}), x^{(t)})$$
$$= g_{i+1}(Tx^{(t)}, x^{(t)}) = (Tx)_{i+1},$$

completing the proof of the Theorem.

Remarks. 1. Theorem 13 extends a result given in [13] for special iterative methods in interval analysis.

2. Evidently Theorem 13 also holds under the assumptions $f(x^{(i0)}) \leqslant_I x^{(i0)}$ and $x^{(i0)} \leqslant_I y^{(0)} \leqslant_I x^{(i0+1)}$.

References

[1] Braess, D.: Die Konstruktion monotoner Iterationsverfahren zur Lösungseinschließung bei linearen Gleichungssystemen. Arch. Rat. Mech. Anal. **9**, 97–106 (1962).

[2] Collatz, L.: Funktionalanalysis und numerische Mathematik. Berlin-Göttingen-Heidelberg-New York: Springer 1964.

[3] Kaucher, E.: Über metrische und algebraische Eigenschaften einiger beim numerischen Rechnen auftretender Räume. Dissertation, Universität Karlsruhe, 1973.

[4] Klatte, R.: Zyklisches Enden bei Iterationsverfahren. Dissertation, Universität Karlsruhe, 1975.

[5] Klatte, R., Ullrich, Ch.: Consequences of a properly implemented computer arithmetic for periodicities of iterative methods. Report on the 3rd IEEE Symposium on Computer Arithmetic, Nr. 75CH 1017-3C, Dallas, Texas (1975), pp. 24–32.

[6] Klatte, R., Ullrich, Ch.: Zur mathematischen Struktur von Rechnerarithmetiken. MNU **30**, 1–8 (1977).

[7] Kulisch, U.: Grundlagen des numerischen Rechnens – Mathematische Begründung der Rechnerarithmetik. Mannheim: Bibliographisches Institut 1976.

[8] Kulisch, U., Miranker, W. L.: Arithmetic operations in interval spaces. (This volume.)

[9] Ortega, J. M., Rheinboldt, W. C.: Iterative solution of nonlinear equations of several variables. New York-London: Academic Press 1970.

[10] Ullrich, Ch.: Rundungsinvariante Strukturen mit äußeren Verknüpfungen. Dissertation, Universität Karlsruhe, 1972.

[11] Ullrich, Ch.: Blockiterationsverfahren bei schwach zyklischen Abbildungen. ZAMM **59** (1979).

[12] Ullrich, Ch.: Über schwach zyklische Abbildungen in nichtlinearen Produkträumen und einige Monotonieaussagen. Aplikace Matematiky **2**, 209–234 (1979).

[13] Wißkirchen, P.: Vergleich intervallarithmetischer Iterationsverfahren. Computing **14**, 45–49 (1975).

Priv.-Doz. Dr. Ch. Ullrich
Institut für Angewandte Mathematik
Universität Karlsruhe
Postfach 6380
D-7500 Karlsruhe
Federal Republic of Germany

Computing, Suppl. 2, 211–229 (1980)

Portable Software for Interval Arithmetic*

J. M. Yohe, Eau Claire, Wisconsin

Abstract

A portable multiple precision interval arithmetic package for FORTRAN has been developed by substituting the multiple precision arithmetic package of Richard P. Brent for the underlying arithmetic in the author's earlier single precision interval arithmetic package. This package, like the earlier version, offers a complete range of operations and functions for interval calculations.

In this paper, we outline the design philosophy of the earlier package and show how this design facilitated the incorporation of the Brent package. We discuss several desirable host system features and possible adaptations of the interval package, and explain how the design of the package would allow it to serve in differing environments with only relatively minor changes.

Since the package may be of direct use to many individuals, we also discuss its installation on other host systems and its use via the AUGMENT precompiler for FORTRAN.

1. Introduction

Numerical experimentation with interval mathematics ([14], [16]) requires the availability of a computing system that supports interval functions and operations. However, even today, the widely-available hardware and software systems do not support interval mathematics, and one must often turn to ad hoc software to obtain this facility.

In order to create interval software that blends naturally with the real arithmetic on a given host computer, it is necessary to take the architecture of that computer into account. In most cases, the directed roundings necessary for interval arithmetic must be implemented by writing software to perform floating point operations, and in many cases this software is most efficiently written in machine language. The result is that the interval arithmetic package written for one machine is not readily transportable to another host system having different architecture.

In [24], we described a FORTRAN interval arithmetic package that is semi-portable, in the sense that the architecture-dependent parts are concentrated in a relatively few modules. The implementation of the package on a different host system requires only that one write these few segments of the package, since the majority of the package is written in an extended dialect of ANSI Standard FORTRAN [1], and can thus be compiled (with the aid of an appropriate precompiler) without significant change.

* Sponsored in part by the U. S. Army under contract numbers DAAG29-75-C-0024 and DACA39-78-M-0023, and in part by the University of Wisconsin – Eau Claire.

Even so, the implementation of that package on a new host system is not a trivial task. One is still obliged to write machine-dependent routines for the four standard arithmetic operations, supply a spate of constants, and, perhaps most bothersome, obtain error bounds for the standard double precision mathematical functions used to calculate the interval equivalents. Even then, the package is capable only of producing single-precision interval results.

Although that package has been successfully implemented in whole or in part on several host computer systems (including the UNIVAC 1100 series, the IBM 360/370 series, Honeywell 600/6000, DEC-10/20, PDP-11, and CDC Cyber series machines), there have been requests for transportable primitives and higher precision versions of the package.

When the transportable multiple precision arithmetic package of Richard P. Brent [5] became available, it was natural to consider the implementation of a new version of the interval arithmetic package, based on Brent's arithmetic. From the point of view of the existing interval arithmetic package, Brent's package could be viewed as simply another "machine", and the implementation could therefore be accomplished by writing the appropriate primitives and supplying the required constants. One of the interesting features of Brent's package is that the precision may be varied by the user and, although the maximum allowable precision is fixed at compile time by specifying the lengths of the arrays containing the multiple precision numbers, the operating precision is determined at run time and, indeed, may be varied at will within the predetermined bounds. This facility is extended to the multiple precision version of the interval arithmetic package, although the user must take certain precautions when changing the precision of interval numbers.

In this paper, we outline the design philosophy of the original interval arithmetic package and indicate how this design facilitated the modifications necessary to implement the multiple precision version. This is done in Sect. 2.

In Sect. 3, we discuss hardware, firmware, and software features that would enhance the efficiency and utility of the interval arithmetic package, and show how its design allows for adaptation to take advantage of these features.

Since the design philosophy of the package should be applicable to higher-level languages other than FORTRAN, Sects. 2 and 3 should be useful to the individual wishing to implement interval arithmetic in another language.

The portable multiple precision FORTRAN version of the interval arithmetic package may be of use to a number of people in its own right. Therefore, we devote Sect. 4 to the use of this package. This discussion examines the steps necessary to install the package on a different host system, the changing of the precision of the package, both statically and dynamically, and the actual use of the package via the AUGMENT precompiler for FORTRAN [9].

2. Package Design

Both the interval arithmetic package [24] and Brent's multiple precision arithmetic package [5] were designed to maximize flexibility and transportability, although neither one was designed with the other one in mind. The design features of both packages, however, made their integration a routine matter. In this section, we briefly describe the design of both packages and the nature of the interface between them.

One of the major factors in making the design of the interval arithmetic package amenable to the incorporation of the multiple precision arithmetic package was the AUGMENT precompiler for FORTRAN. We therefore begin our discussion with a brief summary of AUGMENT's capabilities.

The AUGMENT Precompiler

The need for various special arithmetics at the Mathematics Research Center, University of Wisconsin – Madison, had, over a period of several years, culminated in the development of the AUGMENT precompiler which enabled one to write programs involving nonstandard data types in an extended dialect of FORTRAN. The nonstandard variables are declared in TYPE statements just as though the nonstandard types were in fact standard; AUGMENT, using a description of the nonstandard package(s), generates a standard FORTRAN program in which operations and functions on the nonstandard data elements are replaced by CALLs to appropriate modules of the supporting package(s).

AUGMENT itself was designed for transportability, due both to changes of equipment at the University of Wisconsin – Madison and to the desire of some visitors to the Mathematics Research Center to implement the precompiler capability at their home institutions. Previous Mathematics Research Center precompilers had been dependent on host system architecture and on a specific dialect of FORTRAN, and were therefore not readily adaptable to other host systems. AUGMENT was conceived at about the time the University of Wisconsin – Madison was switching from one vendor's hardware to another, and it was decided to attempt to minimize the system-dependent portions of the precompiler. Most of AUGMENT was written in ANSI standard FORTRAN, and the portions that were critical enough to warrant writing in assembly language were collected in a few primitive modules.

The predecessors of AUGMENT had been restricted to the use of a single nonstandard data type. AUGMENT, however, was designed to accommodate multiple nonstandard data types, since the requirement for this was foreseen.

Thus, in 1975, when the U. S. Army Corps of Engineers Waterways Experiment Station expressed an interest in a transportable version of the interval arithmetic package, the full power of the AUGMENT precompiler was available to assist in the design and implementation of a new interval arithmetic package.

The Interval Arithmetic Package

There were five major considerations in the design of the new interval arithmetic package; the first four of them had been designed into the previous machine-dependent interval arithmetic package. Like the old package, the new package was to be complete, in the sense that both the ANSI standard FORTRAN operations and functions and those preculiar to interval arithmetic, including those defined by Ris in [19], were to be included. The package was to provide the maximum possible accuracy consistent with the number representation and a reasonable execution speed. It was to detect and provide special handling for any exceptions to correct computation, such as out-of-range results, division by an interval containing zero, etc. And, still following the lead of the previous package, it was to be as convenient to use as possible. Unlike the earlier version, however, the revised package was to be designed for maximum transportability.

The transportability criterion probably had the major effect on the design of the package, although this was followed closely in importance by the requirement that the package be easy to use.

Some interesting problems arose from the transportability requirement. Directed roundings ([12]), both in various operations and functions and on input/output conversion, were necessary for the implementation of interval arithmetic, but were not available on any production computer, much less on the range of computers for which the package was being designed. It would be necessary to perform these roundings via software, using the algorithms given in [21]. Even then, however, the development of a package which would be truly transportable was frustrated by the variety of architectures involved: storage of interval numbers is already twice as expensive as the storage of reals, and to develop a data representation which would be applicable to all of the host systems involved in the Waterways Experiment Station project would mean that, for most of the target computers, the representation of the interval endpoints would be both incompatible with the internal representation of real numbers and wasteful of storage space.

Another problem was the evaluation of interval functions such as SIN, EXP, and SQRT. In most cases, these cannot be calculated with acceptable accuracy merely by using the interval analog of the real algorithms; the dependency problem causes the intervals to expand too much too rapidly. In the earlier version of the interval package, we had used double precision functions in the evaluation of the interval functions, but we had some rather comprehensive information on the accuracy of the double precision functions, and they were, of course, available. In striving for transportability, however, we did not want to tie the package to single and double precision; it seemed likely that there might be a need for double precision interval arithmetic, for example, or that double precision functions might not be available in certain environments, so the existence of functions with higher accuracy than the interval endpoints could not, a fortiori, be presumed.

Thus, in order to preserve flexibility and maximize transportability, the interval arithmetic package was designed to be modular, and the data types upon which it was based were nonstandard so they could be left undefined throughout much of the package.

There are three nonstandard data types employed in the interval arithmetic package. The fundamental one is the data type of the interval endpoints, which we have called BPA (mnemonic for Best Possible Answer). It is this data type which supports directed roundings; for the most part, operations on real numbers of the

```
      SUBROUTINE BPASQT(A, R)
C         R = SQRT(A)
C         NO ERRORS ARE POSSIBLE EXCEPT SQUARE ROOT OF A NEGATIVE
C         NUMBER, WHICH RESULTS IN BPAFLT BEING SET TO 18
      COMMON /BPACOM/ OPTION, BPAFLT, ACC, RDU, RDL, RDT, RDN, RDA
      INTEGER OPTION, BPAFLT, ACC, RDU, RDL, RDT, RDN, RDA
      COMMON /BPACON/ ZRO, ONE, ONM, TWO, PIO2, PI, BPAMNB, BPAMXB
      BPA             ZRO, ONE, ONM, TWO, PIO2, PI, BPAMNB, BPAMXB
      COMMON /BPAACC/ IACC(24), LNACC, LGACC, SNHACC, TNHACC, EXPMXA,
     1                EXPMNA, FRACBD, MAXINT
      INTEGER IACC
      BPA LNACC, LGACC, SNHACC, TNHACC, EXPMXA, EXPMNA, FRACBD, MAXINT
      BPA A, R
      EXTENDED EA, ER
      IF(A .LT. ZRO) GO TO 900
      EA = A
      ER = SQRT(EA)
      ACC = IACC(17)
      R = ER
      RETURN
  900 BPAFLT = 18
      RETURN
      END
```

Fig. 2.1. The BPA Square Root Routine

```
      SUBROUTINE INTSQT(A, R)
C         R = SQUARE ROOT OF A          (MONOTONE FUNCTION)
C         THE ONLY FAULT WHICH CAN OCCUR IS THE ATTEMPT TO TAKE THE
C         SQUARE ROOT OF A NEGATIVE NUMBER. THIS WILL BE DETECTED
C         IN BPASQT.
      COMMON /BPACOM/ OPTION, BPAFLT, ACC, RDU, RDL, RDT, RDN, RDA
      INTEGER OTION, BPAFLT, ACC, RDU, RDL, RDT, RDN, RDA
      COMMON /INTCOM/ INTFLT, ID, TA, TB, TR
      INTEGER INTFLT, ID
      INTERVAL TA, TB, TR
      EXTENDED TAE, TBE, TRE
      EQUIVALENCE (TA, TAE), (TB, TBE), (TR, TRE)
      INTERVAL A, R
      TA = A
      OPTION = RDU
      SUP(TR) = SQRT(SUP(TA))
      OPTION = RDL
      INF(TR) = SQRT(INF)TA))
      INTFLT = BPAFLT
      ID = 19
      CALL INTRAP
      R = TR
      RETURN
      END
```

Fig. 2.2. The interval square root routine

precision of the interval endpoints are performed using BPA arithmetic. BPA numbers are bound to a specific format in only a few modules of the package; in most portions of the package, they are declared as type BPA and, as far as the source code is concerned, left unbound.

The algorithm for evaluating BPA intrinsic functions such as SIN, EXP, SQRT, etc. involves evaluating the functions in higher precision, then bounding the result to allow for possible error of approximation, and finally rounding the result in the specified direction to arrive at a BPA number. In implementing this, we introduced the nonstandard data type EXTENDED. Although in practice the EXTENDED data type was declared to be identical with DOUBLE PRECISION, the nonstandard type designation is used in the source code throughout the package. This use of EXTENDED is illustrated by the BPA square root routine (Fig. 2.1).

The third nonstandard type is, of course, INTERVAL. This is declared to be a BPA array of length 2 (as far as AUGMENT is concerned), but again is left unbound throughout most of the source code. Fig. 2.2 illustrates this concept using the interval square root routine.

Thus, the majority of the source code was written using data types which are, as far as the source code is concerned, undefined. The binding of these data types is accomplished by means of a few primitive modules, which AUGMENT invokes at the appropriate places throughout the package. The structure of the package and its dependency on AUGMENT is shown in Fig. 2.3.

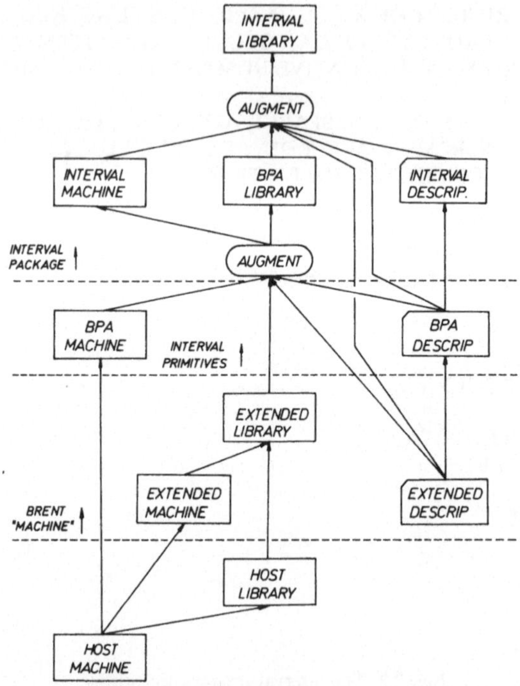

Fig. 2.3. Structure and dependency of the interval package

The task of preparing supporting packages for these three nonstandard data types was then undertaken. Existing higher-precision function routines were utilized by declaring EXTENDED to be the same as DOUBLE PRECISION, so little work was required to produce the supporting package for that data type.

The supporting packages for the other two data types required more extensive programming. For each supporting package, AUGMENT requires that service routines be available to perform elementary operations such as data transfer and negation. In addition, of course, AUGMENT requires that routines be available for each "visible" operation or function for which it is to generate code. Even though the initial plan was to have the BPA numbers represented by a single computer word, so that existing REAL functions could have been used in some contexts, it was decided to implement all of the AUGMENT-required routines explicitly in order to lend an extra measure of flexibility.

Again in the interest of flexibility, it was decided that each individual function should be cast as a separate module. While this would increase the package overhead by introducing subroutine calls for even the simplest operations (see Fig. 2.4), it meant that most package modules would be entirely independent of the actual representations of the data types, and thus different representations could be used for BPA numbers, triplex numbers could be more easily substituted for intervals represented in endpoint form, etc.

Communication between modules could have been implemented by passing all relevant information via calling sequences. However, this would have resulted

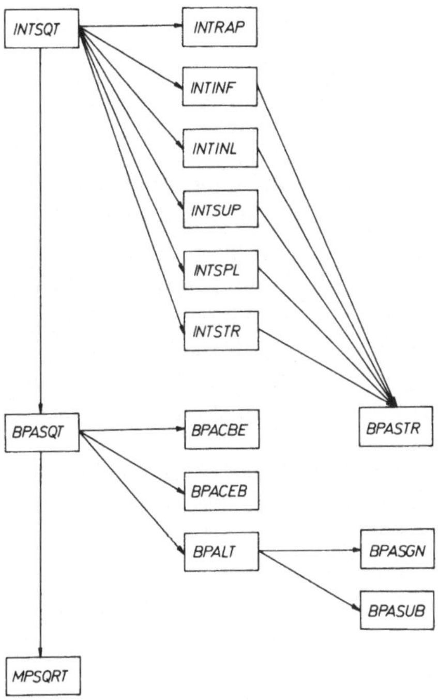

Fig. 2.4. Dependency structure of the interval square root routine

in unwieldy calling sequences, imcompatibility with AUGMENT, and, in some cases, repeating the definitions of constant information in many different modules, to the detriment of transportability. We opted to keep the calling sequences simple and consistent, and use COMMON blocks for the storage of constants and for passing most of the miscellaneous information.

Once the basic design of the package was determined, the BPA portion was written using a "bootstrap" technique. Those modules that were primitive in the sense that they depended directly on the actual representation of BPA numbers were written first. These included the data transmission routines, conversion routines, and the elementary arithmetic operations. The modules which could be written entirely in terms of the primitive BPA modules and the EXTENDED data type were then written, followed by the modules which depended on this second level, and so on. Then the AUGMENT precompiler was used to generate the links between the modules, thereby extending the data type bindings through the BPA package.

Finally, the interval portion of the package was written using the same technique. Here, the number of primitive modules was quite small; such functions as designating which array element contained the INF (or left endpoint) and SUP (right endpoint) of the interval were primitives, but there were few others.

A comprehensive discussion of the design and implementation of the interval arithmetic package may be found in [22], and a summary of this information, together with some additional comments, appears in [23]. We will not go into further detail here; we have presented that information which has a direct bearing on the adaptation of the package to use Brent's multiple precision arithmetic.

Brent's Multiple Precision Arithmetic Package

When this package became available, we decided to investigate the possibility of using it in place of the BPA and EXTENDED portions of the original interval arithmetic package.

The Brent package was an ideal candidate for this purpose, for several reasons. First, it is highly transportable, since it is written in ANSI standard FORTRAN and employs only integer arithmetic in its computations. Brent designed the package specifically for transportability, and the basic design is such that the package can be accommodated on machines of nearly any word length with only changes of a very minor nature. Thus, if the interval arithmetic package could be modified to use the Brent package, the result would be not only a multiple precision arithmetic package, but one that would be extremely transportable.

The second feature of the Brent package that made it attractive is that, unlike most of the multiple precision arithmetic packages we had seen previously, the Brent package includes a complete set of elementary functions, provisions for generating constants, and even a number of functions that are not in the ANSI standard list, but are nonetheless useful. All of these functions are implemented with the same flexibility that the basic operations enjoy.

Third, the Brent package is completely modular, and, indeed, shares many of the interval arithmetic package's basic design characteristics. This, together with the fact that the code in the Brent package is straightforward, lucid, and well-documented, meant that the necessary modifications could be accomplished without undue difficulty.

Finally, although the Brent package was not written with the aid of AUGMENT, or even designed specifically for use with AUGMENT, its basic design is quite compatible with AUGMENT. This implied that the mechanics of incorporating the Brent package into the interval package would not be complicated.

Again, we have not covered the design of the Brent package in great detail, but have only summarized those points which bear upon the interface. The reader who is interested in further details of the Brent package may consult [4].

The Interface

As we have said, the multiple precision version of the interval arithmetic package [25] was the result of interfacing the Brent package with the earlier single precision version of the interval package. The first step in creating this interface was to write an AUGMENT interface for the Brent package. This was a simple procedure; we accomplished it in a short time in collaboration with Brent ([6]).

Next, we considered the question of how to preserve the flexibility of the Brent package and extend it to the interval package. The Brent package relies on an initialization routine to set such parameters as base, precision, and exponent range. We decided to expand the initialization routine to provide, in addition, for run-time generation of the various constants used by the interval arithmetic package; this would lend the same degree of machine independence to the interval package that was already enjoyed by the Brent package. While this does require that the user add a statement to initialize the interval package, such a statement is normally required by the Brent package anyhow, and therefore does not represent additional effort on the user's part.

The major considerations were two problems that turned out to be companions: how to provide for directed roundings, and how to provide for two precisions – the basic precision of the interval endpoints and the higher precision used for function evaluations. The solution in this case turned out to be setting a basic precision for the Brent package and substituting the multiple precision package for type EXTENDED; then writing new BPA primitive modules, patterned after the Brent routines, but using a lower precision than the basic precision chosen for the Brent package, and, of course, including the capability of directed roundings.

The remaining modifications to the interval package proceeded without much difficulty. Most of the modifications were of the housekeeping variety; conversion arrays had to be expanded to accommodate the higher precision and greater range of the multiple precision numbers; the constants had to be computed by the initialization routine instead of being set in DATA statements since the precision is indeterminate until run time; and various primitives which depend on the actual representation of BPA numbers had to be modified. But the majority of the package routines were unchanged, since they had never depended directly on the actual representations of the numbers.

In the interval portion of the package, the only routine requiring major modification was the error-handling routine, which required substantial changes to the output formats due to the significantly different representation of the numbers. There were, of course, a few other modifications, but they were entirely housekeeping changes.

Some modifications were, of course, required to the description file which communicates the characteristics of the nonstandard packages to the AUGMENT precompiler. This file is what enables AUGMENT to extend the data representation bindings thoughout the interval package, and it also enables the precompiler to apply those bindings to the user's program. Even here, due to the flexibility of the description syntax, the changes were relatively minor.

After these changes were made, the revised package was reprocessed by the AUGMENT precompiler, then compiled with the standard FORTRAN compiler, and checked out. The entire process was probably accomplished in less than two months of part-time effort; although no precise records were kept, three or four man-weeks would probably be a reasonable estimate.

3. Other Adaptations

We turn now to the consideration of other possible adaptations of the interval arithmetic package. These might be modifications to accommodate improvements in the host environment, or alterations which would adapt the package to special requirements.

Host Environment Improvements

There are several possibilities for improvements in the host environment which would enhance the performance or versatility of the interval arithmetic package. We mention some of these, and indicate how the package would allow for incorporation of these features.

a) *Directed roundings* in hardware or firmware will probably begin to appear in the computers of the 1980's. Although present-day computers do not include this capability, and thus require software implementation of directed roundings, proposed standards [17] provide for its implementation, and even more accommodating arithmetic schemes have been proposed ([13], [18]).

Modification of the package to take advantage of this feature would be relatively simple. The few modules which perform endpoint arithmetic would need to be rewritten to utilize the hardware/firmware operations, but otherwise, the package should need little or no modification.

b) *Multiple precision arithmetic* in hardware of firmware is another feature which is likely to become available, at least on some computers.

If directed roundings are supplied, the adaptation of the package would be essentially similar to the modification outlined above. Otherwise, the software for endpoint arithmetic would need to be written, but the availability of hardware multiple precision operations should simplify the required code.

c) *Variable precision arithmetic*, where the precision could be set and reset by the user at run time, could have significant advantages for certain applications. We envision a list structure for real numbers, with each variable containing information about its own precision. It is assumed that the variable precision implementation would include the necessary storage management functions, directed roundings, and the capability to convert numbers from one precision to another.

Even for such a radical departure from traditional concepts as this, the design of the package should allow for reasonably uncomplicated adaptation of the package. The basic design of the package implies that subroutines will be called to load and store interval endpoints, and these subroutines need only appeal to the appropriate functions in the variable precision package to incorporate the new arithmetic capability.

d) *Storage management* capabilities may likewise become available for such applications as storing extremely sparse arrays or handling nontraditional data structures.

As above, the design of the package will allow one to rewrite those portions of the package that interface with the storage management functions, leaving the majority of the package intact.

e) *Improved handling of input/output* may become a feature of FORTRAN compilers or precompilers. At present, the process of reading and (especially) writing interval data is rather tedious. It is reasonable to expect that future developments in language processors will provide some relief from the tedium.

At best, improved input/output handling could be designed so that the compiler or precompiler would need only generate a call to one of the routines in the present package each time a conversion was required; in that case, the present package would need little or no adaptation. At worst, the user would need to write additional modules to interface the capabilities of the present package with the protocol required by the compiler/precompiler. In either case, the task should not be onerous.

Other Forms of Interval Arithmetic

For some applications, the endpoint representation of intervals may not be the most desirable form. We mention a couple of alternatives, and indicate the kinds of modifications that would be necessary to implement them.

a) *Triplex arithmetic* is a variant of interval arithmetic which represents an interval as an ordered triple consisting of the left endpoint, the *main* (or most probable) value, and the right endpoint (see e.g. [15]). This form of interval arithmetic is probably more widely used in Europe than is the endpoint representation; indeed, the TRIPLEX-ALGOL-60 compiler at the University of Karlsruhe [20] was one of the first comprehensive interval arithmetic systems developed anywhere.

The interval arithmetic package lends itself to adaptation to Triplex representation with a minimum of tedium. In fact, Klaus Böhmer and this author produced both a single precision and a multiple precision Triplex version of the package in a comparatively short time [3]. The major tasks were changing INTERVAL to TRIPLEX throughout the package (this was done for cosmetic reasons), modifying the interval routines to handle triples, rather than pairs, of numbers, adding functions to store and retrieve the main values, and redefining certain constants to be triplex numbers rather than intervals in endpoint representation. A more detailed discussion of the procedure can be found in [3] or [10].

b) *Midpoint/half-length* representation may be more useful in certain contexts than the standard endpoint representation of intervals. One advantage of this representation is storage economy in certain applications (see Aberth [2]), especially when multiple precision arithmetic is used and the intervals are rather narrow.

The present interval package could be adapted to this representation in one of two ways: a superstructure could be built on the present package to convert the centered form to the traditional endpoint representation, and the present package could be used intact; alternatively, the routines in the present package could be modified to handle the centered form directly. The latter modification would, in all likelihood, be considerably more extensive than the modification to produce a triplex package, but the present interval arithmetic package would nonetheless provide a reasonable starting point.

c) *Complex interval arithmetic* may be useful in some applications. Several variants of complex interval arithmetic exist, including the rectangular form, where the lower left and upper right corners of the rectangle are specified, and the centered form, where the center point and radius of a circular disk are given.

Adaptation of the present package to the rectangular form would not be inordinately difficult, although it would require that all of the interval functions be rewritten to handle complex numbers. Adaptation to the centered form would require an effort comparable to adapting the real version to centered form and adapting the resulting package to complex numbers.

d) *Extended interval arithmetic*, wherein division by zero is allowed to form exterior intervals, is advocated by some (see, e.g., [11]).

The present package can be adapted to this form of interval arithmetic quite easily; one need only modify the package to allow for exterior intervals, and alter certain routines (e.g., division, tangent, etc.) to produce them. (Note that, for an exterior interval, the left endpoint is greater than the right endpoint, but one still must round the left endpoint downward and the right endpoint upward.)

Adding Operations and Functions

Occasionally, a need may arise for some interval operation or function that is not included in the package. Addition of these components is not difficult; one merely writes the new module(s) in the same manner as the existing ones were written, compiles them using the AUGMENT precompiler, and adds them to the interval library. If errors are possible in the new modules, appropriate information should be added to the tables in the error-handling routine; and in special cases, it may be necessary to add code to the error-handling routine to provide an appropriate response in case of failure. Instructions may be found in [22].

Modifying Algorithms

In some circumstances, the user may decide that another algorithm for calculating the result of some operation or function is preferable to the one used in the package. In this case, it is necessary only to replace the unsatisfactory module with one which implements the desired algorithm. The procedure for writing these modules is the same as that outlined above. There need be no concern about unwanted side effects in the remainder of the package; so long as the module performs the correct function, the change will have no adverse effect on the other modules of the package.

The Role of Package Design in Ease of Adaptation

As we have mentioned above, the interval arithmetic package lends itself rather well to adaptations and extensions. Let us look at some of the design features of the package in the context of their impact on adaptability:

a) *Hierarchical structure* makes it a rather simple matter to replace the underlying endpoint arithmetic with a minimum of inconvenience. This allows one to substitute other arithmetic schemes such as hardware-implemented directed roundings, multiple precision arithmetic, and list-type represented numbers.

b) *Modularity* makes it possible to change number representations without having to rewrite the entire package. For example, since all interval storage and retrieval functions are handled by a common subroutine, if one wants to implement triplex rather than endpoint representation, changing one subroutine will take care of all of the data transfer functions in the package. The modularity of the package also makes it a simple matter to add other functions and operations to the interval package, and to employ different algorithms to implement existing functions.

c) *Expression in terms of nonstandard data types* serves two important functions: first, it allows one to write package components in terms of intervals and interval endpoints, without having to define or specify the exact nature of these quantities; second, it provides the means by which bindings to specific data representations are automatically extended through the package.

Thus, we see that the design of the package bestows upon it a rather unusual degree of adaptability.

4. Installing and Using the Multiple Precision Interval Package

Installing the Package

The software: The first step in installing the multiple precision interval package on another host system is, of course, to obtain copies of the software.

The multiple precision interval package must be processed with the AUGMENT precompiler before it can be used, and it depends upon the Brent multiple precision arithmetic package. Thus, three items of software are necessary.

The multiple precision version of the interval arithmetic package, with its documentation ([22], [25]), and the AUGMENT precompiler, with its documentation ([7], [8]), may be obtained from

Mathematics Research Center
University of Wisconsin-Madison
610 Walnut Street
Madison, WI 53706, U.S.A.

Brent's multiple precision arithmetic package, with its documentation [4], may be obtained from

Prof. Richard P. Brent
Computer Science Department
The Australian National University
Canberra, Australia

The primitives: The multiple precision interval package requires that the user provide two primitive routines which depend on the host environment.

The most critical of these is the Hollerith unpacking routine INTUPK. This routine has the calling sequence

CALL INTUPK (PH, UH, MAX, LENGTH)

where PH is the address of the packed Hollerith string to be unpacked, UH is the beginning address of the array where the unpacked string is to be stored, one character per word, MAX is the maximum number of characters to be scanned in the source string, and LENGTH is the number of characters actually unpacked. The multiple precision interval package allows an array length of 132 for the unpacked string, and if the string would exceed this limit, the INTUPK routine must return a value of 133 in LENGTH. The interval arithmetic package recognizes three special characters as string terminators: "=", "#", and "$", and the INTUPK routine should be coded to terminate the scan whenever one of these three characters is encountered. For consistency with the rest of the package, the test for these three characters may be accomplished by a comparison with ICHR (6), ICHR (7), and ICHR (8), where ICHR is an alpha array located in the COMMON block INTCCM (this allows the user to alter the sentinel characters, if desired). The COMMON declaration for this block may be found in Appendix 4 of [22]. If the host system provides an internal sentinel to mark the end of a Hollerith string, INTUPK may be coded to recognize this sentinel in addition to the special characters. Clearly, this routine depends on the storage structure (word size) and character packing of the host system, and cannot, therefore, be made portable.

The other routine, INTWLK, is related to the error handling function. This routine has two arguments. The first of these is an alpha string of six or less characters which names the routine that called INTWLK; the second is an action code IRESP. If IRESP is zero, INTWLK does not terminate the computation; if it is nonzero, INTWLK subtracts 1 from the count specified by the INTERS routine and terminates the computation if the result is zero or negative. On the UNIVAC 1100 system at the University of Wisconsin, INTWLK also causes a trace-back of the subroutine calls that eventually resulted in an error being detected to be printed, followed by a row of asterisks (*). The trace need not be implemented, but for aesthetic reasons, the row of asterisks should be printed in any case.

The UNIVAC 1100 version of the multiple precision interval package contains the UNIVAC versions of these routines; they may be used as models for constructing the routines for the new host system.

Changing the (Static) Precision of the Package

By static precision, we mean the maximum allowable precision for a multiple precision interval number. This precision is governed by the maximum number of storage locations allowed for each BPA number, and is set by the initialization routine. In the standard version of the package, 9 words are allowed for each BPA number, and 12 are allowed for each EXTENDED number. Lower precision may be used dynamically, although the unused storage locations will be idle.

Higher or lower static precision may be specified at the time the package is compiled. This may be done as follows:

1. Alter the precision of the Brent package. This involves changing the number of storage locations reserved for each multiple precision (EXTENDED) number by altering the specification in the AUGMENT description deck for this package as appropriate. The initialization routine, MPINIT (a part of the INTERVAL package), must likewise be altered to reflect this change.

2. Alter the precision of BPA numbers in the description deck for the interval package. The only requirement here is that the precision of a BPA number should not exceed three less than the precision of an EXTENDED number. If the precision of BPA numbers is less than this, a minor change is MPINIT will be required.

3. Alter the COMMON statements in the interval package to provide enough work space for the multiple precision package. Specific instructions for doing this are included in [25], and we will not repeat them here.

Preparing the Package for Use

Once the primitive routines have been provided, the entire package must be processed using the AUGMENT precompiler. The description decks for Brent's package and the multiple precision interval package must be supplied as input to AUGMENT, and we found it necessary to increase AUGMENT's work space to accommodate these files. A work space allocation of 5000 words proved to be more than adequate.

After the package has been processed with AUGMENT, the output of AUGMENT, together with the source code for the Brent package, should be compiled using the standard FORTRAN compiler, and the resulting object code modules should be placed in a library file. The package is then ready for use.

Using the Package

The multiple precision interval package is most easily used via the AUGMENT precompiler; since the precompiler is essential for implementing the package, it is safe to presume that AUGMENT is available.

The applications program is written just as though multiple precision interval were a standard data type in FORTRAN (see Fig. 4.1). Indeed, the basic multiple precision arithmetic of Brent may also be treated as a standard data type in the applications program, too.

Interval and multiple precision variables are declared in TYPE statements in the usual way (multiple precision variables being type EXTENDED):

```
INTERVAL X, Y(20), Z(10, 20)
EXTENDED U, V(2, 5)
```

If variables of the same data type as the interval endpoints are desired, they may be declared in an analogous fashion:

```
BPA B, C
```

Note, however, that the BPA arithmetic and function routines generally require that rounding options be set prior to the call on the desired routines. These rounding options must be set explicitly by storing the appropriate designator in the OPTION variable in the COMMON block BPACOM. If this is not done, the rounding used will be that used by the last BPA operation which did set this option; in general, that will be unpredictable.

In order to set the precision of the multiple precision package and calculate the constants needed by the interval package, the statement

```
C       THIS IS A PROGRAM TO FIND THE ROOTS OF A QUADRATIC
C         EQUATION. SINCE IT IS AN EXAMPLE OF THE USE
C         OF THE INTERVAL PACKAGE, IT IS RATHER SIMPLE
C         AND UNSOPHISTICATED. IT IS ASSUMED THAT THE
C         ROOTS ARE REAL AND THAT THE EQUATION IS NONDEGEN-
C         ERATE. ONLY ONE EQUATION IS SOLVED.
C
C       INTERVAL DECLARATIONS
      INTERVAL A, B, C, X(2)
C       FORMAT INFORMATION FOR INTERVAL WRITE ROUTINE
      INTEGER FMT(4)
      DATA FMT/2, 5, 59, 1H0/
C       INITIALIZATION OF MP INTERVAL PACKAGE
      INITIALIZE MP
C       READ COEFFICIENTS (FREE FORMAT)
      CALL INTRDF (5, A)
      CALL INTRDF (5, B)
      CALL INTRDF (5, C)
C       CALCULATE ROOTS
      X(1) = (-B + SQRT(B**2 - 4.0*A*C))/(2.0*A)
      X(2) = (-B - SQRT(B**2 - 4.0*A*C))/(2.0*A)
C       WRITE RESULTS
      CALL INTWR(6, FMT, X, 2)
C       TERMINATE PROGRAM
      STOP
      END
```

Fig. 4.1. Program using the multiple precision interval package

INITIALIZE MP

must appear in the program before any executable statement involving INTERVAL, BPA, or EXTENDED variables. It is safest to place this statement first in the program. This will cause AUGMENT to generate a call to the initialization routine.

Because AUGMENT does not process input/output statements, it is necessary to accomplish the reading and writing of any nonstandard variables by making explicit calls to the input/output routines. The procedure for doing this is exactly the same as for the single precision version of the package, and the interested reader is referred to [22] and [25] for this information as well as for other details on the use of the package.

Changing Precision Dynamically

One unusual feature of the Brent multiple precision arithmetic package is that the precision may be altered dynamically during a computation. Naturally, any such alteration must be within the limits imposed by the static precision. The user of the multiple precision interval package has this same flexibility, although we reiterate that if the usual precision of the computation is significantly less than the static precision, the space allocated for an interval number will be no different, and the remaining storage locations in each number will not be accessible for any other purpose.

To change the precision of an EXTENDED number, change the number T in blank COMMON to the desired number of multiple precision digits. The documentation of the Brent package [4] will provide further details.

To change the precision of the BPA numbers, change the corresponding value in COMMON BPAMCM to the appropriate value.

Warning: *Making these changes does not affect the precision of previously computed interval numbers, nor of the constants associated with the package.*

The following rules must be observed in order to preserve the integrity of the results:

1. If precision is being increased, make sure that the digits that were previously unused are set to zero. This applies to both EXTENDED and BPA numbers, and to the constants as well as to variables.

2. Make certain that the precision of BPA numbers does not exceed three less digits than the precision of the EXTENDED numbers.

3. If precision is to be decreased, it will be necessary to round all constants and previously computed variables to the lower precision in order to ensure that the intervals you are using contain the higher precision values of those intervals. Again, this applies to the constants used in the package as well as to the previously-computed variables. It is suggested that the user write a short routine to perform the following steps, since no such routine currently exists in the package:

a) Convert the left endpoint of the interval to EXTENDED.

b) Decrease the precision of the BPA numbers to the new precision.

c) Set OPTION in COMMON block BPACOM to the value of RDL (also in BPACOM), to enable downward directed rounding.

d) Invoke BPACEB to round the left endpoint of the interval.

e) Set unused digits of the left endpoint to zero.

f) Reset the precision of BPA numbers to the original value.

g) Repeat the process for the right endpoint, using the value of RDU on the step corresponding to c).

Once all variables and constants have been rounded to the lower precision, set the precision of BPA numbers to the new precision.

The constants may always be reset to a different precision by invoking MPINIT (note, though, that a minor change to this routine will be necessary if the precision of BPA numbers is to be lower than three less than the precision of EXTENDED numbers). However, MPINIT will never have any effect on variables in your program, and the above algorithm must be applied to them.

5. Conclusion

We have attempted to sketch both the design and the use of the multiple precision arithmetic package for FORTRAN. The latter discussion is intended to assist those intending to use the existing package, while the former exposition should provide guidance for those intending to adapt the current package or create an entirely new one for use with another language.

The multiple precision interval arithmetic package is a portable, flexible tool which should prove to be of value to anyone needing to conduct numerical experimentation in interval mathematics.

Acknowledgements

The author is pleased to acknowledge the contributions of many people to this effort. The package was originally developed at the Mathematics Research Center, University of Wisconsin-Madison, with encouragement and financial assistance from the U. S. Army Corps of Engineers Waterways Experiment Station. Since this package is essentially a modification of the package described in [22], those whose contributions were acknowledged there were likewise contributors to this effort. Those people include T. D. Ladner, Dr. F. D. Crary, and S. T. Jones of the Mathematics Research Center; William Boyt and James Cheek of the Waterways Experiment Station; Prof. Bruce Shriver of the University of Southwest Louisiana; Prof. Ronnie Ward of the University of Texas at Arlington; Prof. Richard Hetherington of the University of Kansas; and Prof. Myron Ginsberg of Southern Methodist University. In addition, Prof. Karl Nickel of the University of Freiburg provided advice, encouragement, and the opportunity to work at the Institut für Angewandte Mathematik at Freiburg in the Spring of 1979, and Prof. Klaus Böhmer of the Institut für Praktische Mathematik at the University of Karlsruhe provided similar opportunities there.

References

[1] ANSI standard FORTRAN. New York: American National Standards Institute 1966.
[2] Aberth, O.: A precise numerical analysis program. Comm. Assoc. Comput. Mach. 17, 509–513 (1974).

[3] Boehmer, K., Yohe, J. M.: Triplex arithmetic for FORTRAN. The University of Wisconsin-Madison, Mathematics Research Center, Technical Summary Report #1901, December 1978.
[4] Brent, R. P.: MP users guide. Australian National University, Canberra, Australia, Computer Centre, Technical Report #54, September 1976 (revised July 1978).
[5] Brent, R. P.: A FORTRAN multiple-precision arithmetic package. Assoc. Comput. Mach. Trans. Math. Software **4**, 57 – 70 (1978).
[6] Brent, R. P., Hooper, J. A., Yohe, J. M.: An AUGMENT interface for Brent's multiple precision arithmetic package. The University of Wisconsin-Madison, Mathematics Research Center, Technical Summary Report #1868, August 1978.
[7] Crary, F. D.: The Augment precompiler I: User information. The University of Wisconsin-Madison, Mathematics Research Center, Technical Summary Report #1469, December 1974 (revised April 1976).
[8] Crary, F. D.: The AUGMENT precompiler II: Technical documentation. The University of Wisconsin-Madison, Mathematics Research Center, Technical Summary Report #1470, October 1975.
[9] Crary, F. D.: A versatile precompiler for nonstandard arithmetics. Assoc. Comput. Mach. Trans. Math. Software **5**, 204 – 217 (1979).
[10] Crary, F. D., Yohe, J. M.: The Augment Precompiler as a tool for the development of nonstandard arithmetic packages. The University of Wisconsin-Madison, Mathematics Research Center, Technical Summary Report #1892, October 1978.
[11] Kahan, W. M.: A more complete interval arithmetic. Lecture notes for an Engineering Summer Course in Numerical Analysis at the University of Michigan 1968.
[12] Kulisch, U.: An axiomatic approach to rounded computations. Numer. Math. **18**, 1 – 17 (1971).
[13] Lang, A. L., Shriver, B. D.: The design of a polymorphic arithmetic unit. Third IEEE-TCCA Symposium on Computer Arithmetic, November 1975, pp. 48 – 55.
[14] Moore, R. E.: Interval analysis. Englewood Cliffs, N. J.: Prentice-Hall 1966.
[15] Nickel, K.: Triplex-ALGOL and applications, in: Topics in interval analysis (Hansen, E., ed.), pp. 10 – 24. Oxford University Press 1969.
[16] Nickel, K.: Interval-analysis, in: Proceedings of the conference on the state of the art in numerical analysis held at the University of York April 12 – 15, 1976 (Jacobs, D., ed.), pp. 193 – 226, 1977.
[17] Palmer, J. F.: A proposed standard for floating-point formats and arithmetic. Presented at the Association for Computing Machinery SIGNUM meeting on Mathematical Software, November 3 – 4, 1977.
[18] Ris, F. N.: A unified decimal floating-point architecture for the support of high-level languages (extended abstract). SIGNUM Newsletter **11**, 18 – 32 (1976).
[19] Ris, F. N.: Tools for the analysis of interval arithmetic, in: Lecture Notes in Computer Science 29: Interval Mathematics (Nickel, K., ed.). Berlin-Heidelberg-New York: Springer 1975.
[20] Wipperman, N.-W., et al.: The algorithmic language Triplex-ALGOL-60. Numer. Math. **11**, 175 – 180 (1968).
[21] Yohe, J. M.: Roundings in floating-point arithmetic. IEEE Trans. Computers **C-22**, 577 – 586 (1973).
[22] Yohe, J. M.: The INTERVAL arithmetic package. The University of Wisconsin-Madison, Mathematics Research Center, Technical Summary Report #1755, June 1977 (revised September 1977).
[23] Yohe, J. M.: A semi-portable interval arithmetic package for FORTRAN. Universität Freiburg i. Br., Freiburger Intervall-Berichte 78/3, May 1978.
[24] Yohe, J. M.: Software for interval arithmetic: a reasonably portable package. Assoc. Comput. Mach. Trans. Math. Software **5**, 50 – 63 (1979).
[25] Yohe, J. M.: The interval arithmetic package – multiple precision version. The University of Wisconsin-Madison, Mathematics Research Center, Technical Summary Report #1908, January 1979.

Director Dr. J. M. Yohe
Academic Computing Services
University of Wisconsin-Eau Claire
Eau Claire, WI 54701, U.S.A.